Videofluoroscopic Studies of Speech in Patients with Cleft Palate

Videofluoroscopic Studies of Speech in Patients with Cleft Palate

M. Leon Skolnick, M.D.
Professor of Radiology
University of Pittsburgh School of Medicine
Pittsburgh, Pennsylvania, USA

Ellen R. Cohn, Ph.D.
Instructor, Department of Communications
University of Pittsburgh
Pittsburgh, Pennsylvania, USA

With 199 Figures in 429 Parts

Springer-Verlag
New York Berlin Heidelberg
London Paris Tokyo Hong Kong

Cover: Sketch in base projection demonstrating centripetal movement of velum and pharyngeal walls to close portal. See p. 42.

Library of Congress Cataloging-in-Publication Data
Skolnick, M. Leon.
Videofluoroscopic studies of speech in patients with cleft palate
/M. Leon Skolnick, Ellen R. Cohn.
p. cm.
Includes index.
ISBN 0-387-96958-6
1. Speech disorders—Diagnosis. 2. Cleft palate—Imaging.
3. Videofluoroscopy. I. Cohn, Ellen R. II. Title
[DNLM: 1. Cleft Palate—physiopathology. 2. Fluoroscopy.
3. Speech Disorders—diagnosis. 4. Videotape Recording. WV 440
S628v]
RC429.S57 1989
617.5'225—dc20
DNLM/DLC 89-11588

Media conversion by Precision Graphics, Champaign, Illinois.
Printed and bound by Arcata Graphics/Halliday, West Hanover, Massachusetts.

9 8 7 6 5 4 3 2 1

ISBN-13: 978-1-4613-8876-0 e-ISBN-13: 978-1-4613-8874-6
DOI: 10.1007/978-1-4613-8874-6

We dedicate this book to

Jerry Stone, PhD, medical editor of Springer-Verlag and initiator of this project, whose untimely death robbed us of a good friend and advisor

and to

the patients and staff of the University of Pittsburgh Cleft Palate Center under the leadership of Betty Jane McWilliams, PhD

Many thanks to

Pam Sorensen for the organizing and typing of the manuscript and for preparing the majority of the artwork

Diane Toth for typing the index

Ilene M. Stamps and Jon Coulter for preparing the remainder of the artwork within the text, and to Maria Klein for illustrating the Appendix

To Bob and Irene (our respective spouses) for their patience, understanding and support

Preface

In writing this book, we set out to codify our personal experiences with multiview videofluoroscopy over the past 14 years and to present this material in a practical form so that readers could easily apply it to their clinical needs.

Because the book is mainly directed to speech-language pathologists and radiologists, some material will be primarily of interest to one or the other group, although we hope that most of it will be of interest to both. Certain technical aspects of radiology have been simplified for the benefit of speech-language pathologists, just as speech characteristics have been simplified for the radiologist so as to emphasize basic concepts without losing the reader in details.

Key concepts have been repeated within different chapters to assist in the learning process and minimize the reader's need to flip back to prior chapters to recall important points.

Sketches and line drawings of videotape images rather than still photographs of the actual video frames have been the primary modes of illustration so as to convey our concepts more clearly and rapidly. Often, even photographs of high quality and clarity fail to illustrate all the clinical situations that we have discussed because they lack the information provided by evaluating structures during motion and with simultaneous speech.

We wish to caution the readers that the techniques, observations, and results described in this book are based on our personal experiences and our reading of the literature. We give no assurances that others will achieve the same results that we have achieved, as the conduct and interpretation of videofluoroscopy is as much an art as it is a science.

Squirrel Hill
Pittsburgh, Pennsylvania

M. Leon Skolnick
Ellen R. Cohn

Contents

1
Why Image the Velopharyngeal Portal

Purpose of Book

This book deals mainly with the applications and interpretations of a radiographic technique, multiview videofluoroscopy of the velopharyngeal (VP) portal. This procedure is used to evaluate the effectiveness of velopharyngeal closure in patients with speech evidence of velopharyngeal insuffiency[1] (VPI), and in so doing, provides precise anatomic and physiologic information as to how the VP portal functions during phonation. The data derived by this technique are useful in assisting clinicians to determine the most appropriate course of therapy for each patient.

Determination of Velopharyngeal Insufficiency

Before we even discuss radiologic-imaging procedures, or for that matter before we even consider any imaging procedure or physiologic test for evaluating the function of the VP portal during speech, we first must establish the presence of a speech abnormality, a determination made by the speech-language pathologist. The reader should never lose sight of the essential purpose for all instrumental assessments of VPI: to furnish information that will help the clinician improve the patient's speech. The quality of the perceived speech should be the prime consideration.

One of the most commonly detected speech symptoms of VPI is hypernasal speech. Hypernasality is a speech abnormality in which sounds normally perceived as being non-nasal are produced with an excessively nasal quality. This can occur when the sound wave is diverted into the nose instead of the mouth because of pathologic communication between the oro- and nasopharynx. The site of VPI is usually in the velopharyngeal portal, a result of the incomplete closure of soft palate or velum against the pharyngeal walls (Fig. 1.1). However, VPI may also result from a fistula in the hard or soft palate.

There are also other stigmata of VPI:

1. Reduced intensity in the subject's speech from a loss of some air through the VP portal or palatal fistula.
2. The presence of facial grimacing, which is an attempt to constrict the external nares so as to reduce the escape of air from the nasopharynx because of an incompetent VP portal.
3. Aberrant production of speech sounds that distort the patient's speech and impair intelligibility.

These aberrant sounds are produced by valving (narrowing) the airway at sites other than the velopharyngeal portal. The valving may occur at the oropharyngeal level between the tongue and posterior pharyngeal wall (to produce pharyngeal fricatives) or through the use of momentary closure of the vocal cords (to produce glottal stops) (Fig. 1.2).

---

[1]The term "velopharyngeal insufficiency" and the acronym "VPI" will be used to designate impaired velopharyngeal closure during speech, regardless of the etiology.

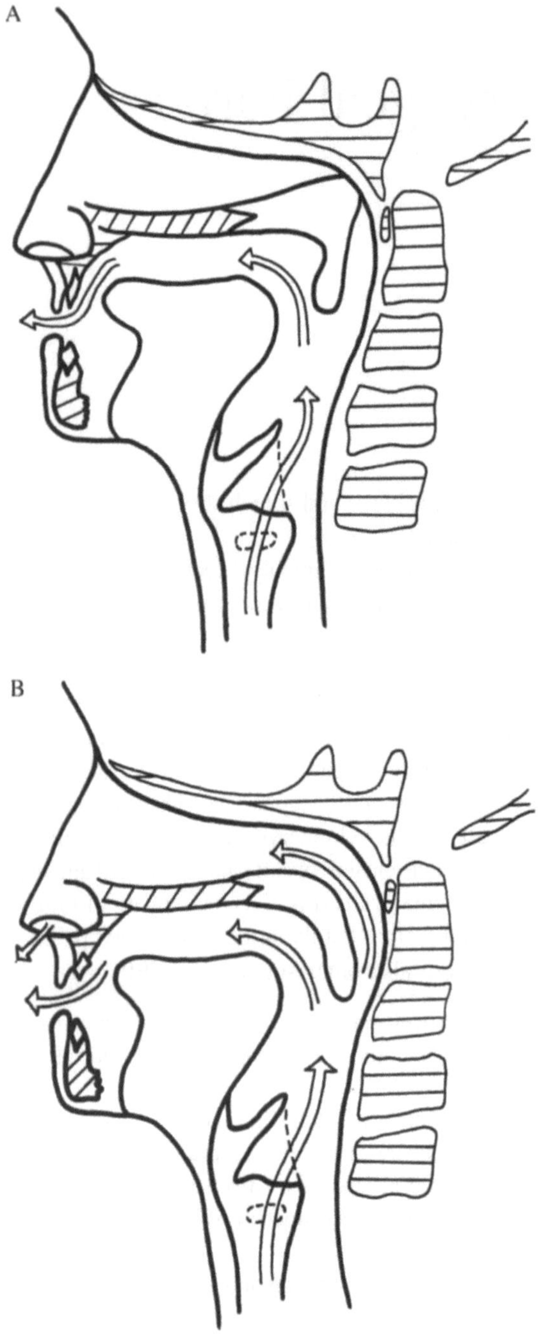

FIGURE 1.1. Path(s) of pharyngeal air flow in subjects with velopharyngeal closure and velopharyngeal insufficiency. (A) During closure air flows only through the oropharynx and out the mouth because the velum presses against the pharyngeal walls. (B) With velopharyngeal insufficiency air flows through both the naso- and oropharynx because the velum does not close the velopharyngeal portal.

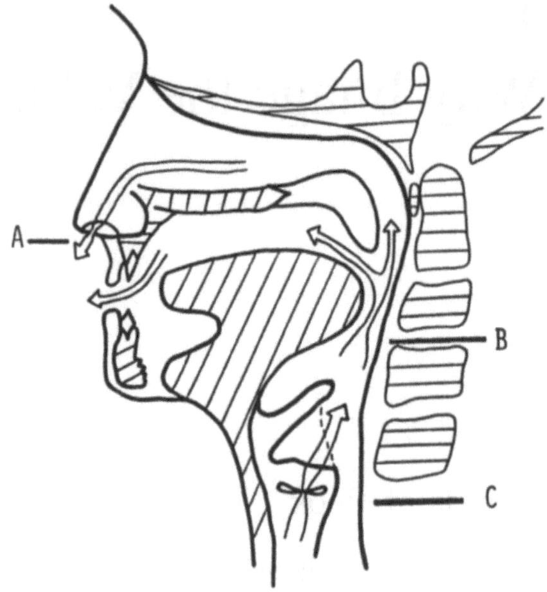

FIGURE 1.2. Maneuvers used to modulate the flow of air through the pharynx at levels other than the velopharyngeal portal in patients with velopharyngeal insufficiency so as to produce consonant sounds, albeit distorted ones. (A) Narrowing of the external nares with nasal grimacing to reduce nasal escape of air. (B) Oropharyngeal narrowing (tongue against oropharyngeal walls) that momentarily builds up air pressure, then suddenly releases it. (C) Transitory narrowing or closure of the vocal cords in an attempt to improve air flow momentarily at the laryngeal level.

These and other speech stigmata of VPI are described in greater detail in Chapter 12.

While typical stigmata of VPI are readily detected by the speech-language pathologist's clinical examination, other patients may present with more complex problems. Speech may, in addition to the stigmata of VPI, contain articulation errors, and perhaps even hyponasal or denasal characteristics because of anatomical deformities within the nasal cavity that restrict the flow of air. However, it may be difficult for the speech-language pathologist to assess the specific contribution of VPI to the abnormal speech pattern of such patients. Under these circumstances, it is vital that the speech-language pathologist have at his disposal diagnostic tools that can precisely determine the presence, extent, and location of the abnormality producing incompetence of the VP portal during speech.

Role of Instrumentation for Assessing Velopharyngeal Insufficiency

There are two broad categories of instrumentation: those that assess VPI indirectly by detecting abnormal flow of air through the nose during the utterance of sounds in which there is normally no nasal air flow (1), and those, like videofluoroscopy, that assess VPI directly by visualizing the velopharyngeal portal during phonation. (Other types of instrumentation for directly viewing the VP portal are discussed in Chapter 2.)

Keep in mind that there can be objective evidence of VPI by instrumental observation with little or no speech evidence of such insufficiency. Warren (2), using air-flow instruments, has observed that, when a VP opening is under 10 mm^2, speech may be perceived within the normal range. Using multiview videofluoroscopy we have also documented small openings in the VP portal, especially with vowels, when there is no perceived hypernasality. (See Chapter 8 for further details.)

When there are discrepancies between the degree of VPI as detected by an instrument [whether it be a physiologic device that measures air flow through the VP portal (1) or an imaging device such as videofluoroscopy (2, 3), or nasoendoscopy (4, 5)], and the severity of hypernasal-speech abnormality as perceived by the speech-language pathologist and by the patient, then the clinical assessment and the patient's wishes should take precedence over the data provided by instruments. After all, it is how the patient sounds to others and himself that is the prime consideration, not what an instrument shows. Instruments are simply aids used to interpret clinical observations more precisely.

Air-Flow Assessment

The air-flow instrumentation can demonstrate the presence and quantify the severity of VPI. This information is useful for confirming the speech-language pathologist's clinical impressions of VPI. However, air-flow instrumentation cannot demonstrate the anatomic deficiencies and/or abnormal movements of the palate and pharynx that cause the VPI.

Videofluoroscopy

If the decision has been made that the patient's hypernasal speech requires treatment, then it is important to directly visualize the anatomic and/or physiologic abnormality causing the VPI using videofluoroscopy. Although nasopharyngoscopy can also directly visualize the velopharyngeal portal, we feel the videofluoroscopy provides more information and can be used on younger subjects.

Because there is a wide spectrum of anatomic and physiologic abnormalities that can produce VPI, it is important to specifically define the cause of VPI prior to deciding upon therapy. At one end of the spectrum is the repaired cleft palate patient who shows good movement of the palate and pharyngeal walls, but who lacks adequate length or sufficient soft tissue mass within the palate to close the VP portal effectively. The patient has a consistent movement pattern, and no amount of speech therapy can increase the movement so as to close the gap in the VP portal. Such a patient will require either a surgical procedure or a prosthetic appliance to close the gap in the portal.

Midway within the spectrum are patients with an inconsistent movement pattern of the palate and pharyngeal walls. Certain non-nasal sounds produce complete or almost complete closure of the VP portal, whereas others result in much poorer palatal or pharyngeal wall movements.

At the other end of the spectrum is the patient who produces almost no palatal and pharyngeal wall motion with any speech task and has severe and consistent VPI. Regardless of the type of therapy, the results will be less than optimal.

Clinical Usefulness of Information Provided by Videofluoroscopy of Velopharyngeal Portal

When the time comes to decide upon the type of surgical or prosthetic device necessary to close a gap in the VP portal, it is important to know exactly where the gap is, how it is created, if there is abnormal movement, and in what manner the various surfaces of the portal move in an attempt to close the portal. All of these factors are crucial in determining the optimal surgical or prosthetic treatment for a specific patient's VPI.

Furthermore, by understanding the deranged anatomy and physiology producing VPI in a specific patient, we can help predict the outcome of a specific type of therapy, thereby giving both the patient and his family an idea of what to expect in the future. For example, when contemplating a pharyngeal flap, the chance of the flap eliminating VPI is much better when there is good pharyngeal wall movement prior to surgery than when there is little or no pharyngeal wall movement, as success of the flap operation depends upon medial movement of the lateral pharyngeal walls against the edges of the flap. Therefore, it is necessary to be able to observe and document the degree of pharyngeal wall movement prior to the surgical procedure. Furthermore, it is helpful that the surgeon knows at what level the maximum pharyngeal wall movement occurs so that he can attempt to place the flap at this same level. A flap placed at a level below maximum pharyngeal wall movement may be much less effective (6).

References

1. Warren DW: Nasal emission of air and velopharyngeal function. Cleft Palate J 4:148, 1967
2. Skolnick ML: Video velopharyngography in patients with nasal speech with emphasis on lateral pharyngeal motion in velopharyngeal closure. Radiology 93:747, 1969
3. Skolnick ML: Videofluoroscopic examination of the velo-pharyngeal portal during phonation in lateral and base projections—A new technique for studying the mechanics of closure. Cleft Palate J 7:803, 1970
4. Matsuya T, Yamaoka M, Miyasaki, T: A fiberscopic study of velopharyngeal closure in patients with operated cleft palates. Plast Reconstr Surg 63:497, 1979
5. Pigott RW, Bensen JF, White FD: Nasendoscopy in the diagnosis of velopharyngeal incompetence. Plast Reconstr Surg 43:141, 1969
6. Shprintzen R, Lewin M, Croft C, Danniller A, Argamaso R, Ship A, Strauch B: A comprehensive study of pharyngeal flap surgery: Tailor made flaps. Cleft Palate J 16: 46, 1979

2
How to Image the Velopharyngeal Portal

Analysis of Problem

In order to determine the best method of imaging the velopharyngeal portal so as to appreciate the interrelationships of palatal and pharyngeal wall movements during speech, one must consider the purposes of the examination: First is the desire to image the velopharyngeal (VP) portal; second, the desire to image the portal in motion; third, the desire to image it while simultaneously recording the patient's speech. If the method of imaging the VP portal does not satisfy all three of these requirements, the study will give less than adequate diagnostic information and inappropriate therapeutic decisions may result.

Now let us examine each of these three requirements. First, when we talk about imaging the VP portal, we must realize that the portal is a three-dimensional structure — a volume. Except for three-dimensional reconstruction of CT scans, which is not clinically practical for imaging the VP portal, all imaging techniques are capable of producing only planar or two-dimensional images. Therefore, to image a volume structure adequately, one must examine it in three mutually perpendicular planes so as to fully describe that volume. Only if the volume is symmetric in two planes can the number of views be reduced to two instead of three. However, the VP portal is not symmetric and, therefore, three planes are optimally required (Fig. 2.1).

Physiology of VP Portal During Speech

In order to appreciate what these planes are, let us briefly review the physiology of the VP portal during speech. The portal consists of two components: a nasopharyngeal component and a velar component. The nasopharyngeal component during quiet breathing is in the shape of an incomplete muscular cylinder with an anterior opening for insertion of the velum. The velar component, a muscular tissue, is attached on its anterior end to the hard palate and on its posterior or free margin into the two sides of the pharyngeal component (Fig. 2.1A). During phonation, the velum elevates and moves posteriorly while the nasopharyngeal component constricts along an approximately horizontal plane, causing the pharyngeal walls to encase the free edges of the velum (Fig. 2.2B-E) in a sphincteric manner (1).

Rationale for Choice of Imaging Techniques

To demonstrate the elevation and posterior movement of the velum toward the posterior pharyngeal wall a lateral view (Fig. 2.1A, 2.2C, 2.3A) is required. This view shows the sides of the lateral pharyngeal walls as they move medially against the margins of the velum and the vertical extent of pharyngeal wall motion. A frontal view is necessary (Fig. 2.1B, 2.2D, 2.3B) to appreciate how the

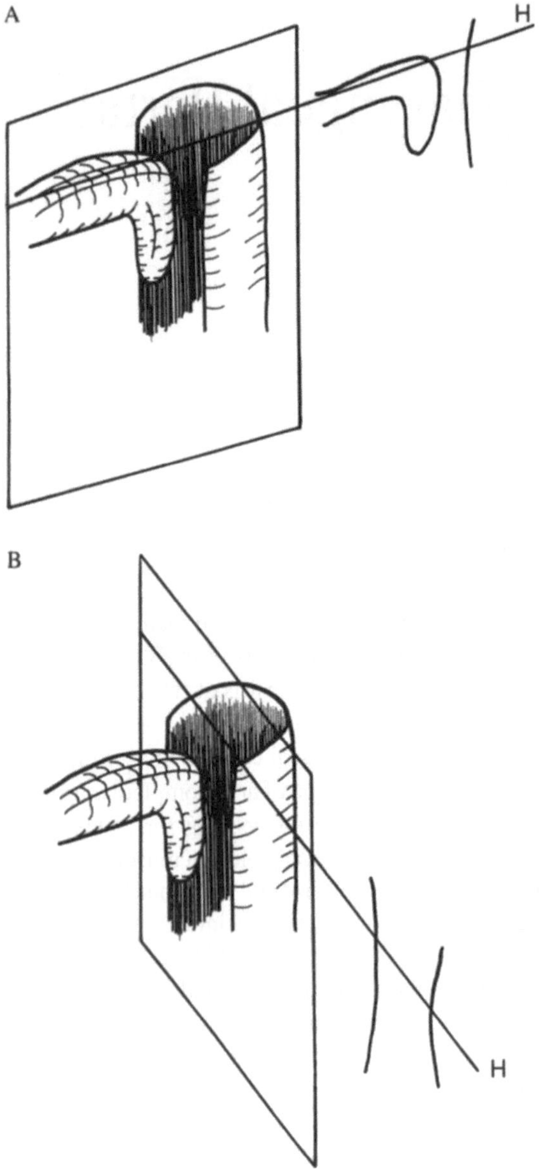

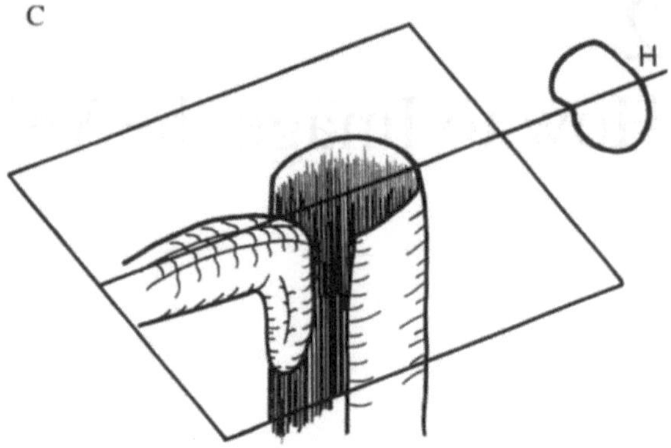

FIGURE 2.1. Velopharyngeal portal during nasal breathing. Portal is open. Three-dimensional representation. (A) Projection through midsagittal plane showing relationships between velum and posterior pharygneal wall (lateral view). (B) Projection through coronal plane showing lateral pharyngeal walls (frontal view). (C) Projection in transverse plane showing en face view of velopharyngeal portal (base view). Line (*H*) is to help the reader appreciate the plane through the VP portal that corresponds to the projected image.

centripetal movement of the pharyngeal walls relates to the palate. To fully appreciate the sphincteric activity, a base or en face view looking down the velopharyngeal portal is needed (Fig. 2.1C, 2.2E, 2.3C). In order to understand adequately the mechanics of VP closure in a particular patient, images must be obtained in each of these three planes — lateral, frontal, and base.

However, it must be recognized that the contribution of each of these component movements to closure of the VP portal can vary widely among different individuals. In some cases closure mainly results from movement of the velar component or from the pharyngeal component, whereas in others it is caused by a relatively equal movement of both components (1, 2) (Fig. 2.4). (Details of the various patterns of closure will be discussed in Chapter 8.)

The second requirement for imaging the portal is that the structures must be imaged during motion because motion occurs during speech. Specifically, we are interested in observing the behavior of the VP portal during the patient's usual speech activities.

The third criterion is that the patient's speech should be recorded simultaneously with the movements of the VP portal.

Thus, we have the need to record the VP portal in multiple planes, each in motion, and simultaneously with the patient's speech. Any other method of imaging the VP portal that does not satisfy these three requirements will not provide an adequate evaluation of the portal when the purpose of the study is to evaluate the dynamics of the portal during speech.

Furthermore, it would be very beneficial if the examination is one that does not disturb the patient's speech in the process of recording data, does not require insertion of an instrument in the pharynx, does not require any topical anesthesia to be

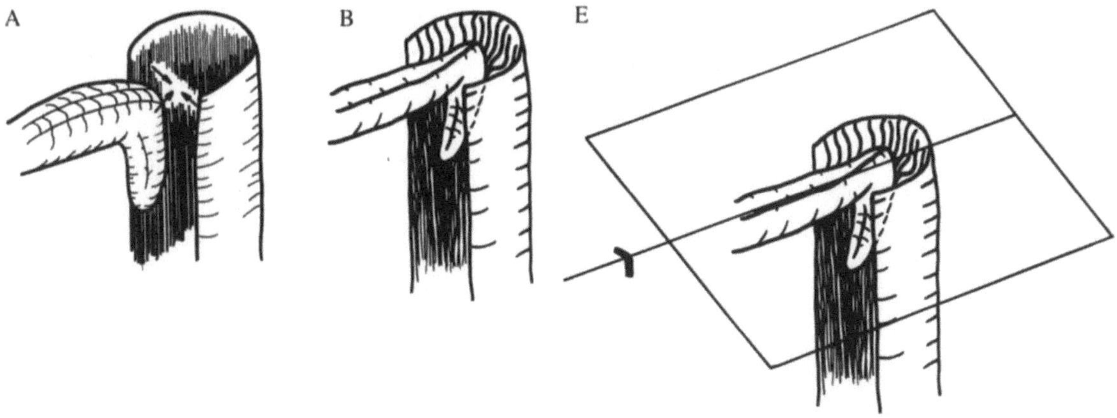

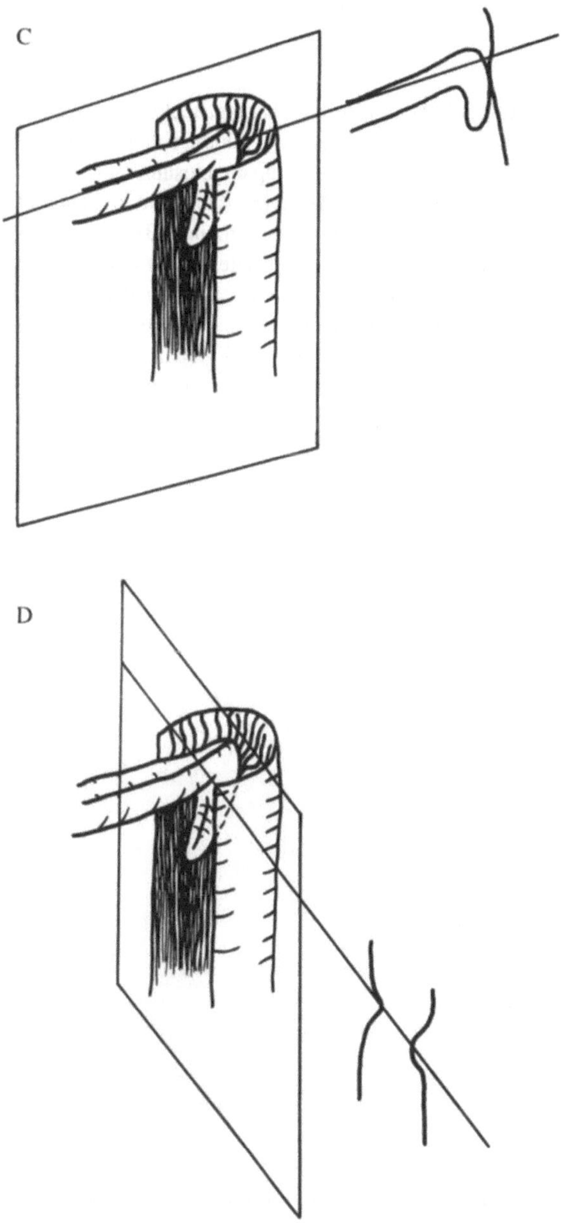

FIGURE 2.2. Velopharyngeal closure during phonation. (A) Direction of velar and lateral pharyngeal wall movements as portal closes is indicated by arrows. (B) Three-dimensional representation showing closure of velopharyngeal portal. (C) Projection through midsagittal plane (lateral view) showing velopharyngeal contact. (D) Projection through coronal plane (frontal view) showing localized medial movement of the lateral pharyngeal walls. (E) Projection through transverse plane (base view) showing en face view of velopharyngeal closure.

applied to the pharyngeal mucosa (because of the possibility of drug reaction from the anesthetic agent), and does not require any complicated patient preparation or premedication.

The study should be simple enough to be performed on any patient cooperative enough to speak to the examiner. In our experience we have been able to perform this examination occasionally on children as young as age 3. By age 4 most children can cooperate well enough for the radiologist to obtain a good study.

Practical Choices

Now let us examine the various choices we have for visualization of the velopharyngeal portal and see how closely they approach these ideal characteristics.

Endoscopy

Our choices fall basically into two categories: endoscopic and radiographic. Among the endoscopic methods are oral endoscopy and nasoendoscopy.

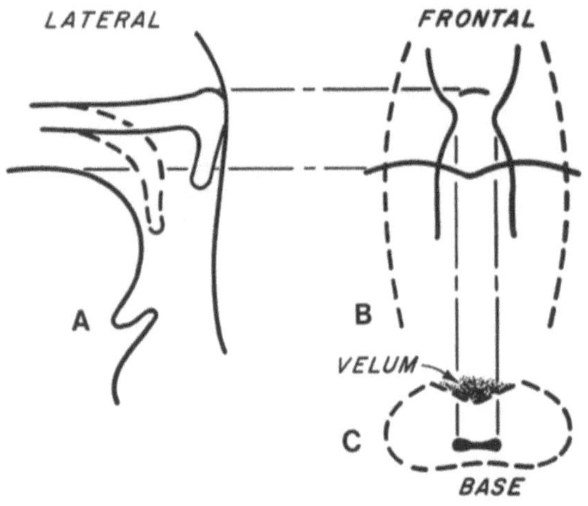

FIGURE 2.3. Velopharyngeal portal during rest (dotted lines) and closure (solid lines) in (A) lateral, (B) frontal, and (C) base views.

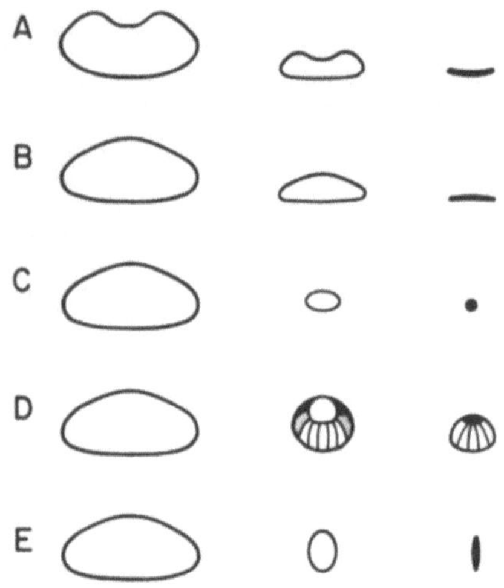

FIGURE 2.4. Patterns of velopharyngeal closure as seen on base view. Left column — rest; middle column — partial closure; right column — complete closure. (A) Coronal closure pattern in normal subject (convexity in center of velum represents uvula muscle contour). (B) Coronal closure pattern in repaired cleft palate subject (concavity in center of velum represents site of primary palatal repair). (C) Circular closure pattern in repaired-cleft-palate subject. (D) Circular closure pattern in repaired cleft-palate subject with Passavant's ridge. (E) Sagittal closure pattern in repaired cleft palate subject. Reproduced with permission from: Skolnick ML, McCall GN, Barnes M: The sphincteric mechanism of velopharyngeal closure. Cleft Palate J 10:286, 1973

Oral Endoscopy

We can dismiss the oral endoscopic approach with a few words because the information gained by oral endoscopy is so limited. The procedure is as follows: A rigid or flexible endoscope is inserted into the mouth. The tip of the endoscope has an angled mirror, which is pointed cranially to look at the VP portal. The presence of the instrument in the mouth limits the patient's utterances to a few isolated phonemes. No meaningful observations can be made during connected speech as the endoscope interferes with free oral movements. The phonemes that can be observed are usually vowels and these are not the phonemes at which maximum VP closure occurs (3). Thus, although one can obtain an en face or base view of the VP portal during rest, this technique provides a less than optimal attempt at closure during phonation. At times the uvula may obscure the observer's view of the VP portal itself, further limiting the usefulness of oral endoscopy.

Nasal Endoscopy

Nasal endoscopy is much better than oral endoscopy for visualizing the VP portal. Two types of instrumentation can be used: either a rigid or flexible endoscope. The rigid endoscope (4, 5) Fig. 2.5) has the advantage of the larger field of view and a better-quality image, but has the disadvantage that the diameter of the tube is greater than that of the flexible endoscope, making it more difficult to in-

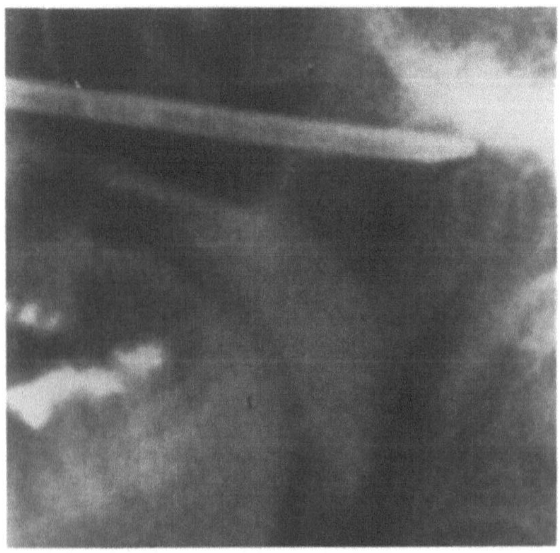

FIGURE 2.5. Rigid nasendoscope in nasopharynx examining velopharyngeal portal. Lateral x-ray with endoscope in nasal cavity viewing velopharyngeal portal during nasal breathing. Illustration courtesy of R. Pigott.

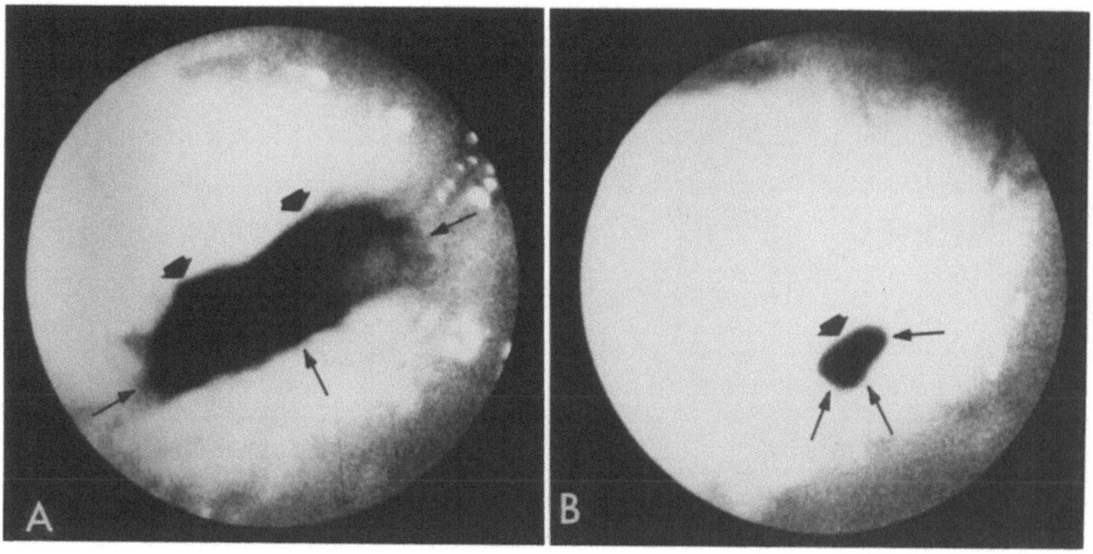

FIGURE 2.6. Endoscopic view of patient with circular closure pattern and small central gap. (A) Quiet breathing. Velum — arrowheads; pharyngeal walls — arrows. (B) Maximum closure effort during phonation. Central gap remains. Illustration courtesy of R. Pigott.

sert in the nose. In addition, it is more difficult to learn how to use. Potentially, the rigid scope can seriously injure the patient if the patient makes a sudden movement while the endoscope is in the nasal cavity.

The flexible endoscope (6) uses a fiberoptic viewing system instead of the lenses and prisms incorporated within the rigid endoscope. The field of view of the flexible endoscope is smaller. Even though this instrument is more flexible than the rigid scope, it is sometimes difficult to fully bend the tip of the endoscope caudally to view the VP portal. However, the flexible endoscope is easier for the operator to learn how to use and possesses a significantly lower chance of injuring the patient if the patient makes a sudden movement. It can be inserted through a narrower-diameter nasal cavity, thus facilitating its use in younger patients. In addition, the instrument can be inserted partially or completely through the VP portal to look at lower levels of pharyngeal movement and even to see laryngeal and vocal cord movements.

The rigid and flexible endoscope do share several features in common. The patient can engage in normal conversational speech with either rigid or flexible endoscope while the instrument is in the nasopharynx, thereby allowing the observer to see the dynamics of the VP portal unimpeded by the observing instrumentation. Both the rigid and flexible endoscopes require the application of topical anesthesia to the nasal passage and nasopharynx. Although nasal endoscopy is usually innocuous, occasional reactions to the topical anesthesia can occur that can be quite serious, and, on extremely rare occasions, even fatal. Finally, it may be difficult to perform an endoscopic study on young children when either their nasal cavity is too small to admit the diameter of the instrument or the patient would not cooperate for the procedure.

Limitations of nasal endoscopy are as follows. The procedure provides only one view of the VP portal: that of the en face view showing the margins of the pharyngeal walls and palate as seen from above. One cannot see the full vertical extent of the lateral pharyngeal walls as they move medially so one cannot ascertain the level of maximum medial movement of the pharyngeal walls relative to the edges of the palate (Fig. 2.3B). In addition, information concerning the length and thickness of the palate, the region and vertical extent of maximal palatal-pharyngeal wall contact, and the relationship of the Passavant's ridge to the palate usually cannot be obtained from endoscopic observations.

However, endoscopic studies can provide very useful information. They can demonstrate the contribution of palatal and pharyngeal wall movements to the total closure of the VP portal, and show regions of the portal where incomplete closure occurs (Fig. 2.6). Endoscopy can visualize a defect in

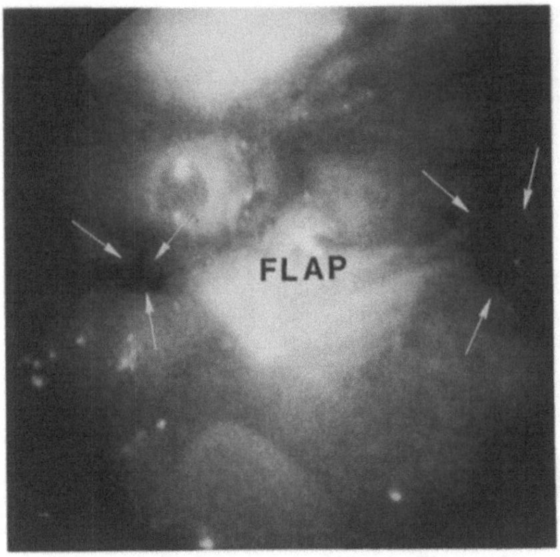

FIGURE 2.7. Endoscopic view of pharyngeal flap with bilateral portal inadequacy (arrows) during maximum phonation effort. Illustration courtesy of R. Pigott.

A

A'

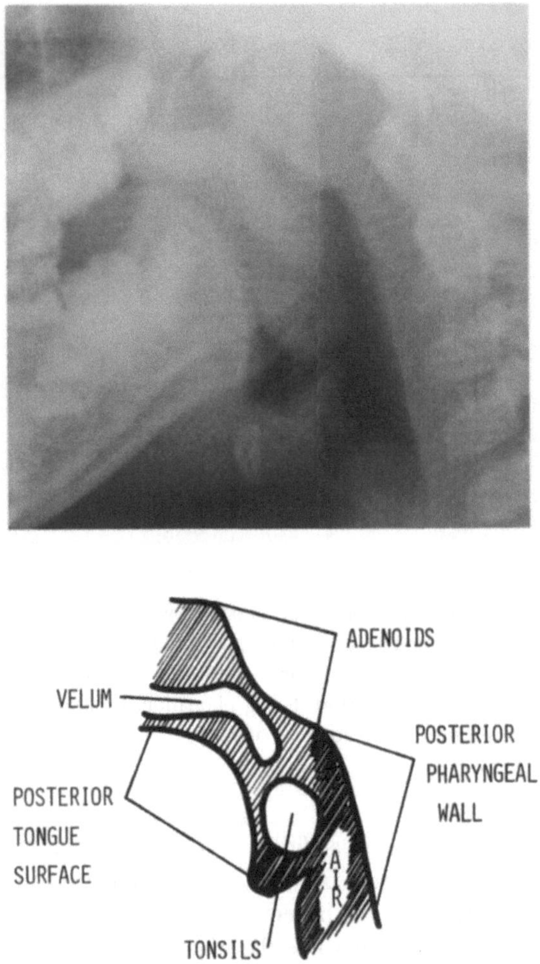

FIGURE 2.8. Lateral cephalometric views (A, A'). Phonation of "e." Thin column of air between velum and pharyngeal wall indicates that these two structures do NOT touch.

the posterior portion of the palate whether such a defect results from an unrepaired submucous cleft palate or scarring at the site where the margins of the cleft palate were sutured together. Endoscopy can also demonstrate incomplete closure of the lateral portals in patients who have had a pharyngeal flap (see Chapter 10 for details) (Fig. 2.7). Thus, nasal endoscopy can substitute for the base view if that view cannot be fluoroscopically obtained. However, to fully appreciate the dynamics of the portal, the portal must also be imaged in the frontal and lateral views. These views can only be obtained using radiographic techniques at present.

Radiological Choices

Now let us examine our radiographic options. Basically, there are two: radiography or fluoroscopy. Radiography produces still images; fluoroscopy allows the recording of moving structures.

Lateral Cephalographs

A frequently used radiographic examination for imaging the VP portal is a lateral x-ray usually referred to as a lateral cephalometric film (7). This x-ray must be obtained during quiet breathing, during the phonation of a sustained sound, or both. It reveals static anatomy including the length and configuration of the velum, the shape and depth of the nasopharynx, and the region of maximal VP contact or approximation during the maintenance of a static sound (Fig. 2.8).

However, this technique has several major shortcomings. First, because of the static nature of the cephalometric x-ray there is no way of recording the sound actually uttered by the patient as the x-ray was taken. We only know what sound the examiner asked the patient to say. What is asked versus what is uttered may not be similar in a patient with abnormal speech. Furthermore, the movements of the VP portal during the utterance of an isolated sound may not be related to the type of movement that the palate and pharyngeal walls undergo during

A

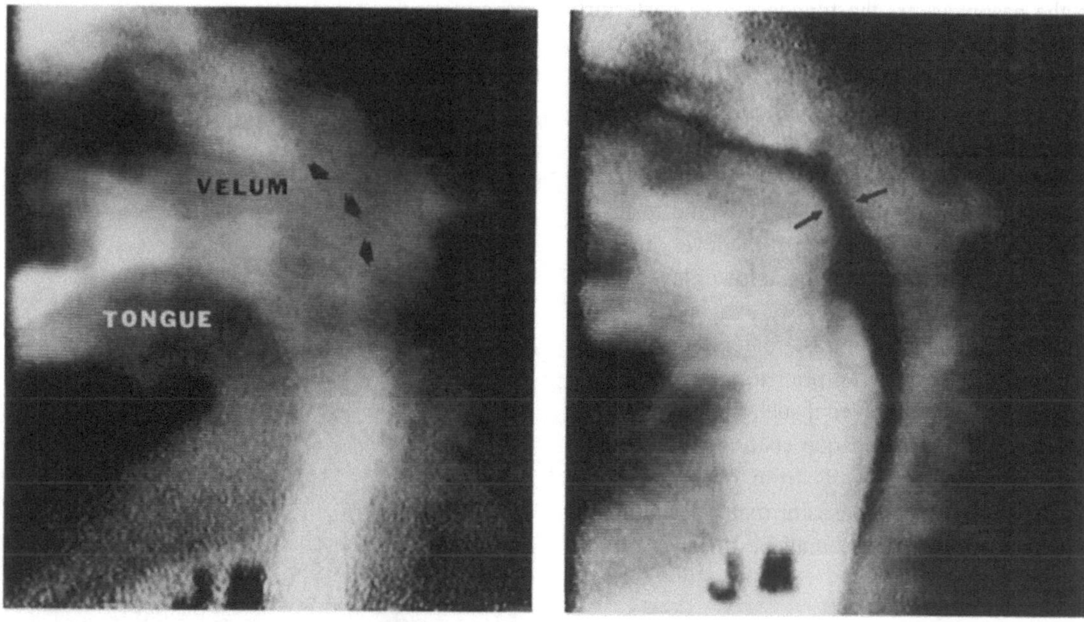

B

FIGURE 2.9. Comparison of lateral videofluoroscopic views during phonation without and with barium coating velum and pharyngeal walls. (A) Without barium. Velum appears to touch posterior pharyngeal wall (arrowheads) because no definite column of air is seen between these two structures. (B) Same speech task with barium. Thickness of barium column (distance between arrows) indicates actual gap between the velum and pharyngeal wall. Gap not seen on image without barium because amount of air in gap is too little to produce enough x-ray contrast to demark surfaces of velum and pharyngeal walls. Barium, being so absorptive of x-rays as compared with air and soft tissue, is readily seen even as a fine column.

connected speech, especially during normal conversational speech. Therefore, the effectiveness of closure or degree of VP gap seen on a static x-ray may not be characteristic of what the patient can do during connected speech, and the former should not be substituted for the latter.

Second, the ability to visualize the soft palate on the lateral view depends upon its being surrounded by air within the nasopharynx. Air, which is much less absorptive of x-rays than soft tissues, provides an excellent contrast material by which to delineate the margins of the palate (Fig. 2.8). However, as the VP portal narrows during speech because of posterior palatal motion, the amount of air between the palate and pharyngeal walls becomes markedly reduced and finally disappears when the palate touches the pharyngeal walls. Therefore, the ability to distinguish radiographically the posterior surface of the palate from the anterior surface of the posterior pharyngeal wall is greatly reduced as these surfaces approach each other. The two structures may appear to touch when in fact there may be a small air gap between them that is too small to appreciate radiographically. This concept is demonstrated in Fig. 2.9 with the use of frames from a lateral videotape study.

Limitations of Cephalometric X-Rays for Frontal and Base Views

In the frontal and base views of the VP portal, the air in the nasopharynx is insufficient to provide enough contrast to delineate the margins of the palate and pharyngeal walls without the use of barium.

In the frontal view the nasal bones and cervical vertebrae lie in front of and behind the nasopharynx. These bones cast numerous vertical and horizontal shadows that obscure the shadows produced by the interfaces between the pharyngeal walls and nasopharyngeal air column.

For the base view another problem arises. When the VP portal is viewed en face during speech, there may be structures moving at several levels in this cylinder that appear to lie in one plane because they are projected superimposed one upon the other. Thus, although the palate and pharyngeal walls move in the nasopharynx, the tongue moves posteriorly at the oropharyngeal level, and the vocal cords move at the laryngeal level. These structures can appear to be on the same level on the base projection.

To distinguish the pharyngeal walls from surrounding bones on the frontal view and to distinguish the soft tissues moving at the level of the VP portal from structures moving at lower levels in the vocal tract on the base view, one needs an extremely radiopaque contrast material to coat the margins of the VP portal prior to the roentgen examination. However, even if such a material is used, differentiating a radiopaque coating on the velum and nasopharyngeal walls from overlying bones (frontal view) or from the same material coating the oropharynx (base view) may be difficult on still images.

Barium as a Contrast Agent

The radiopaque substance most commonly used to coat the nasopharyngeal surfaces is a suspension of colloidal barium sulphate that is introduced into the nasopharynx (see Chapter 4 for further details). This radiopaque contrast material is vital for imaging the margins of the nasopharynx in the frontal and base views. The barium is also extremely useful when obtaining lateral radiographs because it more precisely delineates the opposing surfaces of the velum and posterior pharyngeal wall than can be seen with the air in the nasopharynx acting as the contrast agent. This delineation is more important when there is a minimum (1-3 mm) gap between the velum and posterior pharyngeal wall because this minimal gap may not be appreciated when a lateral radiograph is taken without intranasal barium contast material (Fig. 2.9).

Tomographic Methods

Besides obtaining static radiographs of the portal in the lateral, frontal, and base projections, one can also obtain tomographic views using a CT scanner. However, CT scanning has a number of limitations. Because the CT scanner is designed basically to slice the body in the plane perpendicular to the couch upon which the patient lies, the base or axial view (Fig. 2.10) is the easiest one to obtain and the one containing the greatest resolution. Images in the lateral (sagittal) or frontal (coronal planes) would have to be electronically reconstructed from a series of closely spaced axial scans and these reconstructions would lack the detail seen in the base (coronal) view.

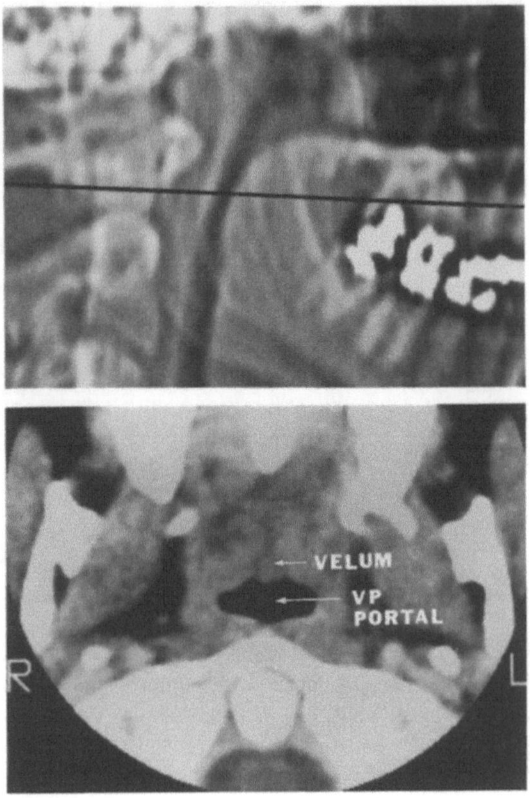

FIGURE 2.10. CT scan in transverse plane through velopharyngeal portal during quiet nasal breathing. (A) Lateral scout view showing plane of CT scan (black line). (B) CT scan. Appearance is similar to videofluoroscopic base view.

Because each CT scan takes 1 to 5 seconds to produce, images can only be obtained during sustained phonemes in isolation ("e," "ah," "s," etc.) except for one CT scanner, the Imatron (Imatron, Inc., 389 Oyster Point, South San Francisco, CA 94080) (8), which can make images every 50 milliseconds. This scanner is just being introduced and there are few installed. Thus, CT scanning usually does not provide information about VP function during continuous speech.

Videofluoroscopy

We believe that videofluoroscopy is at present the method best suited for evaluating the movements of the VP portal during speech (1, 2, 9). First, videofluoroscopy fulfills the function of being able to simultaneously record speech and to show roentgen images of the portal in motion so that the observer can correlate the movement of the palatal and pharyngeal walls with the sounds that the subject utters. Second, it allows one to obtain views of the VP portal in each of the three projections necessary to describe totally the movements of the portal during speech. Third, it is readily acceptable by patients. No special patient preparation aside from barium nose drops is required, no topical anesthesia is used, and no instrument is kept in the patient as the examination is being conducted.

Videofluoroscopy Versus Cinefluoroscopy

We keep referring to the term *videofluoroscopy* as the most appropriate way for recording fluoroscopically obtained movements of the palate and pharyngeal walls synchronous with patient's speech. Videofluoroscopy is not the only method — there is also cinefluoroscopy — but we feel that videofluoroscopy is the best method for several reasons. First, it is easiest to simultaneously record fluoroscopic images and speech using a videotape recorder because that equipment can record video and audio signals together on the same tape. Second, a video recording system is very sensitive to recording images at low light levels. Therefore, it can record fluoroscopic images at much lower light intensities than cinefluoroscopy so as to reduce significantly the radiation required to produce an image.

Cinefluoroscopy we believe is a less satisfactory procedure for studying the dynamics of velopharyngeal function. Although the sharpness of the image and the detail within each image frame are greater with cinefluoroscopy than with videofluoroscopy because of the higher resolution of film, cinefluoroscopy requires a radiation dose approximately ten times greater than videofluoroscopy (10). Furthermore, with cinefluoroscopy systems it is much more difficult to simultaneously synchronize the sound to the cine image than with a video recording system because there are few commercial cine systems that provide simultaneous sound synchronization built within the equipment. Often external synchronizers are required and the system becomes clumsy and quite expensive. In our experience, the improved image quality obtained with cine does not provide significantly greater diagnostic information than does videofluoroscopy. Finally, unlike cinefluoroscopy, the results of videofluoroscopy are immediately available. Because film development is not required, the radiologist can check the adequacy of the examination before the patient leaves the department.

Radiation Exposure

A word about radiation exposure is appropriate at this point. Whereas x-rays are not innocuous and can potentially produce some somatic and genetic damage even at low doses, one must carefully balance the benefits obtained from a radiographic or fluoroscopic procedure against the consequences of deciding upon a course of therapy without having obtained the information that the radiographic or fluoroscopic procedure could have provided. Thus, the preoperative exposure of the patient to several rads of radiation from videofluoroscopy could help the clinician to better decide upon the treatment that is most appropriate (whether surgical, prosthetic, speech therapy, or a combination), and to better inform the patient as to what the results and expectations of treatment would be like. We believe that the minimal chance of producing harmful effects from the radiation are more than compensated for by the potential for improved treatment results and more realistic patient expectations. Those who are concerned about the potential harm of radiation should balance this concern with the potential risks from anesthesia and surgery when an inappropriate operation is performed because of an incomplete diagnostic imaging evaluation.

To put radiation exposure in perspective, let us compare CT, lateral cephalometric x-rays, and videofluoroscopy. For a single CT slice, the radiation exposure is between 1 and 4 rads (depending upon patient size and equipment used). For a single lateral cephalometric film the exposure is about 0.25 rads (to the skin surface closest to the x-ray table). For 1 minute of videofluoroscopy in lateral view the radiation exposure at skin surface can range from 0.5 to 0.025 rad, while for frontal or base views (which require a higher radiation level) the exposure can range from 1 to 0.125 rad. The very low radiation exposures are currently being obtained at Children's Hospital of Pittsburgh using a state-of-the-art General Electric MPX fluoroscopic system with a Plumbicon TV camera tube. A more likely low radiation exposure is about 0.1 rad/minute (lateral view) and 0.45 rad/minute (base view) (10). Using these later figures and assuming 3 minutes for a videofluoroscopic study (1 minute each for lateral, frontal, and base) and the same exposure for frontal and base views, then total radiation exposure for a videofluoroscopic study would equal four lateral cephalometric films. However, using the current fluoroscopic radiation levels from Children's Hospital of Pittsburgh, then a 3-minute videofluoroscopic study would require approximately the same radiation as one lateral cephalometric film.

References

1. Skolnick ML, McCall GN, Barnes M: The sphincteric mechanism of velopharyngeal closure. Cleft Palate J 10:286, 1973
2. Skolnick ML: Videofluoroscopic examination of the velopharyngeal portal during phonation in lateral and basal projections — a new technique for studying the mechanics of closure. Cleft Palate J 7:803, 1970
3. Taub S: The Taub oral panendoscope: a new technique. Cleft Palate J 3:328, 1966
4. Pigott RW: The nasendoscopic appearance of the normal palato-pharyngeal valve. Plast Reconstr Surg 43:19, 1969
5. Pigott RW, Benson JF, White FD: Nasendoscopy in the diagnosis of velopharyngeal incompetence. Plast Reconstr Surg 43:141, 1968
6. Matsuya T, Yamaoka M, Miyasaki T: A fiberscopic study of velopharyngeal closure in patients with operated cleft palates. Plast Reconstr Surg 63:497, 1979
7. Williams WN, Eisenbach CR II: Assessing VP function: the lateral still technique vs. cinefluorography. Cleft Palate J 18(1):45, 1981
8. Moon, JB, Smith WL: Application of cine computed tomography to the assessment of velopharyngeal form and function. Cleft Palate J 24:226, 1987
9. Skolnick ML: Video velopharyngography in patients with nasal speech with emphasis on lateral pharyngeal motion in velopharyngeal closure. Radiology 93:747, 1969
10. Isberg A, Julin P, Henrikson CD: Absorbed doses and energy imparted from radiographic examination of the velopharyngeal function during current speech. Cleft Palate J, 26:105, 1989

3 Equipment for Multiview Videofluoroscopy

This chapter discusses the equipment needed to perform a multiview videofluoroscopic study and briefly describes how the equipment operates. Because our book is written for both the speech-language pathologist and the radiologist, we ask the indulgence of the radiologist while we describe for the benefit of the speech-language pathologist the operation of the fluoroscopic system. In later chapters the situation will be reversed when speech production is discussed for the benefit of the radiologist.

Basic Concepts of Patient Positioning

In order to appreciate equipment requirements, a few comments on patient positioning are in order. (Details of this topic are presented in Chapter 4.)

The three basic views of videofluoroscopy are the lateral, frontal, and base. The lateral represents a midsagittal plane through the VP portal that demonstrates the relationships between the velum and posterior pharyngeal wall. The frontal view represents a coronal plane through the portal that shows the relationships between the lateral pharyngeal walls and velum. The base view is a transverse section through the portal at the level of the velopharyngeal (VP) aperture that projects an en face view showing all of the margins of the velum and pharyngeal walls. To display each of these views, the x-ray beam must be oriented perpendicular to the plane of the particular view (Fig. 3.1).

For the lateral view, the patient is positioned erect, either sitting or standing, in front of a vertically oriented fluoroscope. The side of his face, and therefore the midsagittal plane of the portal, is parallel to the face of the fluoroscopic screen. The x-ray beam is horizontally oriented so as to project the silhouette of the velum and posterior pharyngeal wall on the fluoroscopic screen (Fig. 3.2A, 3.1A).

For the frontal view, the patient is rotated 90° so as to face the fluoroscopic screen in order to orient the coronal plane parallel to the face of the screen. The horizontally oriented x-ray beam perpendicularly intersects the coronal plane and projects the barium-coated margins of the lateral pharyngeal walls and superior surface of the velum on the fluoroscopic screen (Fig. 3.2B, 3.1B).

For the base view, we have found it easiest to rotate the fluoroscope 90° so that the table is now horizontal and the x-ray beam becomes vertically oriented. The transverse plane through the portal is positioned parallel to the surface of the fluoroscopic table by having the subject lie prone on the table with his head hyperextended on his neck (in a sphinxlike position). The vertically oriented x-ray beam intersects the transverse plane to project the margins of the portal on the fluoroscopic screen as a circular or oval shadow that contracts centrally during phonation (Fig. 3.2C, 3.1C).

Equipment Needs for Multiview Videofluoroscopy

The basic equipment needed for multiview videofluoroscopy can be found in almost any hospital radiographic department and in many private radiographic offices as well. It consists of a standard

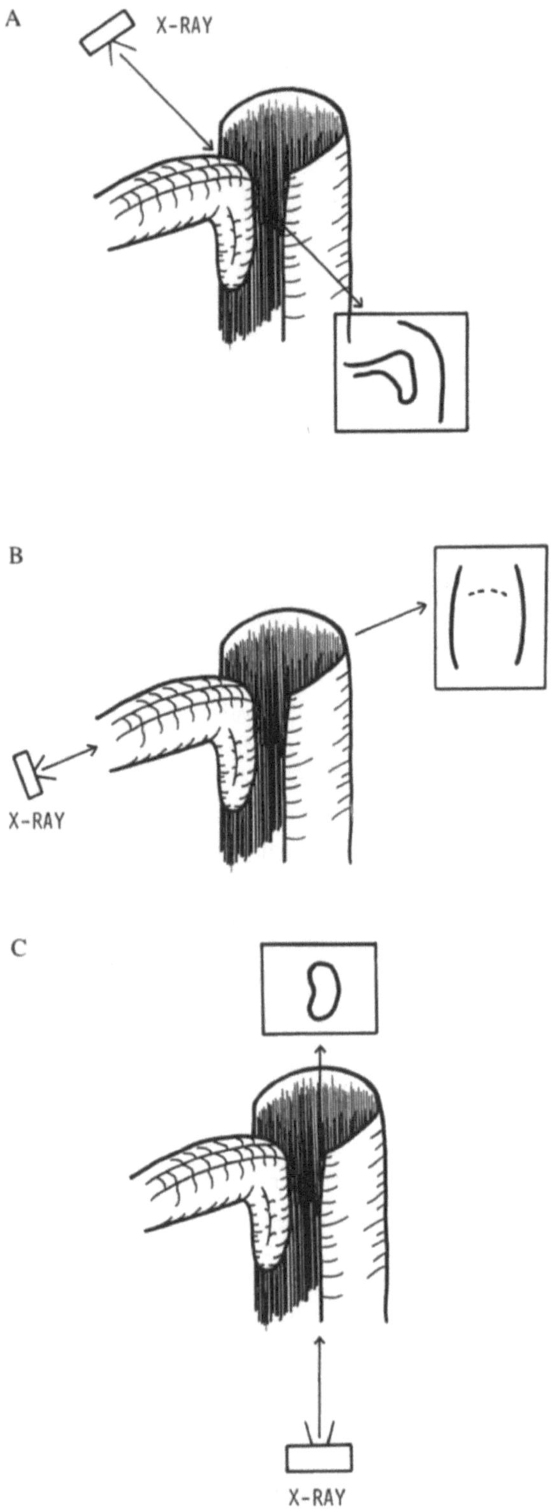

FIGURE 3.1. Orientation of x-ray beam to velopharyngeal portal for imaging in (A) lateral, (B) frontal, and (C) base views.

fluoroscopic/radiographic system that is used for the routine examination of the gastrointestinal tract (i.e., for the performance of a GI series or a barium enema). This system contains a patient-examining table that can move from horizontal to vertical position. Attached to the table is a fluoroscopic carriage that remains perpendicular to the table as the table is moved from horizontal to vertical position. The fluoroscopic carriage can be freely moved as the patient is being fluoroscoped so images of various parts of the body can be seen without shifting the patient's position. The fluoroscopic system must display the image in TV format so that it can be recorded using a conventional videotape recorder.

Operation of a Fluoroscopic System

A few comments are in order about the operation of the fluoroscopic system so that the non-radiologist who reads this book will appreciate some of the factors that are involved in producing an optimal fluoroscopic image.

Production and Characteristics of X-Rays

To produce x-rays, a beam of electrons must be rapidly accelerated through a vacuum into a material of some very heavy atomic weight and high melting temperature, which is called the target. The interaction of the electron beam with the atoms of the target results in the emission of x-rays.

The electron beam is generated by heating a coil of wire called the filament, with an electric current. The greater the amount of current rated in milliamperes, or MA, that flows through the filament, the greater the number of electrons that are available to be accelerated into the target to produce x-ray photons. The filament (cathode) and target (anode), usually both composed of tungsten, are enclosed in a glass tube containing a high vacuum. To produce the necessary acceleration of the electrons released by the heated filament, a high voltage, ranging from 50,000 to 120,000 volts (referred to as kilovolts or KV for short), is connected between the filament and target. The target side is positive so that the negatively charged electron would be attracted to it (Fig. 3.3). The penetrating power in tissue of the

A

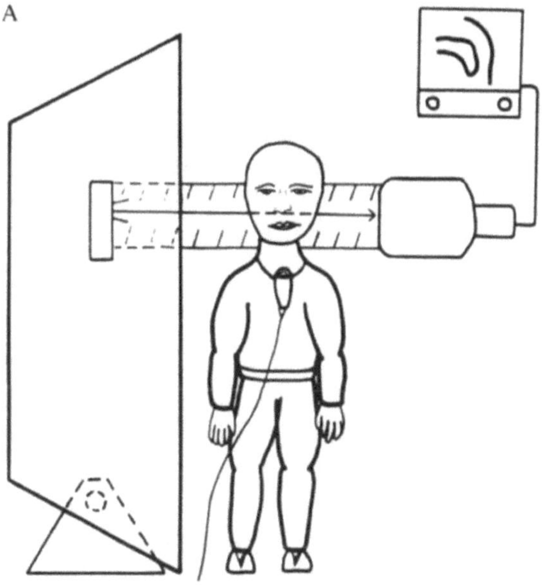

C

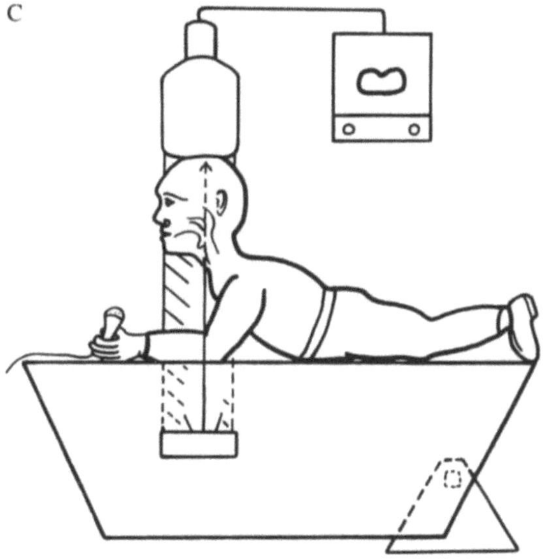

FIGURE 3.2. Patient positioning in fluoroscope for (A) lateral, (B) frontal, and (C) base views.

B

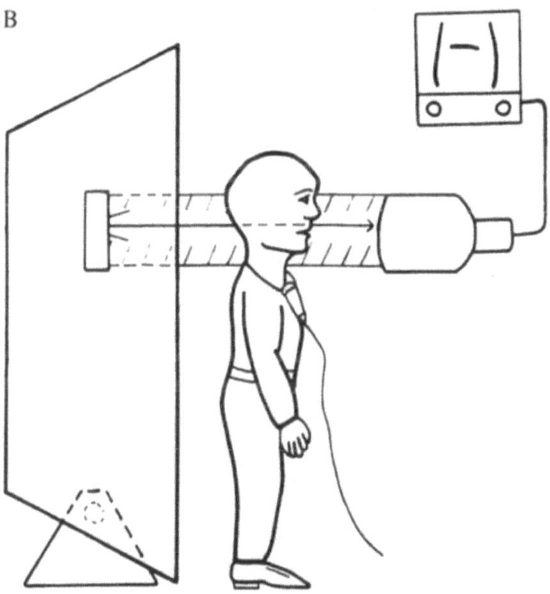

produced x-ray beam is proportional to the velocity of the electron beam that hits the target (anode), which is proportional to the voltage accelerating the electrons into the anode. Therefore the penetrating power of an x-ray beam is rated in terms of KV (1).

The thicker the portion of the body being examined, the greater the penetrating power of the x-ray beam required to go through that region in order for some of the x-rays to reach the fluoroscopic screen. In the frontal and base projections, there is a greater thickness of tissue in front of and behind the nasopharynx than in the lateral view. Much of this tissue is composed of bone, which absorbs and stops x-rays to a greater degree than the muscles, air, and fat that comprise the tissues in the region of the VP portal on the lateral view. Thus, for the frontal and base views a higher degree of x-ray beam penetration or kilovoltage is required than for the lateral view.

To produce an x-ray image, two requirements must be met. The energy (KV) of the x-ray beam must be high enough to penetrate through the subject to reach the film or fluoroscopic screen, and there must be a sufficient quantity of x-ray photons to form an image. The number of photons is proportional to the filament current (MA). The greater the number of photons (higher the MA) the clearer the image detail but the greater the radiation dosage.

Image Intensified Fluoroscopy

If one would look at a fluoroscopic screen directly, one would see a dim image produced where the x-ray photons hit atoms of fluorescent material that coat the fluoroscopic screen. The fluorescent material is a chemical that produces light when hit by x-rays. The images on the fluoroscopic screen are so dim that they cannot usually be perceived unless the

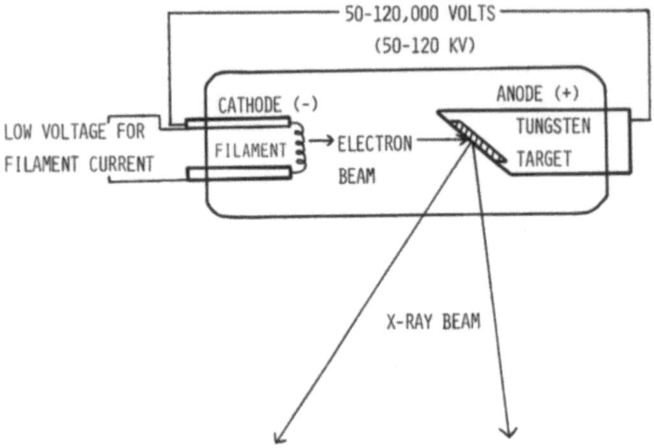

FIGURE 3.3. Diagram of x-ray tube showing production of x-rays. Filament (cathode), heated by low-voltage electric current, releases electrons that are accelerated into target (anode) by high-voltage difference between cathode and anode. Interaction of electron beam with tungsten target produces x-rays.

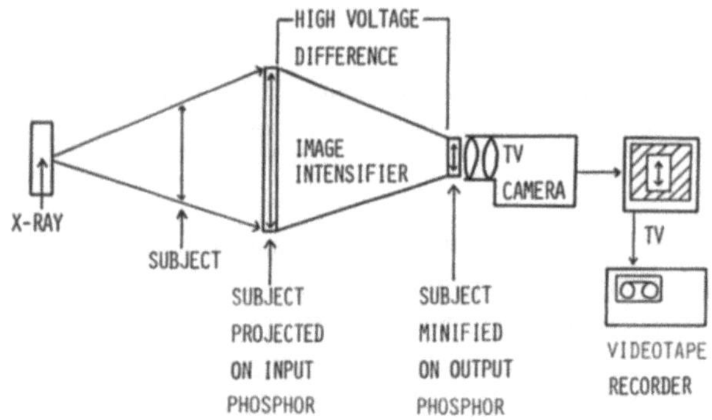

FIGURE 3.4. Operation of image intensifier system. X-ray beam projects image of subject onto input phosphor of image-intensifier tube. Input phosphor absorbs x-rays and emits light that interacts with an adjacent coating (photocathode) to release electrons in proportion to the intensity of recieved light. Electrons are focused and accelerated by high voltage to the output phosphor, which absorbs electrons and emits light. A minified but much more intense image is produced that is viewed by a TV camera and displayed on a TV monitor. The image can be simultaneously recorded on a videotape recorder.

room is darkened and one's eyes have dark adapted by having the pupils dilate to increase the amount of light reaching the retina. In order to view the image in room light without having to dark adapt, an image-intensifier TV system, which electronically intensifies the dim x-ray image, replaces the fluoroscopic screen (2, 3).

The image intensifier is a special vacuum tube that increases the brightness but minifies the size of the fluoroscopic image. A TV camera that is attached to the image intensifier converts the minified image into video format for display on TV monitors and for videotaping (Fig. 3.4).

Collimating the X-Ray Beam

One other component of the fluoroscopic system deserves a brief comment. That component is the shutter for collimating or limiting the field of the x-ray beam. Beam collimation is important for two reasons. First, because x-rays are not innocuous, we wish to limit the x-ray beam to the region of interest within the patient that we wish to examine. We do not want the beam to hit other areas of the face and neck or for that matter to hit areas lower in the body, especially the gonads. The two regions most sensitive to radiation in the face and neck are the lens of the eye and the thyroid gland in the neck. Whenever possible one should avoid radiating these regions.

To control the x-ray beam, these collimating shutters, which are usually plates composed of steel and lead, are moved toward each other to limit the size of the beam going through the opening between the metal plates (Fig. 3.5). In modern fluoroscopic systems there are several series of shutters so that any x-ray beam that penetrates through the first set of shutters is stopped by the second and possibly

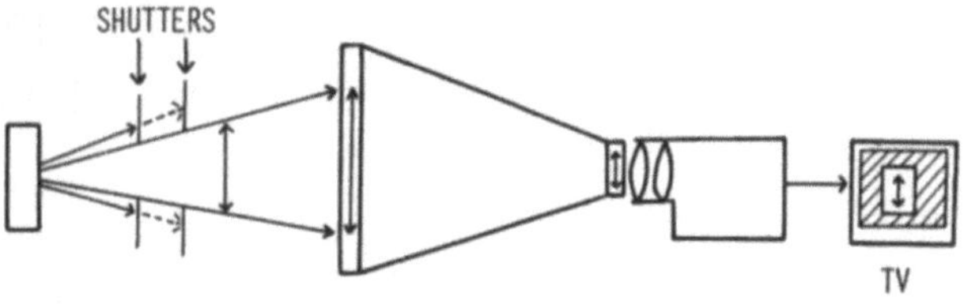

FIGURE 3.5. Collimation of x-ray beam by shutters so as to limit size of beam.

third set of shutters. Therefore, there is almost no x-ray beam going beyond the confines of the area delineated by the shutters, which can be manually set by the operator to precisely demarcate the area of interest.

Another benefit of reducing the x-ray beam is that one reduces the scatter radiation within the region being x-rayed. Scatter radiation occurs when an x-ray beam does not pass in a straight line through the subject to the fluoroscopic screen but rather enters in the subject at one angle and is deflected within the tissue to exit at a slightly different angle. This deflected or scattered x-ray beam does not provide any diagnostic information but only adds additional light to the fluoroscopic screen at the point of its interaction with the phosphor. This additional light from the scattered x-ray beam reduces the overall contrast of the image and degrades the image quality (4).

Besides reducing scatter and preventing tissues beyond the region of interest from being irradiated, a tightly collimated x-ray beam can prevent x-rays from striking the fluoroscopic screen when there is no or very little intervening patient tissue and thus can eliminate areas of extreme brightness on the fluoroscopic screen that can distract the viewer's attention. For example, if the vertical shutters are too far apart when the patient is positioned for the lateral image, then the x-ray beam may go through the thin soft tissues below the mandible or perhaps even go through the air adjacent to the skin underneath the mandible (Fig. 3.6A). Narrowing the distance between the vertical shutters so as to limit the x-ray beam to the area of interest (Fig. 3.6B) eliminates a distracting region of brightness below the mandible on the TV monitor and also improves overall image quality.

A tightly collimated x-ray beam is even more important on the frontal and base views because the beam must go through thicker tissues than for the lateral view, that increases the degrading effects of scatter radiation.

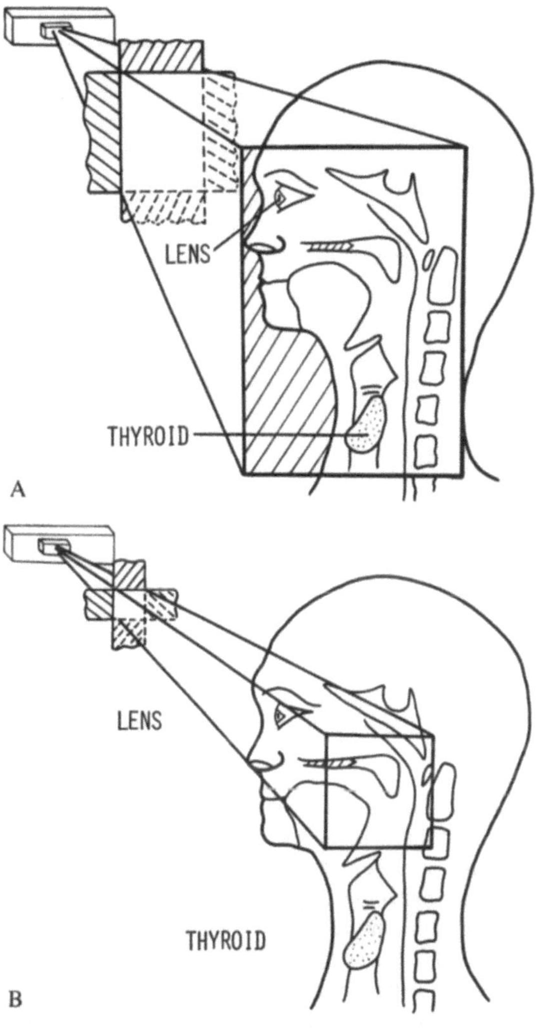

FIGURE 3.6. Use of shutters to limit x-ray beam to region of interest. (A) Poorly collimated beam. Lens, thyroid, and air around face and neck are included in image. Lens and thyroid (the regions most sensitive to radiation) are unnecessarily irradiated. The air included within the x-ray beam produces a very bright region on the image that is distracting to the viewer. (B) Beam collimated tightly to region of velopharyngeal portal eliminates radiation exposure to the eye and thyroid, and improves image contrast by reducing extraneous radiation that adds distracting brightness to the image.

A

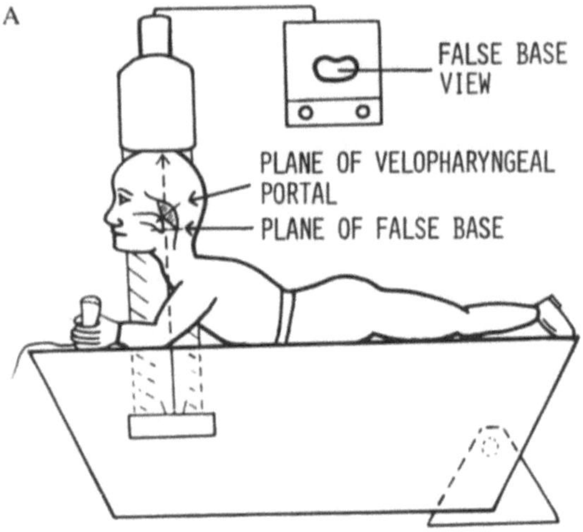

B

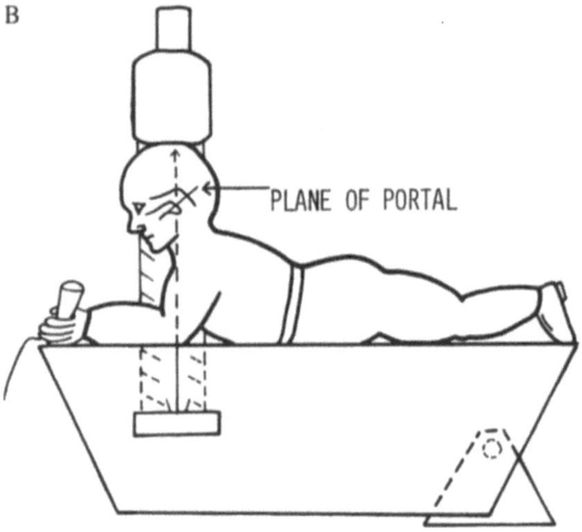

FIGURE 3.7. Limitations of conventional fluoroscopic system for obtaining base view. (A) Patient with enlarged adenoids. The plane of the velopharyngeal portal is shifted from a horizontal to a diagonal position by the enlarged adenoids so that the vertically oriented x-ray beam is not perpendicular to the portal and thus cannot produce an en face image of the portal. Base view displayed on TV monitor is a false base representing the en face projection of the nasopharynx below the level of the adenoid mass. (B) Large adult subject. Because of distance limitations between the image intensifier unit and x-ray table, the subject cannot elevate his head enough so as to bring the plane of velopharyngeal portal perpendicular to x-ray beam.

Limitations of the Conventional Fluoroscopic Tilt Table

There are two serious limitations when a conventional tilting fluoroscopic table is used for performing multiview videofluoroscopy. First, the patient must change his position from erect for the lateral and frontal views to prone for the base view. Thus, the orientation of the head and neck are not the same for the base compared with the lateral and frontal views. For the base view the head is usually more hyperextended than for the lateral or frontal views so that the midsagittal velopharyngeal relationships are slightly different when the patient is in base view, compared with lateral views (see Chapter 4 for details). Second, when performing the base view in certain patients the head cannot be hyperextended enough to bring the plane of the VP portal perpendicular to the x-ray beam so that the portal can be seen en face. Usually this problem either occurs in young children with large adenoids (Fig. 3.7A) or in older subjects, who are either too inflexible, too large, or both, to adequately hyperextend the neck within the limited space between the x-ray table and fluoroscopic carriage (Fig. 3.7B).

The "C" Arm Fluoroscopic System

The limitations of the conventional fluoroscopic table can be overcome by using a "C" arm fluoroscopic system. Such systems may be found in an angiographic suite or the cardiac catheterization laboratory (which may not even be in the main x-ray department). In the "C" arm system the image intensifier and x-ray tube are mounted at opposite ends of a "C"-shaped arm. The arm is suspended from the ceiling and the x-ray beam can be rotated in any plane around a subject without changing the position of the subject. The patient has to be positioned only once because the "C" arm is rotated around him for different views. It is usually most convenient to position the patient prone on the x-ray table, as for the base view (Fig. 3.8). Provided that the subject does not move his head between views, the relationships of the pharyngeal anatomy can be more precisely correlated among the three views than could be done with the conventional method of performing videofluoroscopy that entails moving the patient for different views. For routine use such precise correlations are rarely necessary. However, if one is doing a research project, it may be helpful to be able to fix the nasopharynx in space and rotate the fluoroscopic system around it.

Another advantage of the "C"-arm system is that the distance between the x-ray tube and image intensifier is about 40 inches or approximately twice the distance between the x-ray table and fluoroscopic carriage in a conventional system. Thus, in patients who are large, who have inflexible necks, or whose adenoids are enlarged, a properly positioned base view can be obtained. In the subjects with large adenoids, the plane of the x-ray beam can be adjusted to intersect perpendicularly the plane of the portal even if the portal plane cannot be positioned parallel to the horizontal. To do so, the operator determines the inclination of the plane of the portal from the lateral view and then adjusts the inclination of the x-ray beam for the base view to perpendicularly intersect this plane (Fig. 3.8D).

Videotape Recorders

Videotape recorders are similar in operation to audio-frequency tape recorders with one significant difference. In the audio-frequency tape recorders the tape moves across recording and playback heads but the heads themselves are stationary. In a videotape recorder the recording/playback head rotates at several thousand revolutions per minute as the tape moves at several feet per second across the drum on which the rotating heads are located. The use of the rotating heads for the videotape recorder results in a much greater density of recorded information on the tape, compared with that obtained from the audiotape recorder. Such higher density is required because there is much greater information on each TV image than there is in the voice signals recorded with an audiotape recorder. In addition to recording video information, the videotape recorder also records one or sometimes two audio channels.

It is important for the reader to appreciate the difference in the way audio and video signals are recorded. When a video signal is to be recorded, the rotating drum must first get up to its operating speed before the signal can be accurately recorded on the tape, whereas with the audiotape recorder there is no time delay before the signal can be recorded. One simply starts the tape moving and the signal is recorded. This delay time in videotape recorders can vary, but usually is between 5 and 10 seconds. Therefore, to record a nondistorted TV image, the tape recorder should be started between 5 and 10 seconds before one begins to record the desired image on the tape. The need to remember to start the recorder several seconds before turning on the fluoroscopic screen so as to insure a technically good image can be an annoying chore.

In some videotape recorders one can turn on the rotating drum and bring it up to speed before one begins to advance the tape. This feature is called "pause control." With such a piece of equipment, the tape can be turned on and off with no significant intervening delay before the image can be properly recorded. The head continues to rotate before the tape starts moving and after it stops. However, this arrangement has its deficiencies as well. When the drum containing the head revolves without the tape being moved past it, then the heads keep pressing upon the same segment of tape and begin to wear its surface down. The tape can be physically damaged. The iron oxide coating (the material upon which the image is recorded) can be rubbed off and may clog up the video heads, thereby degrading image quality either during recording or playback.

Again, in a manner akin to audiotape recorders, videotape recorders come in cassette models. The advantages of cassette models are that the tape does not have to be threaded when first placed on the machine and the spool does not have to be rewound to the beginning before removing the cassette.

The last characteristic of a tape that is important to consider is the interchangeability of the tape with other tape recorders. There are several different-format recording systems that are now available: 3/4 inch U-matic cassettes, 1/2 inch video cassettes using the Betamax® (Sony Corporation) and VHS formats; and 8-mm video cassette recorders. Cassettes from a one-format system cannot be recorded or played on a different-format system, although the images can be electronically copied from a one-format cassette to another format. No attempt will be made to describe the advantages and disadvantages of specific machines because models are changing too rapidly. However, we have observed that videotape recorders using wider width tape (say 3/4 inch cassette versus 1/2 inch cassette) usually produce better-resolution images.

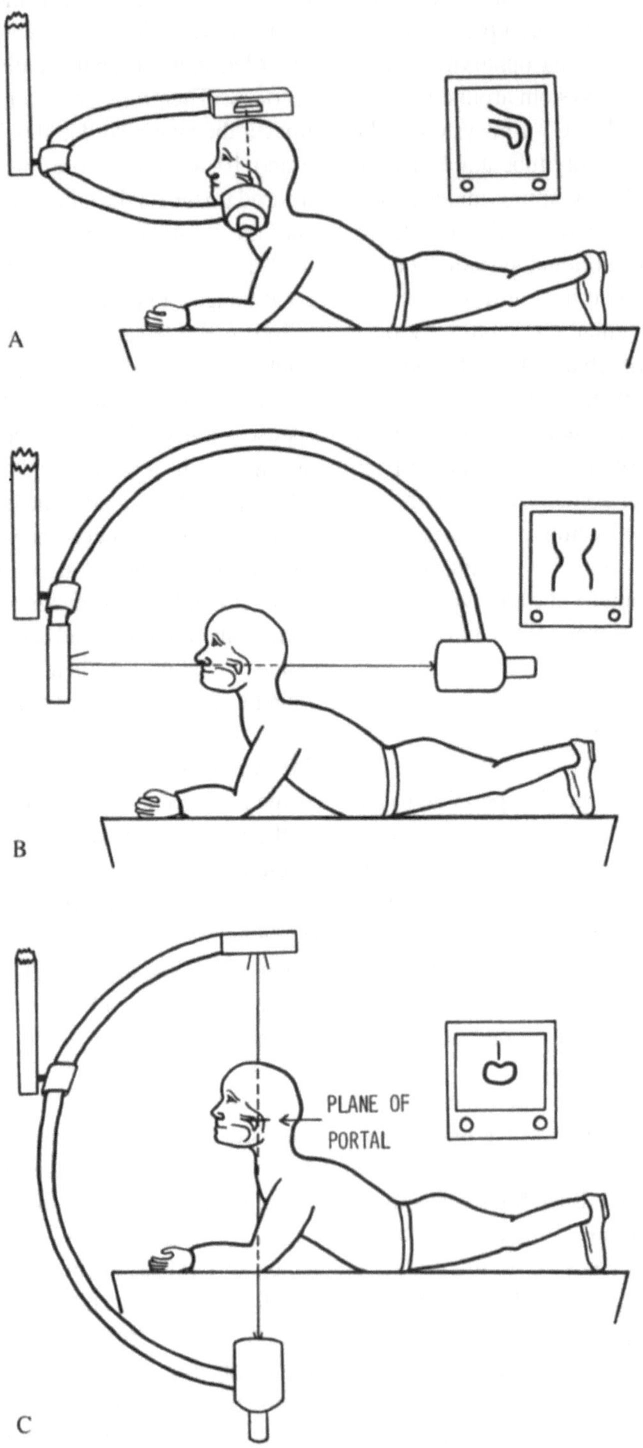

FIGURE 3.8. Use of "C" arm to position patient for (A) lateral, (B) frontal, and (C, D) base views. Patient position remains unchanged while "C" arm rotates around head to produce the three views. To orient "C" arm for base view, plane of velopharyngeal portal is first determined from lateral view (A). In (C) the plane of portal is horizontal so "C" arm is oriented vertically to intersect perpendicularly the plane of the portal. In (D) plane of portal is diagonal because of enlarged adenoids. By rotating "C" arm, x-ray beam perpendicularly intersects plane of portal to produce en face view.

D

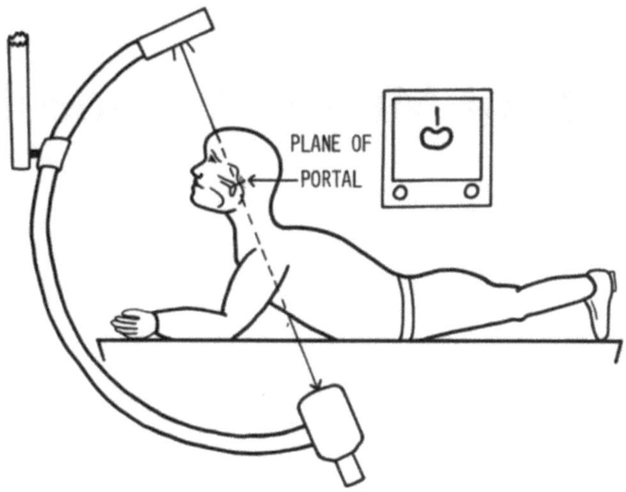

We suggest choosing a popular format video recorder so that hopefully the system will be in production for at least a number of years. Nothing can be more frustrating than the experience of having accumulated a large library of tapes on a particular format and then learning that your video recorder is irreparably broken and a replacement cannot be purchased. Seeking a popular video format is also helpful if you want to exchange tapes with colleagues or easily present video recordings at a scientific meeting.

Finally, a few cautions about the quality and validity of the recording of the patient's speech obtained during the videofluoroscopic examination. In our experience these speech recordings are poor representations of the patient's true speech and should not be used to make judgements about the quality of the patient's speech. Several factors account for the poor-quality speech recordings. There is considerable background noise in the fluoroscopic room from the equipment cooling fans and rotation of the x-ray tube. The quality of the audio recording on the videotape recorder is not high fidelity. There may also be some distortion of the patient's speech from the presence of the nasally introduced barium.

References

1. Christensen EE, Curry TS III, Dowdy JE: Production of x-rays. In: An Introduction to the Physics of Diagnostic Radiology, 2nd Ed. Philadelphia: Lea & Febiger, 1984; p5
2. Christensen EE, Curry TS III, Dowdy JE: X-ray image intensifiers. In: An Introduction to the Physics of Diagnostic Radiology, 2nd Ed. Philadelphia: Lea & Febiger, 1984, p194
3. Christensen EE, Curry TS III, Dowdy JE: Television. In: An Introduction to the Physics of Diagnostic Radiology, 2nd Ed. Philadelphia: Lea & Febiger, 1984, p234
4. Christensen EE, Curry TS III, Dowdy JE: Basic interactions between x-rays and matter. In: An Introduction to the Physics of Diagnostic Radiology, 2nd Ed. Philadelphia: Lea & Febiger, 1984; p49

4
Techniques of Multiview Videofluoroscopy

In Chapter 3 we briefly described the three basic views — the lateral, the frontal, and the base — so that the reader could appreciate the rationale behind the choice of the equipment with which to perform this procedure. This chapter will describe information provided by each of the views, methods of performance, and normal anatomy as seen in each of the views.

Prepare the Patient

It is a good idea to explain thoroughly to the patient before the radiographic examination begins exactly what the procedure will consist of. This explanation first allays the patient's fear, which can be quite overwhelming, especially if he* is young and small and sees fluoroscopic and radiographic equipment towering over him. Second, it gives the patient an idea of what the speech tasks are so that with minimal prompting he can then repeat the speech tasks when requested to, and thus reduce the amount of radiation required for the examination. In fact, it is advisable to practice the speech tasks several times with the patient beforehand. If the patient can read, then he may be given a card listing the phrases or the assistant may hold the card, with the phrases typed in large print, several feet away from the patient. (See Chapter 13 for discussion of patient preparation and Chapter 14 for speech-task selection.)

*Whenever the masculine pronoun is used, the feminine pronoun is understood to be present as well.

Examination Routine

The routine examination of multiview videofluoroscopy consists of the following:

1. A lateral view without the use of intranasal or intraoral contrast material
2. A barium swallow in lateral projection
3. A lateral view after intranasal barium has been introduced
4. A frontal view with intranasal barium
5. A base view with intranasal barium

As discussed in Chapter 14, for each view except the barium swallow, the patient recites a standard set of speech tasks.

Lateral View

Information Obtained

This view demonstrates the anatomy of the velum and posterior pharyngeal wall in the midsagittal plane, and shows movements of these structures during breathing and phonation (Fig. 4.1). One can appreciate the length and configuration of the velum, the location and height of the velar eminence, and the region of contact or closest approximation between the velum and posterior pharyngeal wall during maximum phonatory effort. If the opposing velar and pharyngeal surfaces touch each other, then the vertical length of the region of contact can be measured (Fig. 4.1B). The forcefulness of contact can be suggested by observing the degree of indentation of the posterior pharyngeal wall by the velum (Fig. 4.1C).

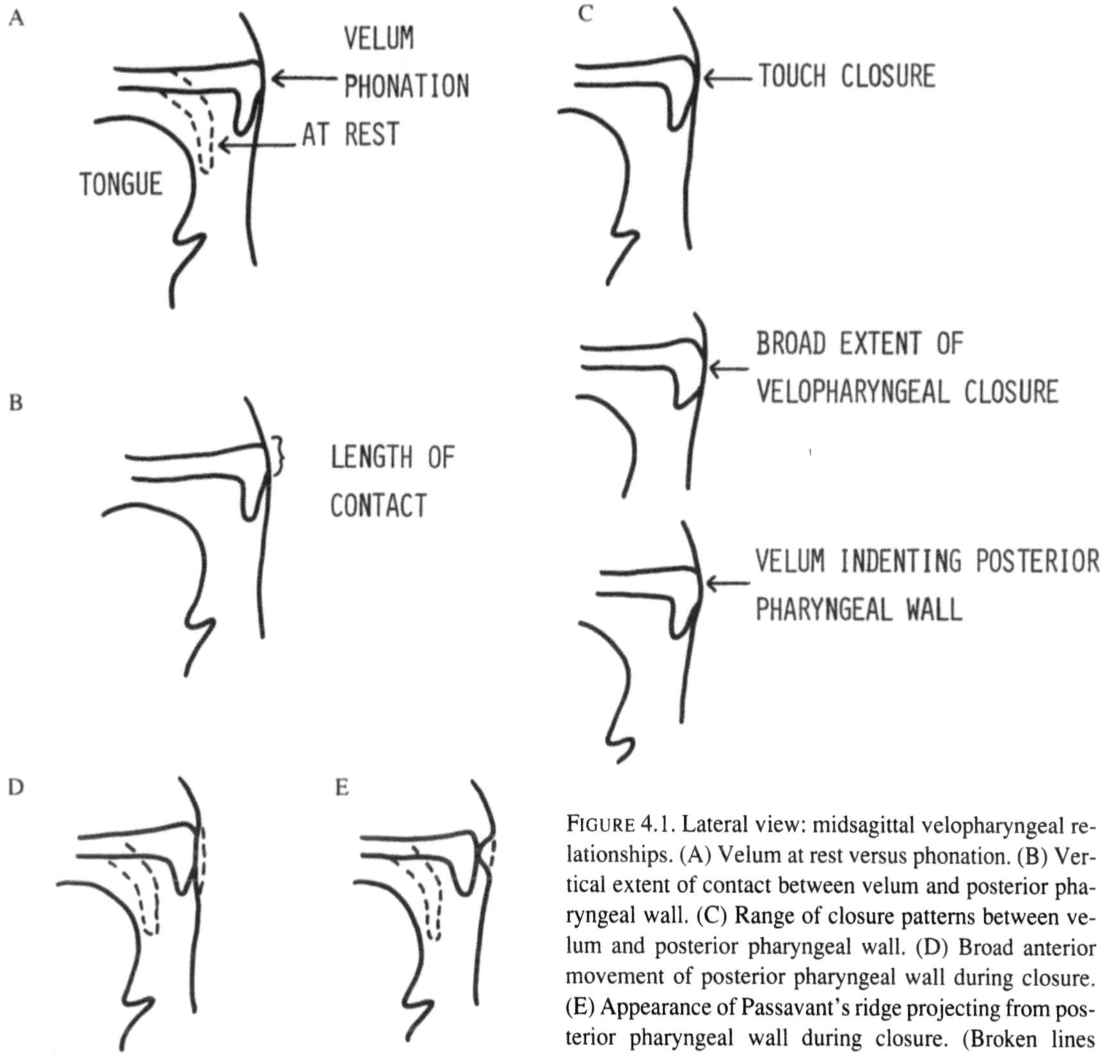

FIGURE 4.1. Lateral view: midsagittal velopharyngeal relationships. (A) Velum at rest versus phonation. (B) Vertical extent of contact between velum and posterior pharyngeal wall. (C) Range of closure patterns between velum and posterior pharyngeal wall. (D) Broad anterior movement of posterior pharyngeal wall during closure. (E) Appearance of Passavant's ridge projecting from posterior pharyngeal wall during closure. (Broken lines indicate structures at rest.)

The posterior pharyngeal wall can be observed for the presence of anterior motion during speech. This motion can be broad (Fig. 4.1D) or localized as when a Passavant's ridge (Fig. 4.1E) is present. The size and level of the ridge in relationship to the velum can be determined (see Chapter 9 for details of Passavant's ridge). If adenoids or tonsils are enlarged, their effects on palatal and tongue motion can be assessed.

One should also evaluate the timing of the movement of the velum, posterior pharyneal wall, and tongue relative to speech so as to assess whether or not the speech mechanism is functioning in a normally coordinated manner. Abnormal speech may also be related to stiffening in the palate resulting from scarring secondary to previous surgery, so one should be alert to observe reduced pliability of the palate during phonation.

If a pharyngeal flap is present (see Chapter 10 for details), one should look for its site of origin on the posterior pharyngeal wall, its region of insertion into the velum, and the midsagittal relationships between the velum and posterior pharyngeal wall during phonation above the level of the flap. Although one usually sees almost no velopharyngeal closure at this level, occasionally there is effective midsagittal closure, which raises the question of what the functional role of the flap is in such a

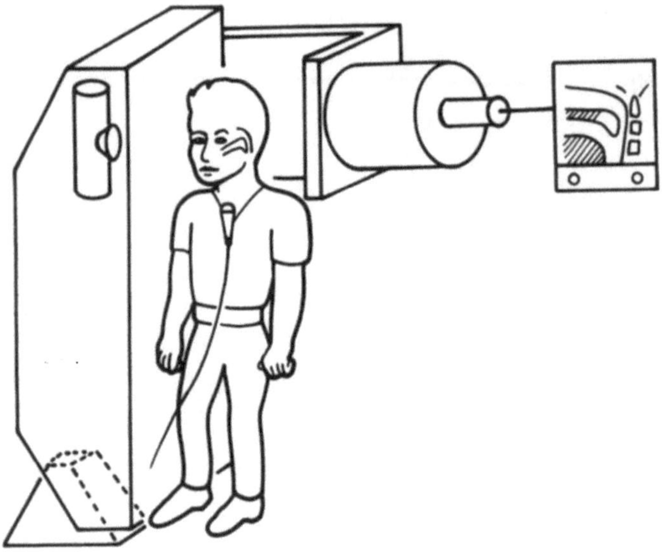

FIGURE 4.2. Patient positioned for lateral view. Plane of hard palate is horizontal. Head must not be tilted or rotated.

A

B

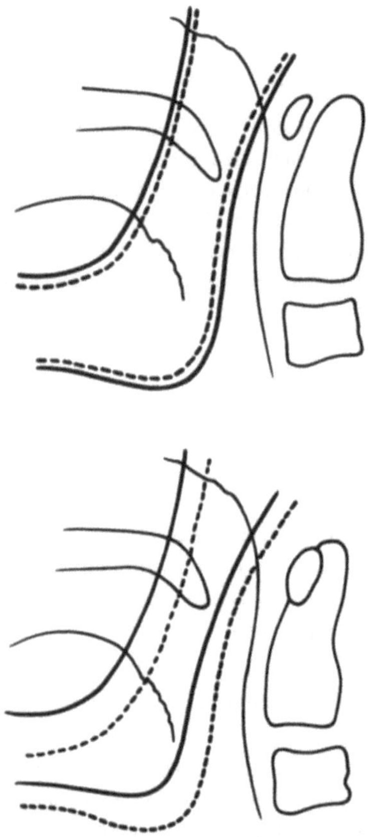

FIGURE 4.3. (A) Head in true lateral view. Both mandibular rami are superimposed. (B) Head rotated and tilted. Mandibular rami do not superimpose.

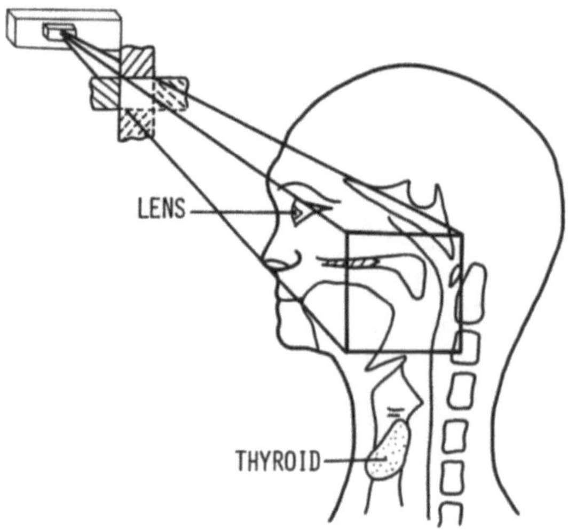

FIGURE 4.4. X-ray beam tightly collimated to region of velopharyngeal portal eliminates radiation exposure to eye and thyroid.

patient. Except for the presence of a pharyngeal flap, all the above are better seen after nasal instillation of barium.

Patient Positioning

The fluoroscopic table is vertically positioned for the lateral examination. The patient stands in the gap between table and the fluoroscopic screen with his right side adjacent to the table. The patient looks straight ahead (Fig. 4.2) with head in neutral position (plane of hard palate parallel to floor). The cervical vertebrae and head are in lateral projection with reference to the x-ray beam. To insure that the head is truly in lateral position (neither tilted nor rotated), the rami of both sides of the mandible must appear to be superimposed (Fig. 4.3). The beam is centered at the level of the nasopharynx and the size of the beam reduced so that it just covers the naso- and oropharynx, the anterior portion of the vertebral column, and the posterior portion of the tongue (Fig. 4.4). If the beam is kept to this area, then the thickness of the tissues through which the beam penetrates is relatively uniform. In addition, the distractions produced by bright areas on the television monitor caused by the x-ray beam going through thin tissues or through air below the chin are avoided. (Refer to Chapter 3 for discussion of beam collimation.) No immobilization device is routinely used to prevent the patient from moving his head while the examination is being performed. Therefore, it is important for the patient to focus on a person or object during the procedure to help him maintain a constant position. A good approach is to have an assistant, wearing a lead apron, stand several feet in front of the patient, and have the patient fix his gaze at one specific level of the assistant's body, at a small toy or bright light (such as a penlight). Besides being useful in fixing the patient's gaze, the assistant can remind the patient of the sequence of standard speech tasks that he must utter during the procedure. Alternately, a mark on the wall can be used as a visual fixation point, thereby eliminating the need for the assistant.

We prefer to have the patient stand rather than sit because children, especially young ones, are more likely to fidget when sitting than when standing.

Noncontrast Lateral View

Although we believe the best information can be obtained after the intranasal instillation of contrast material (described below), we nonetheless obtain a videorecording without contrast material first. Sometimes children between 3 and 7 years of age may become frightened and do not permit the examiner to instill contrast material in the nasopharynx. If such is the case, we can obtain a significant amount of information just from the noncontrast lateral views, especially if the patient shows gross VP incompetence. A second reason for using the noncontrast lateral view first is that a pharyngeal flap may be better appreciated in a noncontrast study than after nasal instillation of contrast (see Chapters 6 and 10).

Barium Swallow in Lateral Projection

The barium swallow is primarily performed (1) to detect fistulas in the hard and soft palate or pharyngeal flap that may produce hypernasal speech by providing an abnormal flow of air from the oral to the nasopharynx; and (2) to demonstrate reflux through the VP portal, which is a secondary sign of VPI. The barium swallow can also be used to evaluate patients with respiratory symptoms associated with swallowing to see if laryngeal aspiration is present.

A fistula through the soft palate or pharyngeal flap may produce hypernasal speech because the fistula provides a path for the flow of air from the oro- to nasopharynx, similar to that provided by an incompetent portal (Fig. 4.5). A fistula may also produce symptoms of nasal regurgitation of fluid.

The reflux of barium through the velopharyngeal portal during swallowing is, in our experience, indicative of velopharyngeal insufficiency (VPI). In patients who cannot tolerate the nasal instillation of barium, and in whom a small velopharyngeal gap is suspected on the noncontrast lateral view, the visualization of nasal regurgitation of barium will confirm the presence of VPI.

In a study of 195 patients (1), we observed that in approximately 50% of the patients with fluoroscopic evidence of VPI, clinical evidence of hypernasal speech, or a combination of both, barium refluxed through the VP portal during a barium swallow. The presence of barium refluxing through the VP portal during swallowing was significant (Fig. 4.6), although the extent of barium reflux did not correlate with the size of the VP gap during phonation (Fig. 4.7). Therefore the width of the velopharyngeal gap during barium reflux cannot be used as an indication of the width of the velopharyngeal gap during phonation. The absence of barium reflux was of no significance because, in some patients with VPI (Fig. 4.8) and in patients with effective closure (Fig. 4.9) during speech, there was no reflux of barium through the portal. We did not see regurgitation of barium through the VP portal in any subject who had neither hypernasal speech

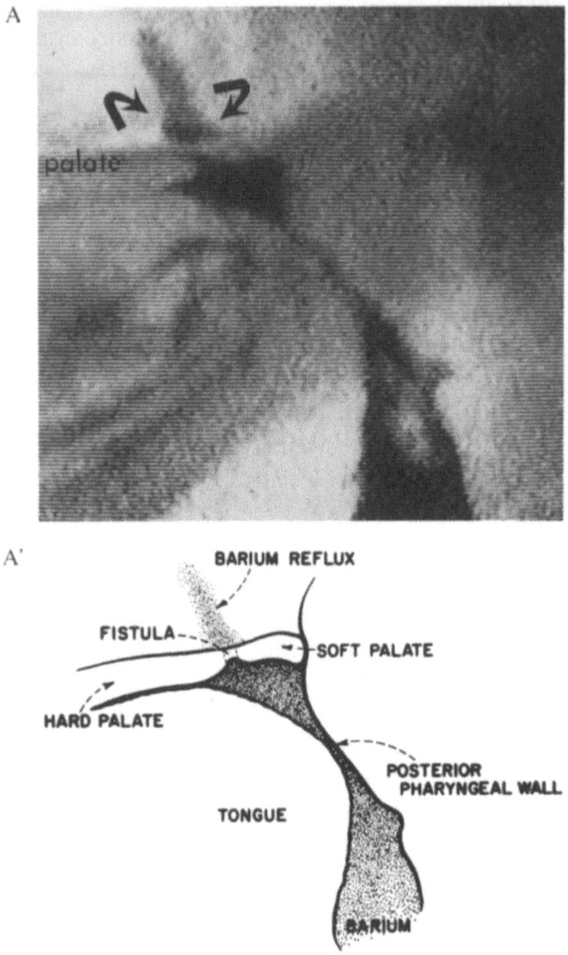

FIGURE 4.5. Barium swallow demonstrating reflux of barium from oro- to nasopharynx through fistula in soft palate. (A) Still frame from videotape (lateral view). (A') Corresponding sketch.

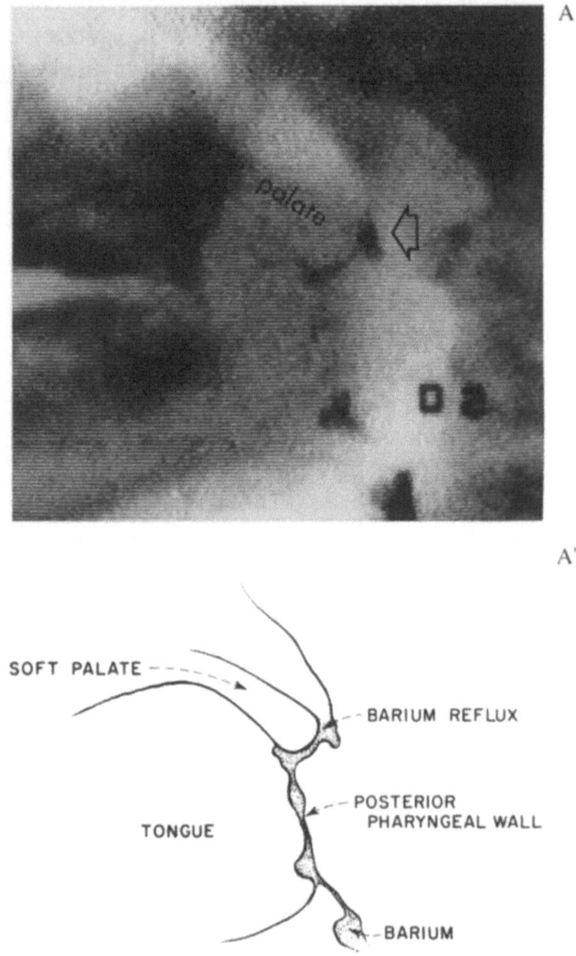

FIGURE 4.6. Barium swallow demonstrating reflux of barium through velopharyngeal portal. Reflux during swallowing predicts velopharyngeal inadequacy during phonation. (A) Still frame from videotape during swallowing (lateral view). (A') Corresponding sketch. (B) Still frame during phonation and (B') corresponding sketch. Barium (B)-filled gap between palate and pharyngeal wall confirms presence of velopharyngeal inadequacy (lateral view after nasal instillation of barium).

nor radiographic or fluoroscopic evidence of VPI (1).

Normally a barium swallow is performed by asking the patient (while still in the same position as the initial lateral speech study) to rapidly swallow several gulps of a barium mixture of a type normally used for the examination of the upper gastrointestinal tract. However, before the barium swallow is performed, the examiner should inquire whether the patient has any difficulties when he drinks or eats. Such difficulties could range from coughing, or perhaps problems in breathing after eating or drinking, to food coming out of the nose during eating or drinking.

If there are respiratory symptoms during eating or drinking, there may be tracheal and laryngeal aspiration of contrast material. Therefore, if such symptoms are elicited, or if the patient's VPI is believed to be caused by neuromotor abnormalities rather than anatomic defects in the palate, the barium swallow exam should be initiated with a single small swallow and the examiner should carefully observe the laryngeal region and upper trachea for aspiration of barium. If none occurs, additional swallows of barium can be cautiously taken while

B

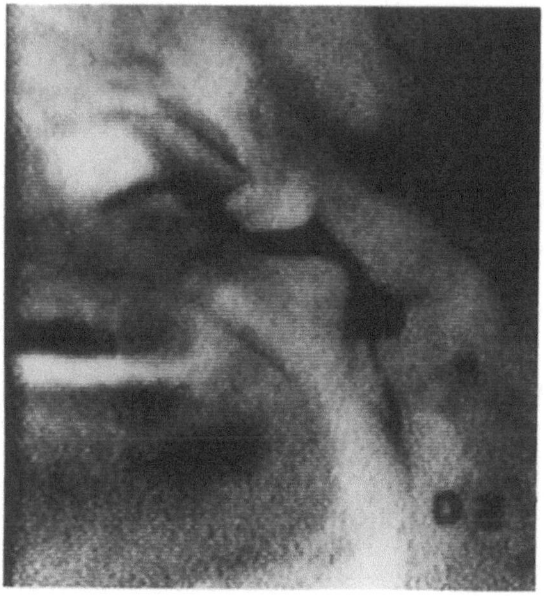

A

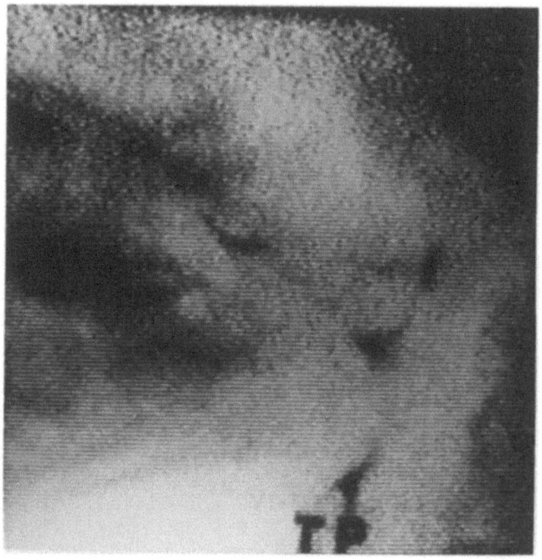

A'

B'

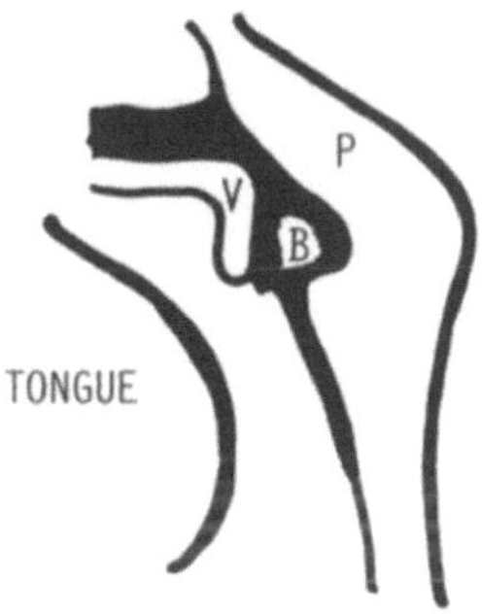

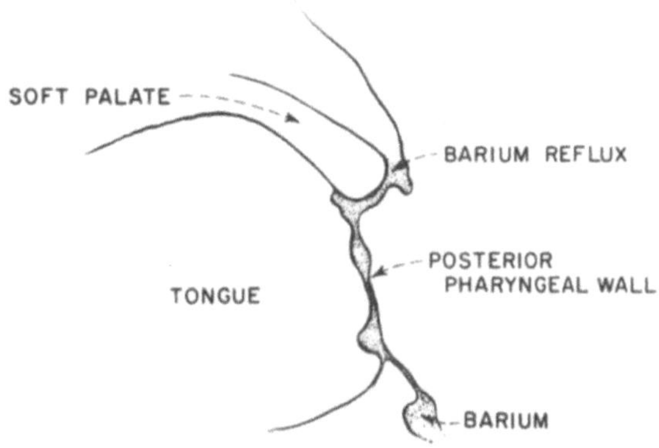

fluoroscoping the pharynx and laryngeal region. The study should be terminated if *any* laryngeal aspiration of contrast is seen. If no tracheal or laryngeal aspiration occurs, then the naso- and oropharynx are examined.

If symptoms of nasal regurgitation are elicited, suggesting a defect in the hard or possibly soft palate, it is important to determine exactly what foods produce it. Sometimes only solids or perhaps even particular types of solids will go through a palatal fistula. For example, we had one particular patient in whom food only regurgitated through the nose when the child ate a particular type of chocolate. The intial swallow of liquid barium showed no abnormalities. Even after barium paste was used, no fistula was demonstrated. Only after the patient chewed a mouth full of the offending chocolate mixed with a tablespoon of barium paste for several minutes did the fistulous tract become visible in the anterior hard palate (Fig. 4.10).

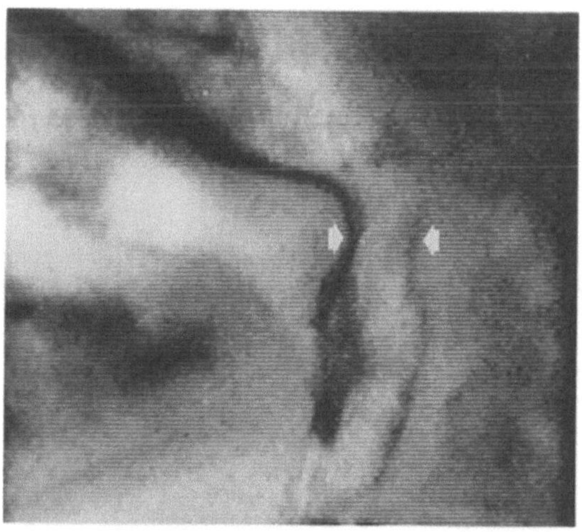

FIGURE 4.7. Lack of correlation between width of velopharyngeal gap during barium reflux (swallowing) and during phonation. (A and A') Barium reflux with minimal gap. (B) During phonation, a much wider VP gap (arrows) is present. (Still frame from videotape.)

A

A'

B

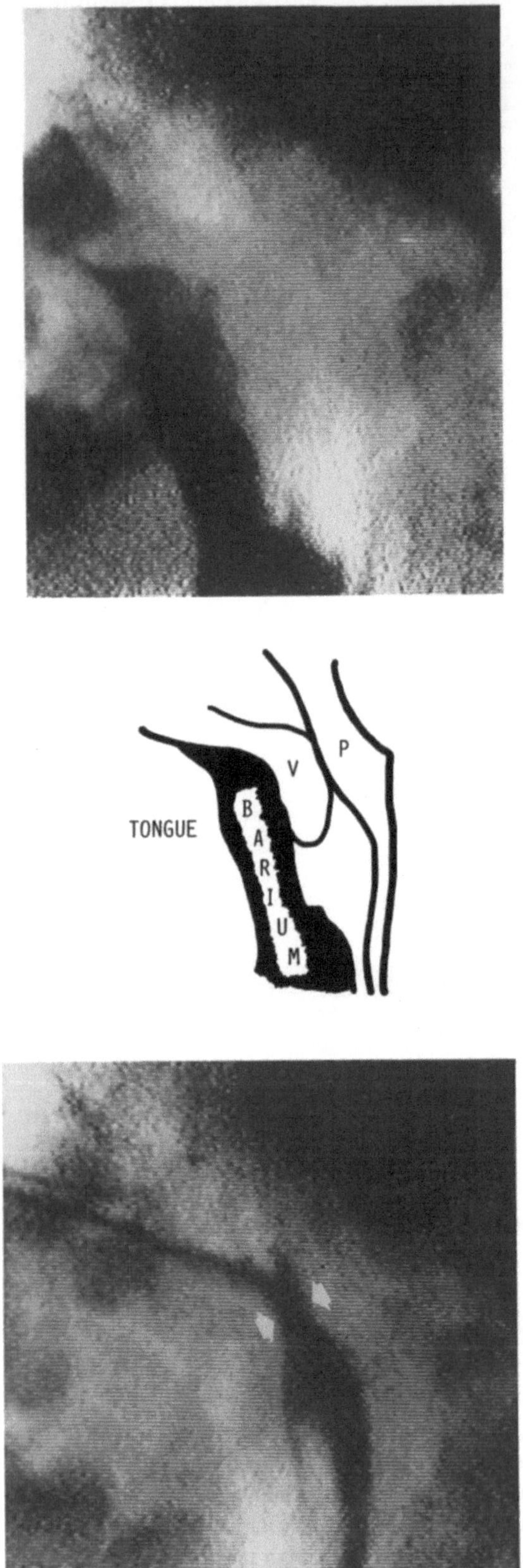

FIGURE 4.8

A

A'

B

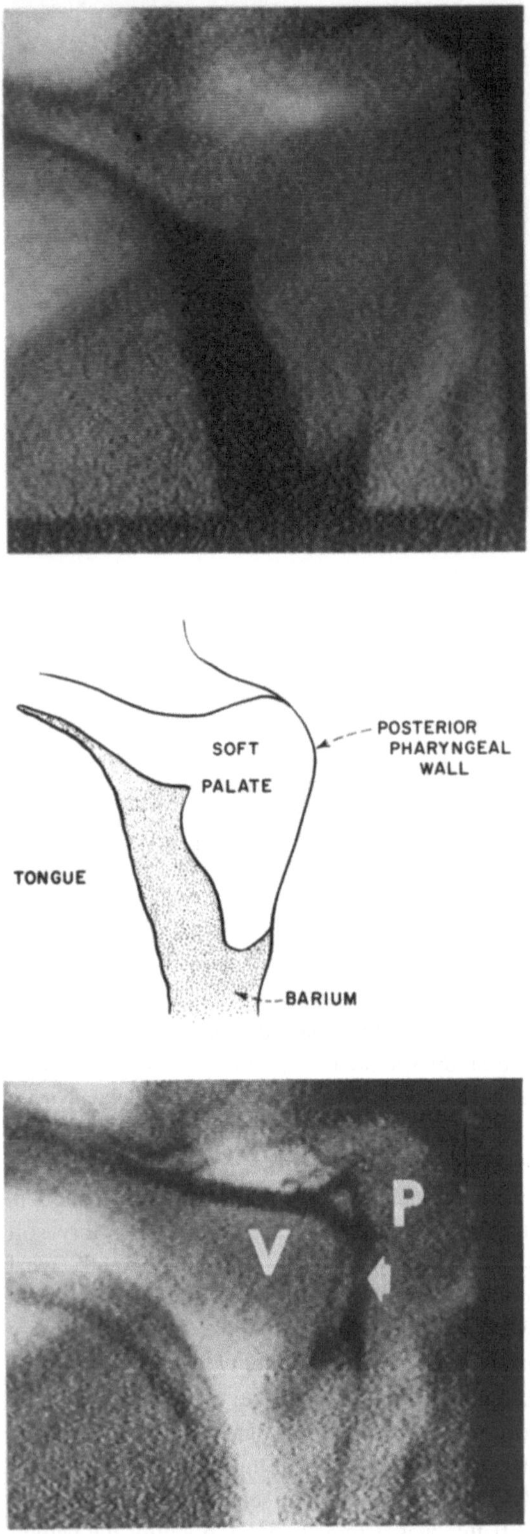

FIGURE 4.9

◁◁ FIGURE 4.8. Normal barium swallow does not exclude velopharyngeal inadequacy during phonation. (A) No reflux through velopharyngeal portal during barium swallow (lateral view). (A') Corresponding sketch. (B) Velopharyngeal gap (arrows) between palate and pharyngeal walls during phonation (lateral view after nasal instillation of barium).

◁ FIGURE 4.9. Normal barium swallow with no velopharyngeal inadequacy during phonation. (A) No reflux of barium through velopharyngeal portal (swallow — lateral view). (A') Corresponding sketch. (B) Same patient, normal velopharyngeal closure during phonation following nasal instillation of barium. The barium (arrow) is squeezed into a fine line (arrow) from contact between velum (*V*) and posterior pharyngeal wall (*P*).

We have found the barium swallow in frontal and base views of no significant clinical use.

Intranasal Contrast Material

In order for the reader to appreciate why intranasal radiopaque contrast material is an improvement over the use of no such material, we will briefly digress to explain why one sees the palate and pharyngeal walls on a lateral x-ray without intranasal contrast material. The normal contrast agent in the nasopharynx is the column of air that is interposed between the palate and pharyngeal walls. Because air is much less absorbant of x-ray photons than the soft tissues of the palate, the pharyngeal walls, or the overlying mandibular rami, the air appears much blacker than the other structures. The greater the thickness of air interposed between the palate and pharyngeal walls, the greater the contrast difference that is appreciated between the air and the surfaces of the palate and pharynx. As the palate approaches or touches the posterior pharyngeal wall during speech, the amount of air interposed between these opposing surfaces is reduced or completely eliminated. Thus it may be difficult to differentiate between actual contact of these opposing surfaces and the existence of a small gap between them because the interposed air may not produce enough contrast differentiation to distinguish between the two situations (Fig. 4.11).

Barium sulfate is an extremely radiopaque material that can be identified radiographically even when seen in thin layers. It is used in a liquid form so as to coat the mucosal surfaces of the palate and pharyngeal walls. Because of the much greater radiopacity of barium, compared with the mucosa, the interface between the two is precisely defined. Even when the palate presses against the posterior pharyngeal wall, if there is a thin coating of barium on the opposing mucosal surfaces, this coating may be visualized and serves to delineate the interface between the two soft-tissue structures (Fig. 4.9B).

Although barium sulfate is quite radiopaque when it coats a mucosal surface, it is best seen when the x-ray beam is tangential to the surface. The barium coating can be likened to a thin piece of paper seen on end. The paper may be several millimeters wide and only 1/10 mm thick. When the x-ray beam goes through it tangentially, it transverses the entire width and, therefore, a considerable amount of the beam is absorbed by the barium. If the x-ray beam hits the barium coating perpendicularly, then the beam only transverses a very thin layer of barium, and the amount of absorption by that very thin layer is minimal and not readily detected fluoroscopically.

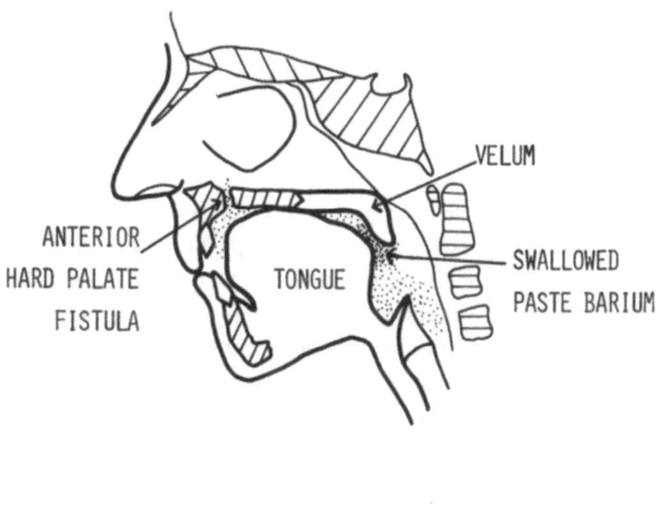

FIGURE 4.10. Patient chewing barium paste mixed with chocolate to demonstrate refluxing through anterior hard-palate fistula.

A

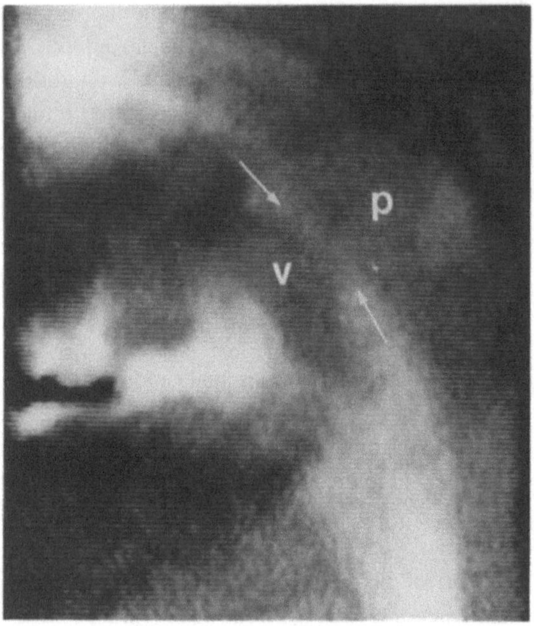

C

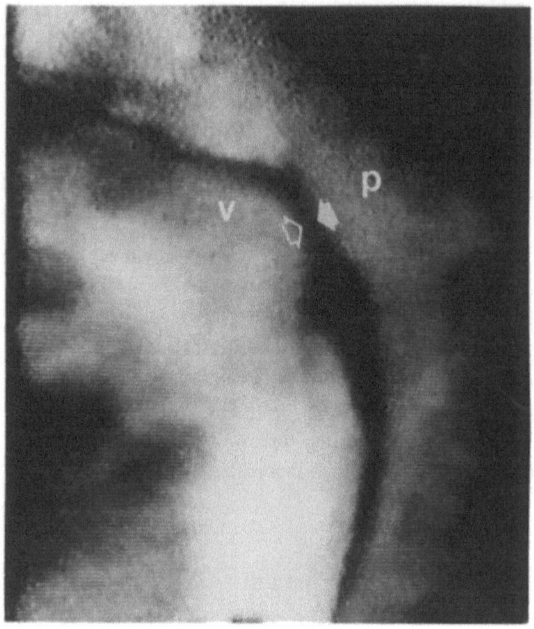

B

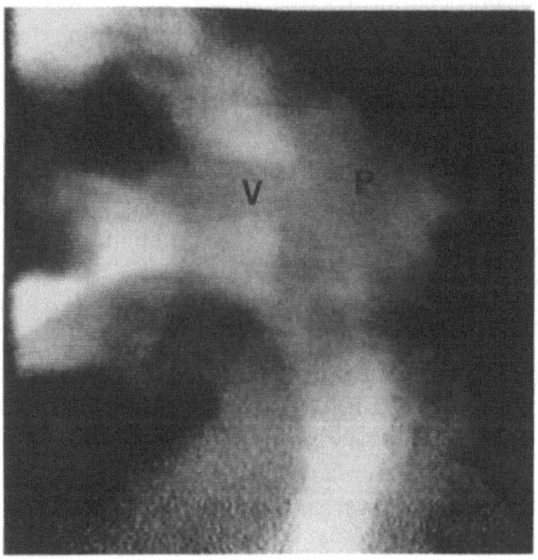

FIGURE 4.11. Air versus barium as contrast agents in the naso- pharynx for determining the presence or absence of velopharyngeal closure (lateral view). (A) Several millimeter velopharyngeal gap (arrows) defined by air contrast. (B) Velum appears to blend with posterior pharyngeal wall. Patient has hypernasal speech but no velopharyngeal gap is seen. (C) Same patient as in (B). Nasal barium demonstrates small velopharyngeal gap (arrows) not appreciated when only intranasal air was the contrast agent.

Therefore, in the lateral projection one can readily appreciate the posterior surface of the palate and anterior surface of the posterior pharyngeal walls because these two structures are tangentially viewed by the x-ray beam, but one does not see the lateral pharyngeal walls because the x-ray beam transverses perpendicular to these surfaces (Fig. 4.12).

Technique of Nasal Instillation of Barium

The barium used is colloidal barium sulfate that comes either as a powder that is mixed with water or as a premixed liquid which is supplied by a number of vendors. In either case, colloidal barium sulfate is used because the barium does not precipitate out of the liquid and therefore provides a uniform coating over the mucosal surfaces.

Whether we start with a powder and add water, or a premixed liquid, we prefer the consistency of the final material to be that of heavy sweet cream. However, it may be necessary to experiment with various viscosities of the final material, depending upon the brand that one uses, as different brands may provide greater or lesser degrees of coating for a given viscosity.

The barium is instilled in the nasopharynx by dripping the liquid through the external nares using

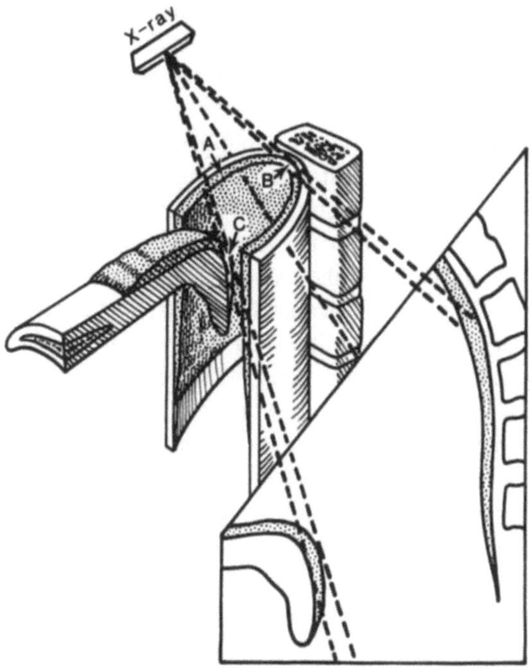

FIGURE 4.12. Physical basis for visualizing barium coating on palate and pharyngeal walls. When the x-ray beam traverses the barium coating tangentially (over the posterior surface of the palate, *C*, and posterior pharyngeal wall, *B*), the x-ray beam goes through a relatively thick amount of barium. This thickness of barium absorbs enough of the x-rays so that the barium coating appears as a distinct black stripe on the fluoroscopic image. When the x-ray beam intersects the barium coating perpendicularly as when it goes through the barium coating on the lateral pharyngeal wall *(A)*, the barium layer that the x-ray goes through is quite thin. This amount of barium does not absorb a significantly greater amount of x-rays so as to change the displayed density of the beam.

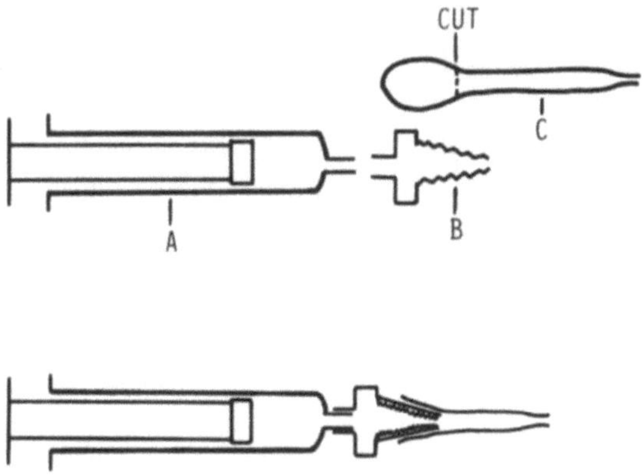

FIGURE 4.13. Device used for nasal instillation of barium. *A* Five- to 10-ml syringe. *B* Plastic adapter for connecting catheter to syringe. *C* One-piece plastic nose dropper. Bulb end cut off and remainder of dropper attached to syringe via adapter.

a modified syringe that serves as a large nose dropper. A 5- or 10-ml syringe is fitted with a short, soft plastic or rubber tip made from a disposable nose dropper (Fig. 4.13), and 1-3 ml of material is instilled into each nostril. Before the material is introduced, the patient blows both sides of his nose to expel any retained mucus that may prevent the barium from flowing freely. The patient stands or sits with his head hyperextended as the material is introduced (Fig. 4.14). As the barium is instilled, the patient is encouraged to sniff and breathe in through his nose to hasten the posterior flow of the barium. He then keeps his head in this hyperex-

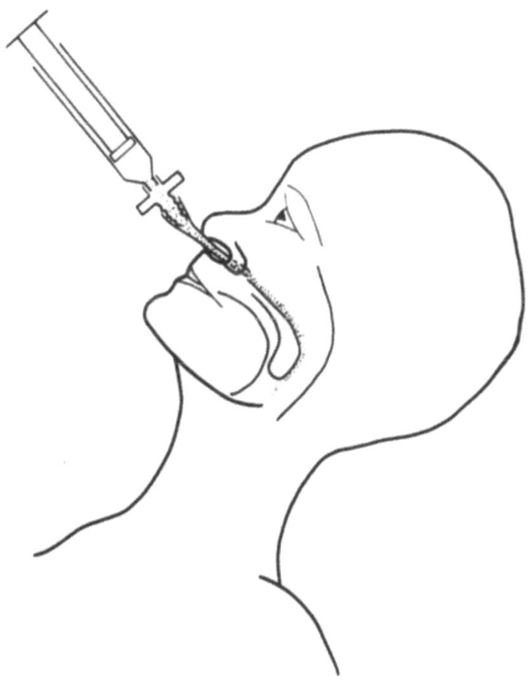

FIGURE 4.14. Procedure for instilling barium into nasopharynx using syringe-nose dropper assembly. Head hyperextended; tip of nose dropper inserted about 1-2 cm beyond external nares into nasal cavity.

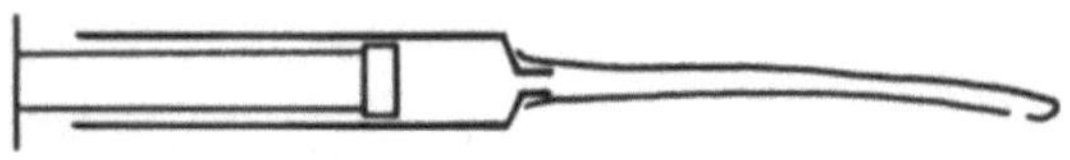

FIGURE 4.15. Alternate device for nasal instillation of barium. Five- to 10-ml syringe (nonleur lock) with #8 to #10 single-side hole catheter.

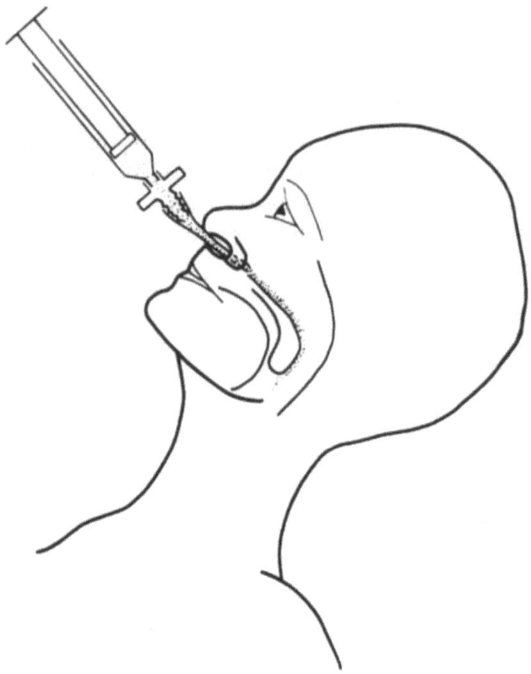

FIGURE 4.16. Procedures for instilling barium into nasopharynx using syringe-catheter assmebly. Catheter inserted about 6 to 8 cm into nostril so as to position end of catheter over soft palate close to velopharyngeal portal.

tended position as he tilts it from side to side for 1 to 2 minutes to allow the barium to coat the velum and nasopharyngeal walls.

In order to gain the patient's cooperation, especially if he is young, we have found it helpful to allow the patient first to taste the barium mixture and to feel the tip of the catheter on the syringe so as to assure himself that it is soft and not painful to touch.

When we initially began our investigations, we introduced the barium into the nasopharynx in a different manner. Instead of dripping it into the external nares, we placed it directly into the nasopharynx by attaching a #8 or #10 French, single-side hole, soft rubber or plastic catheter to the syringe (Fig.4.15). This catheter was introduced into each side of the nose without anesthesia until the catheter hit the posterior pharyngeal wall; it was then withdrawn slightly, and contrast material instilled directly into the nasopharynx (Fig. 4.16). The advantage to the catheter approach is that it directly places the contrast material in the nasopharynx area and eliminates the wait for the contrast material to drip along the floor of the nose. It also prevents any contrast material from remaining along the floor of the nose to obscure the pharyngeal walls on the frontal projection. The disadvantage of using the catheter approach is that it is more uncomfortable and initially quite frightening to the patient when he is told that this long rubber catheter must be inserted into his nose. If the patient cooperates initially, and the coating disappears before the study is completed, the patient is probably less likely to permit a recoating. However, the catheter can be left in place and taped to the nose after the syringe is removed so that if recoating is necessary, the catheter does not have to be disturbed. The presence of the catheter in the nose does not affect the speech and is usually quite readily tolerated by the patient.

As a result of our experience with the two techniques, we prefer to use the dropper technique. The minor inconvenience of barium remaining in the floor of the nasopharynx during the frontal projection can be minimized by having the patient slightly elevate his head, which projects this barium above the level of the VP portal.

Because the barium coating on the palate and pharyngeal mucosal surfaces lasts only for several minutes, it is important that once the barium is instilled nasally, the remainder of the examination be performed rapidly. Because the greatest amount of overlying bones and soft tissues are on the frontal projection, we have found that the pharyngeal walls need the best coating of barium to be identified in this view. Therefore, we prefer to begin with the frontal view. However, for people just getting familiar with multiview videofluoroscopy, it is probably better for them to start in the sequence of lateral, frontal, and then base views, even if it means introducing additional barium between the lateral and frontal views to maintain good coating of the pharyngeal walls. Beginning with the lateral view after the introduction of barium allows the inexperienced examiner to get used to the effects of barium by comparing it with the more familiar nonbarium lateral view.

Cautions in Use of Nasally Instilled Barium

A word of caution is in order concerning the patient's position during the nasal instillation of barium. We keep the patient erect with head hyperextended because in this position gravity insures that the barium will flow along the floor of the nasal cavity and into the nasopharynx. If, on the other hand, the patient is positioned supine and barium is dripped into the nasal cavity, there is a chance that some may flow into the paranasal sinuses. If the method of catheter insertion is used, such a chance filling of the nasal sinuses is less likely. When we were initially trying a variety of positions to determine which was the best for inserting barium into the nasopharynx, we occasionally saw some barium go into the maxillary or ethmoid sinus when the barium was inserted with the patient supine. This barium usually disappeared from the sinus before the patient's examination was completed. Even when it persisted after the exam was terminated, we were aware of no ill effects produced by its presence. Presumably, it was spontaneously cleared within a short time after the patient left our department. However, it is conceivable that if barium persists in a paranasal sinus it could harden in that location and form a foreign body. Therefore, we recommend that the supine position not be used to instill barium into the nose.

Lateral View Following Nasal Instillation of Barium

The patient is positioned in the same manner as he was for the lateral view without the nasal instillation of barium (the plane of the hard palate is approximately parallel to that of the floor). The same speech sample is used. As in the lateral view without barium, no attempt is made to restrain the head mechanically for the post-barium lateral or any other post-barium views. (The pros and cons of head immobilization are discussed later in this chapter.)

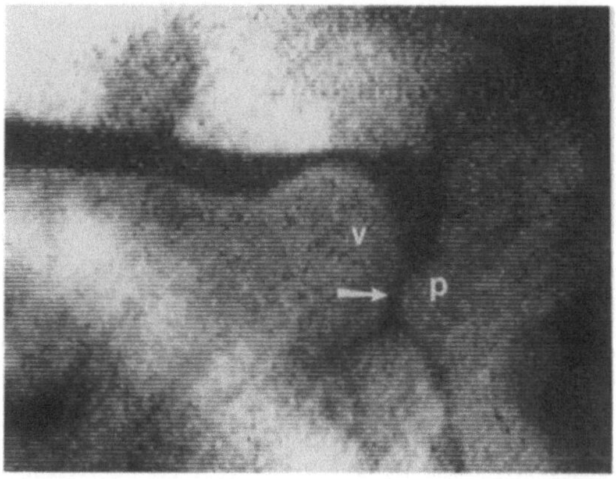

FIGURE 4.17. Appearance of Passavant's ridge during phonation of plosive sound. Interface between velum (*V*) and Passavant's ridge (*P*) delineated by barium coating (arrow). In the absence of barium, the velum and ridge would blend together and the ridge would not have been appreciated.

The barium coating on the velum and posterior pharyngeal wall precisely delineates these opposing surfaces so as to differentiate true contact from close approximation of these surfaces. The barium coating is also very useful in identifying the presence of a Passavant's ridge (Fig. 4.17), which often cannot be seen when just air is used as the intranasal contrast agent. The ridge is most prominent during plosive or sibilant sounds when VP narrowing is maximum (in a subject with VPI) or when VP closure is most forceful (in a competent subject). Under these conditions, there is almost no air in the nasopharynx to delineate the margins of the ridge and distinguish it from surrounding tissues (2). (See Chapter 9 for details.)

If there is a minimal gap between the palate and pharyngeal wall or if these mucosal surfaces just appear to gently touch, the lateral view should be repeated with the patient's head in marked hyperextension. Hyperextension often unearths VP gaps not easily seen with the head in neutral position. In hyperextension the distance between the palate and posterior pharyngeal wall is increased so that a closure mechanism, which is borderline with the head in neutral position, now becomes obviously incompetent (Fig. 4.18) (3). Hyperextension can also exaggerate the presence of a Passavant's ridge

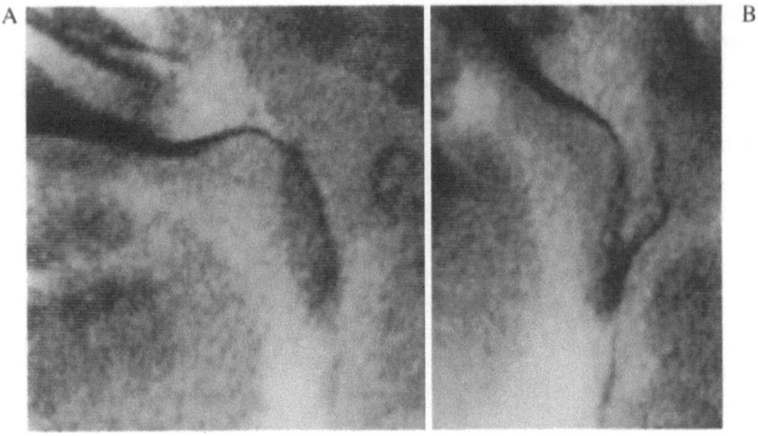

FIGURE 4.18. Lateral view: neutral position versus hyperextension. (A) Neutral position: velum touches posterior pharyngeal wall. (B) Hyperextension position: gap between velum and posterior pharyngeal wall appears.

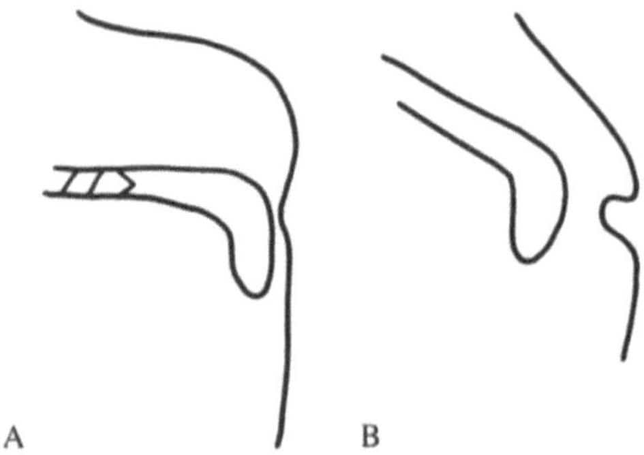

FIGURE 4.19. Patients with Passavant's ridge (*P*). Ridge is minimally (A) or not seen in neutral position. Ridge becomes very prominent when head is in hyperextension (B).

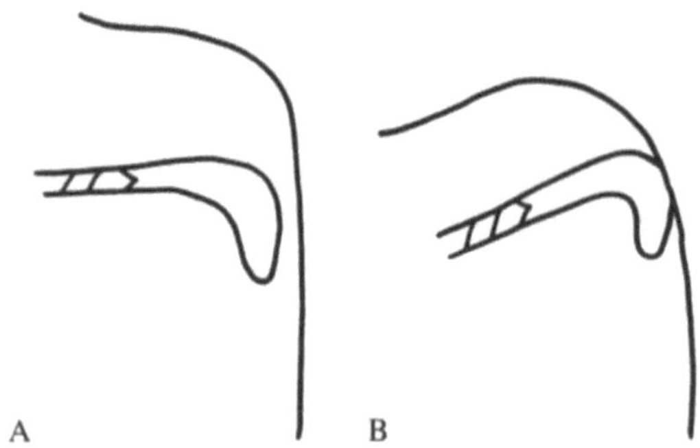

FIGURE 4.20. Decrease in velopharyngeal gap as head moves from (A) neutral position to (B) flexion.

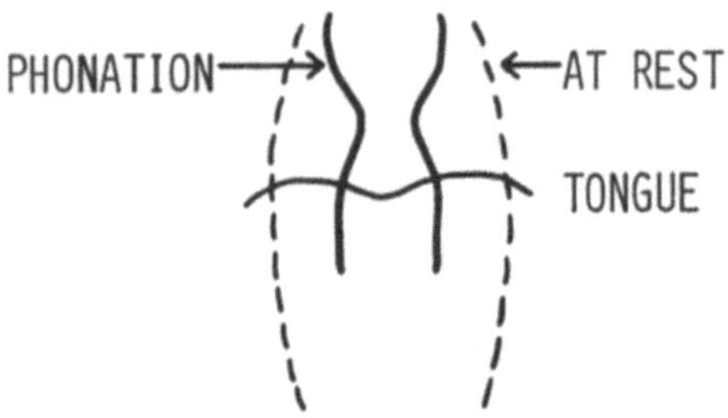

FIGURE 4.21. Frontal view comparing positions of lateral pharyngeal walls at rest and during phonation.

that may only be minimally seen with the head in neutral position (Fig. 4.19) (2). If, however, the head were flexed, a portal that appears minimally incompetent in neutral position may appear competent in flexion (Fig. 4.20) because the distance between the palate and the posterior pharyngeal wall is reduced in flexion. Therefore, lateral views with the head flexed should be avoided.

Frontal View

Information Obtained

This view is best for delineating the movements of the lateral pharyngeal walls during phonation. Besides observing the horizontal direction of the maximum pharyngeal-wall movement during speech (which is usually in the medial direction although occasionally it can be paradoxical and move laterally), one observes the vertical position and length

Figure 4.22. Patient positioning for frontal videofluoroscopic view. To prevent head movement, patient fixes his gaze at tape attached to front of image intensifier.

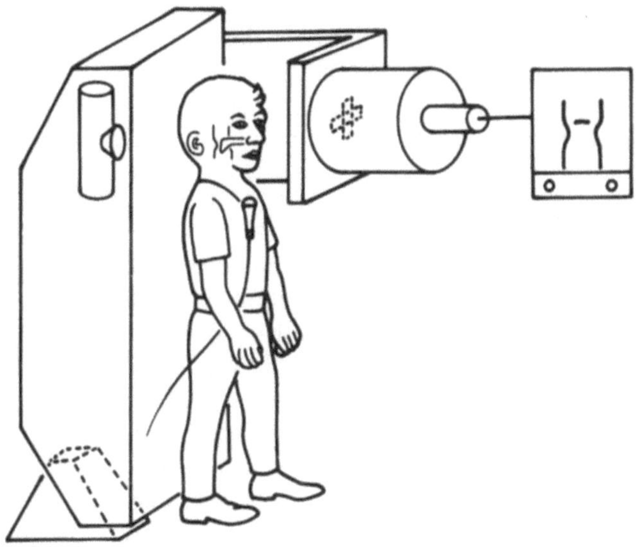

of the segment of moving walls (Fig. 4.21) and the symmetry of the lateral pharyngeal wall movement.

This view also has another purpose — to relate the level of the velum or pharyngeal flap to the region of maximum medial movement of the lateral pharyngeal walls.

Patient Positioning

For the frontal view the patient is rotated 90° from his position in the lateral view. The patient is now looking at the back of the fluoroscopic carriage (Fig. 4.22). To prevent his head from moving, it is advisable to place a small piece of brightly colored tape in the center of the image intensifier at approximately eye level. The patient is instructed to gaze upon that piece of tape and by so doing will not move his head from side to side. If the barium in the floor of the nasal cavity obscures the nasopharynx, the patient is instructed to slightly elevate his head just enough to see the pharyngeal walls moving below the nasal collection of barium.

When the head is in a neutral position or just minimally hyperextended, the plane of the velar eminence or pharyngeal flap is approximately tangential to the x-ray beam and the upper barium-coated surface of the eminence is usually appreciated as an arc between the lateral pharyngeal walls (Fig. 4.23A). If the head is hyperextended, the x-ray beam intersects the velum in a more perpen-

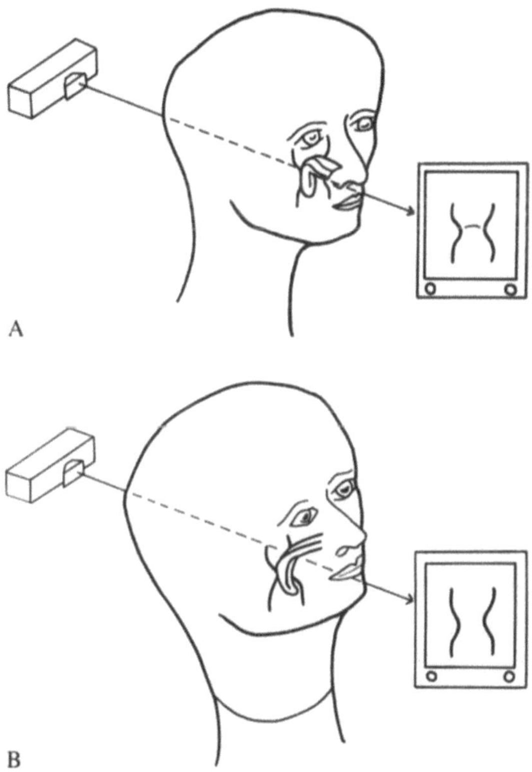

FIGURE 4.23. Frontal view during phonation. (A) Head in neutral position. Superior surface of velum is tangential to x-ray beam and projected as barium-coated arc between lateral pharyngeal walls. (B) Head in hyperextension. Superior surface of velum is almost perpendicular to x-ray beam. Barium-coated velar surface is not detected.

dicular manner so that the barium coating absorbs little radiation. As a result, these structures are not fluoroscopically visible (Fig. 4.23B).

As in the lateral view, we repeat the same speech sample for the frontal view.

Problems with Barium Coating

Unlike the lateral view, essentially no information can be obtained about the movements of the lateral pharyngeal walls and velum from the frontal view unless these surfaces are coated with a radiopaque material. When no radiopaque coating is used, the soft tissue-air interface between the lateral walls and the nasal air column cannot be appreciated. There are too many osseous structures lying anterior and posterior to the plane of the nasopharynx. These are superimposed on the shadows of the nasopharynx and obscure the difference in radiopaque contrast produced by the air-soft tissue interface between the pharyngeal walls and nasopharyngeal air column. If the pharyngeal walls are not well delineated on the frontal view, then the examiner can introduce another 1-3 ml of barium in each nasal orifice to obtain a better quality of coating. After this additional barium is introduced, it may take several minutes for the material to drip down to the pharyngeal walls to coat them effectively. Therefore, the operator should not be disconcerted if, when he fluoroscopes shortly after reinforcing the nasal barium, he does not see well-coated pharyngeal walls. He should wait for a few minutes and then fluoroscope again (Fig.4.24). Having the patient swallow several times will also help transfer some of the barium from the palate to the pharyngeal walls.

If, however, after a second application of nasal barium the pharyngeal walls are still not well coated, it may mean that the patient has an excessive amount of secretions in the nasopharynx, which prevents adherence of the barium to the pharyngeal walls. In such a situation, better visualization of the pharyngeal walls may be achieved if the fluoroscopic table is placed horizontally and the patient positioned supine on it. Then the barium in the nasal cavity will not flow as rapidly into the oropharynx because the effect of gravity has been removed. However, as previously noted, the barium is best instilled in the nose with the patient erect to minimize the passage of barium into the paranasal sinuses.

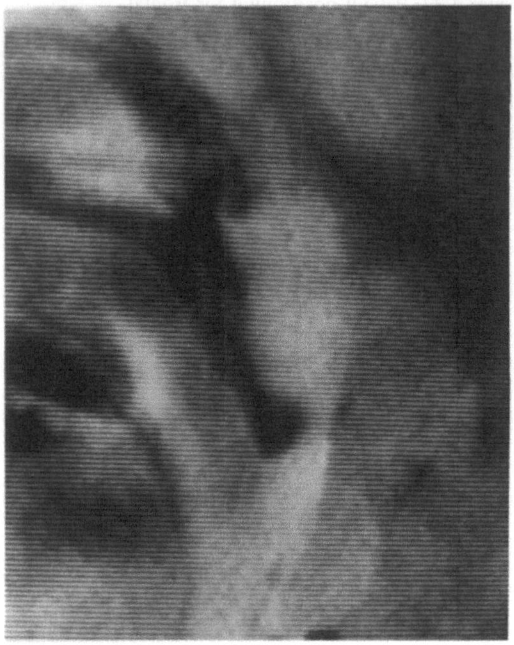

FIGURE 4.24. Lateral view. Barium coating descending over velar surface, but barium has not yet coated pharyngeal walls.

Defining Lateral Pharyngeal Walls with Barium

Even though the pharyngeal walls appear on frontal fluoroscopic images as approximately parallel structures moving toward and away from each other, the reader should remember that the nasopharynx is really a cylindrical structure with the anterior portion of the cylinder removed so as to accommodate the velum. While the barium coats all of the margins of this cylinder, the margins that are seen on any x-ray projection are those margins that are tangential to the x-ray beam. On the frontal view, the lateral pharyngeal walls at the greatest coronal diameter of the pharyngeal cylinder are tangential to the x-ray beam.

However, although it is usual that the maximum amount of pharyngeal wall movement occurs within this region of the lateral pharyngeal walls (Fig. 4.25A), this situation may not always be true. In some patients the greatest degree of lateral pharyngeal wall movement may occur in the plane that is posterior to the greatest coronal diameter of the nasopharynx. This region of lateral pharyngeal wall movement may not be appreciated on the erect

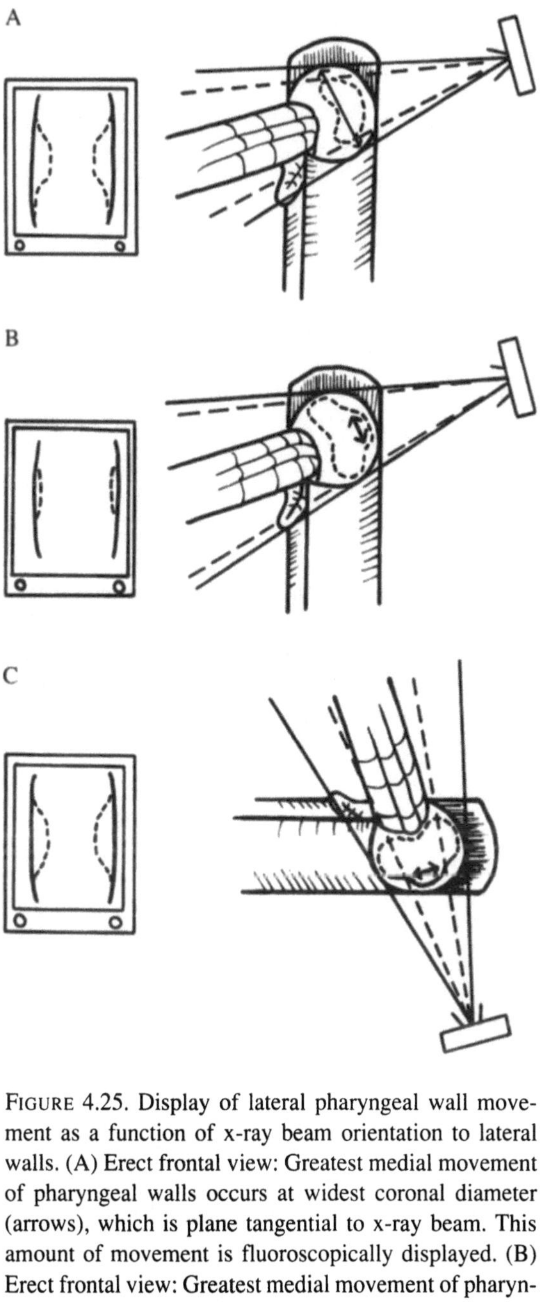

FIGURE 4.25. Display of lateral pharyngeal wall movement as a function of x-ray beam orientation to lateral walls. (A) Erect frontal view: Greatest medial movement of pharyngeal walls occurs at widest coronal diameter (arrows), which is plane tangential to x-ray beam. This amount of movement is fluoroscopically displayed. (B) Erect frontal view: Greatest medial movement of pharyngeal walls occurs at plane posterior (arrows) to widest coronal diameter. Fluoroscopic image does *not* display maximum lateral wall movement at this region but shows only the minimal degree of medial movement that occurs at the widest coronal diameter. (C) Supine frontal view: Barium forms thicker layer over plane of greatest lateral wall movement (arrows), thereby absorbing more radiation as x-ray beam goes through this region of the wall. Fluoroscopic image now displays lateral wall movement at this plane.

frontal projection because the moving pharyngeal walls do not tangentially intersect the x-ray beam (Fig. 4.25B). Movements of the lateral pharyngeal walls in planes other than the plane tangential to the x-ray beam can be readily detected on the base view or when the patient is examined in supine position for the frontal view (Fig. 4.25C). With the patient supine, the barium flows posteriorly, providing a thicker coating over these surfaces and making them more readily detectable. If a base view cannot be obtained in a patient with little or no medial movement of the LPWs on erect frontal, then a supine frontal is recommended.

Centered Versus Rotated Head Position

During quiet breathing the barium-coated pharyngeal walls appear as two vertically oriented and parallel structures approximately equal distance from the midline of the face. The midline is an important anatomic landmark in evaluating lateral pharyngeal wall movement. It is, therefore, important that the head be centered for the frontal projection and not rotated so as to project the midline off to one side or the other. One can determine whether the head is rotated or not by observing whether the nasal septum is equidistant from the lateral margins of the maxillary antra, whether the appearance of the antra are symmetric (usually only the inferior portions are seen if the x-ray beam is well collimated), and whether the incisor teeth are symmetric on either side of the nasal septum. The appearance of the mandibular vertical rami and condyles can also be used to judge symmetry, but the shutters of the collimator should not remain open so wide as to include these structures on the image during the entire frontal examination.

Base View

Information Obtained

This view shows an en face view of the VP portal. The posterior margins of the velum are seen in continuity with the lateral and posterior surfaces of the nasopharyngeal walls. The base view enables one to see how the combined movements of the palate and pharyngeal walls create a sphincterlike narrowing or closure of the VP portal (Fig. 4.26A).

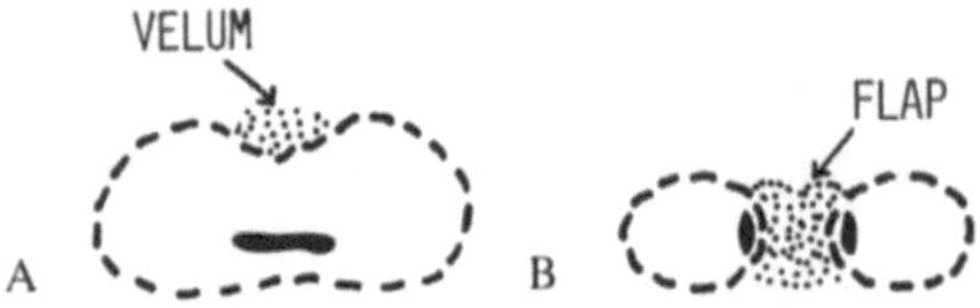

FIGURE 4.26. Base view: centripetal movement of velum and pharyngeal walls to achieve velopharyngeal closure. (A) Nonflap patient. (B) Patient with pharyngeal flap. Quiet breathing, – – –; phonation, —— .

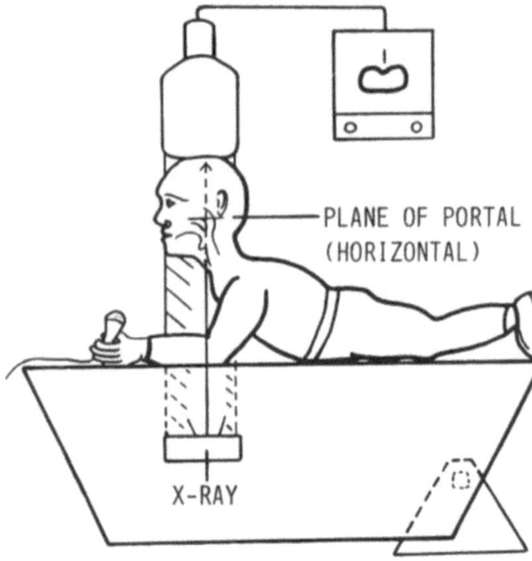

FIGURE 4.27. Base view: patient positioning demonstrated. Vertically directed x-ray beam must perpendicularly intersect plane of velopharyngeal portal in order to visualize completely margins of velopharygneal portal. Therefore the plane of the portal must be horizontal.

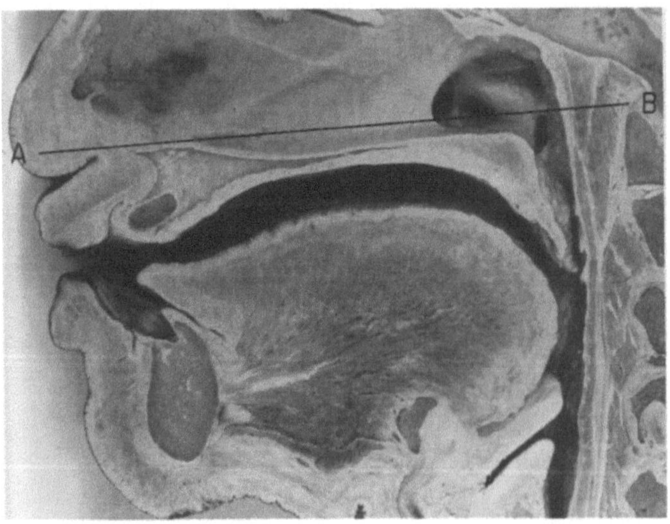

D

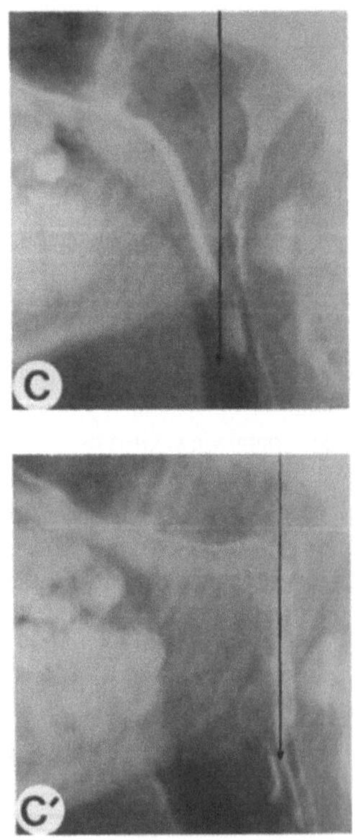

FIGURE 4.28. Pairs of simultaneous lateral-base still films during quiet breathing and phonation of "e" showing proper appearance of velopharyngeal portal on base view and corresponding appearance of lateral cadaver view. (A, A') Base view: Velopharyngeal portal (black) is marginated by barium coating (white line). (B) Transverse anatomic plane through a cadaver head corre-

The base view is also useful in assessing velopharyngeal function in patients with a pharyngeal flap. In this projection, the width of the pharyngeal flap is well delineated and the portals lateral to each side of the flap are clearly visualized (Fig. 4.26B).

Patient Positioning

Positioning for the base view is the most difficult of the three views because the plane of the VP portal must be perpendicular to the direction of the x-ray beam for the portal to be seen correctly.

Positioning for the base view is as follows: The patient lies prone on the horizontally positioned x-ray table supporting himself on his elbows with his hands outstretched. His back is arched and his head hyperextended on his neck (Fig. 4.27). The appearance is akin to that of the Egyptian Sphinx. Because the direction of the x-ray beam is vertical when the fluoroscopic table is horizontally positioned, the plane of the VP portal must be horizontal for it to be optimally seen. This horizontal plane is achieved by hyperextending the head. To further establish optimal positioning, the patient is encouraged to extend his neck forward in a manner similar to that of a swan. Such a maneuver helps project the nasopharynx in front of the cervical vertebrae and reduces the amount of overlying osseous structures. The examiner's assistant (of course wearing a lead apron) is positioned at the end of the fluoroscopic table to prompt the subject with the speech tasks. The subject is encouraged to fix his gaze at the forehead of the assistant so as to maintain proper head position. The microphone is placed on a small stand and positioned on the fluoroscopic table between the patient's hands, or the patient holds the microphone in his hands.

It is usually necessary to make final adjustments of the head position for the base view during fluoroscopic observation of breathing and phonation. If the head is properly positioned, one sees the velopharyngeal portal as an oval or round barium marginated structure (Fig. 4.28). During phonation, the margins of the structure move centrally as the portal closes (Fig. 4.28, 4.29). If the plane of the portal is not perpendicular to the x-ray beam, all of the margins will not be seen at once, resulting in an incorrect representation of the movements of the margins. Instead of seeing a complete barium-margi-

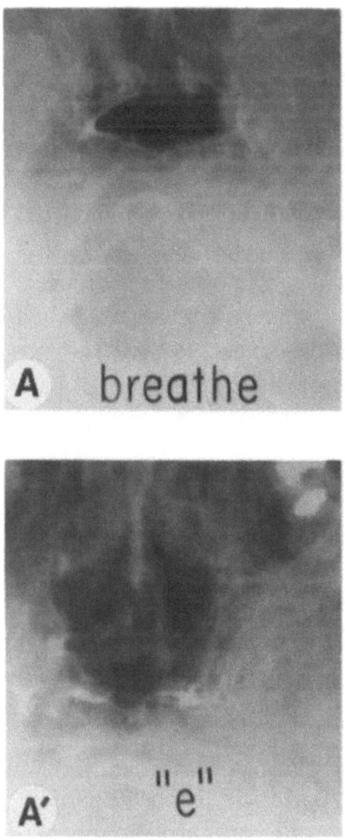

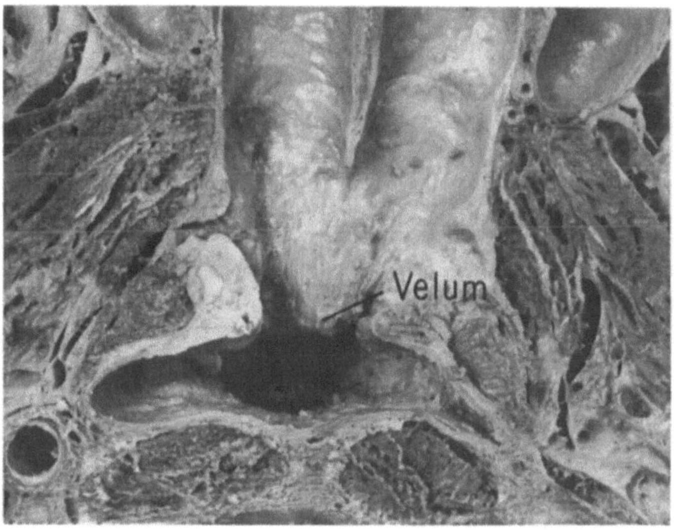

B

sponding to base view in (A). (C, C') Lateral view: Arrow indicates vertically oriented x-ray beam that perpendicularly intersects plane of velopharyngeal portal. (D) Midsagittal anatomic plane through cadaver head corresponding to lateral view in (C). (Line *A-B* represents plane of transverse anatomic section in B.)

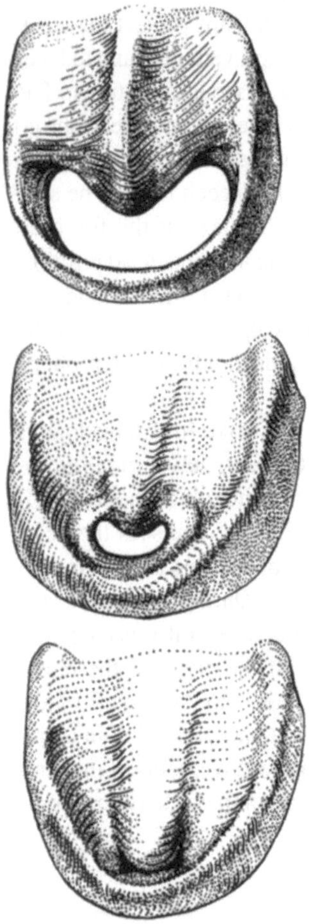

FIGURE 4.29. Sketch in base projection demonstrating centripetal movement of velum and pharyngeal walls to close portal.

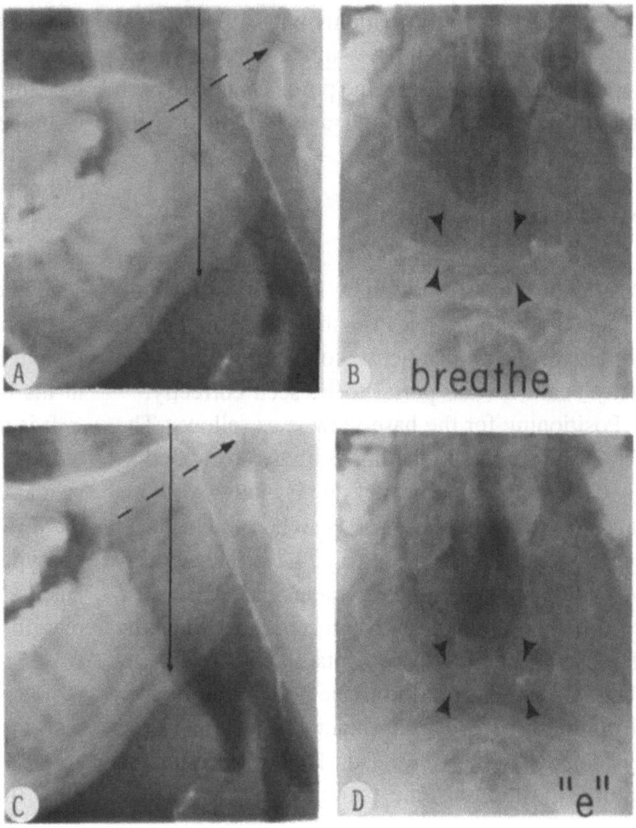

FIGURE 4.30. Pair of simultaneous lateral and base still films during breathing and phonation of "e" with head flexed. (A) and (B) Breathing. (A) Lateral view: Vertical x-ray beam (solid arrow) is not perpendicular to plane of portal (dotted arrow). (B) Complete oval of barium-marginated velopharyngeal portal is not seen. Portions of barium-coated portal seen as lines (arrows). (C) and (D) Phonation of "e." Velopharyngeal closure occurs: Lateral view (C) shows velum touching posterior pharyngeal wall. However, plane of velopharyngeal portal (dotted arrow) is inclined (not perpendicular) to vertical x-ray beam (solid arrow). (D) Base view [similar to (B) base view during breathing] does not show sphincteric closure of VP portal.

nated oval or circle, one sees arcs of barium usually representing portions of the lateral and possibly posterior pharyngeal walls (Fig. 4.30).

The usual cause of improper patient positioning is a failure to hyperextend the head adequately (Fig. 4.31). To achieve optimal head positioning for the base projection, the operator should ask the subject to flex or extend his head slightly during quiet breathing so as to obtain the optimal appearance of the portal.

In order to prove that the proper position of the head for the base view could be determined by observing the fluoroscopic image of the barium-marginated VP portal, we obtained simultaneous biplane base and lateral view in several subjects (Fig. 4.32) when the base view was first introduced by Skolnick in 1970 (4). The lateral view of these simultaneous basal-lateral pairs was used to determine the orientation of the plane of the VP portal for the image of the portal that appeared on the base view. When the portal appeared during breathing as a barium-marginated oval structure on base view, then the portal was properly positioned (portal plane horizon and perpendicular to vertical x-ray beam).

FIGURE 4.31. Head not extended enough for satisfactory imaging of the velopharyngeal portal on base view. Plane of velopharyngeal portal is not perpendicularly intersected by vertical x-ray beam.

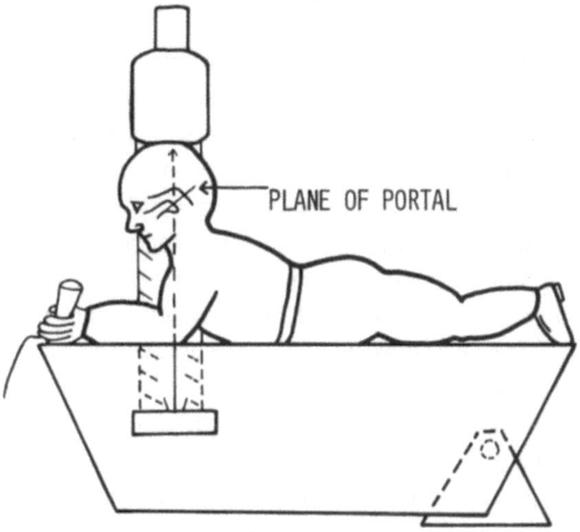

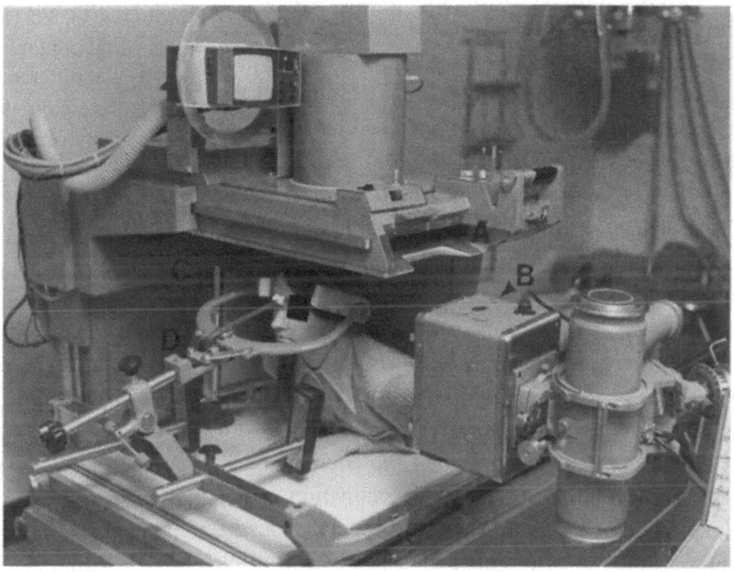

FIGURE 4.32. Technique for performing biplane simultaneous lateral and base still films with subject positioned for base view. (A) Spot film holder for base view x-ray film. (B) Horizontally oriented x-ray tube (from portable x-ray unit) to record lateral view of velopharyngeal portal on (C) vertical x-ray cassette. Head restraint clamp (D) to immobilize head position.

When the portal margins were not well seen on base view, then the lateral view showed that the VP portal was inclined forward and the portal plane did not intersect the vertical x-ray beam in a perpendicular manner.

This simultaneous biplane procedure was performed in the following way. The subject was first positioned for the base view using fluoroscopy to determine the head position that best displayed the barium-coated margins of the velopharyngeal portal during breathing and speech tasks. Then the head was restrained in that position using a clamp that attached to the x-ray table (Fig. 4.32). Because we did not have a biplane fluoroscopic system for simultaneous base and lateral fluoroscopy, we took simultaneous biplane still films in breathing and phonation of "e" (Fig. 4.28). The base still films were obtained with the spot film device of the fluoroscope and the lateral views were obtained by placing a portable x-ray unit next to the patient. This same fluoroscopic positioning procedure (without the need for lateral-base comparison) was then used for all future base views because the accuracy of the technique had already been proven.

Difficulties in Patient Positioning

Inability to position the plane of the VP portal perpendicular to the x-ray beam can result from two causes. First, if the patient has large adenoids the plane of VP closure may become more inclined than if the adenoids are absent (Fig. 4.31). To bring this inclined portal plane to a horizontal position may not be possible because of the patient's inability to extend the head high enough within the limited space between the fluoroscopic carriage and x-ray table. In such situations, there is little one can do with conventional fluoroscopic equipment to obtain a true base projection. The evaluation of the VP portal is then limited to the frontal and lateral views, and perhaps the oblique view, which will be described later in this chapter.

The other problem in obtaining a correctly positioned base view relates to the patient's physical size and spine flexibility. Most adults cannot hyperextend their head and neck to bring the plane of the VP portal parallel to the horizontal and yet fit within the gap between the top of the fluoroscopic table and the undersurface of the fluoroscopic screen. These subjects simply are not flexible enough to arch their back and hyperextend their neck adequately. If, however, a fluoroscopic system were used in which the distance between the table top and the fluoroscopic screen was greater than the 20 inches of conventional equipment, perhaps as great as 40 inches, then true base views could be taken in almost any subject provided the adenoid mass were not too large to distort the inclination of the plane of the VP portal. There are special types of fluoroscopic "C" arm units that can be used for this purpose (see Chapter 3).

Proving Accuracy of Base View

Because the base view looks at the VP portal as if one were looking down the long axis of a tube, centripetal movements of any structures within the pharyngeal tube, whether they lie at the same or different levels, will project to appear as if they were lying at the same level (see Fig. 4.33).

In order to convince ourselves and our readers as well that the palate and pharyngeal walls actually move at one level in the nasopharynx to close the VP portal in a sphincterlike manner, we have used oblique views to confirm that movements of the palate and posterior pharyngeal wall (seen on the

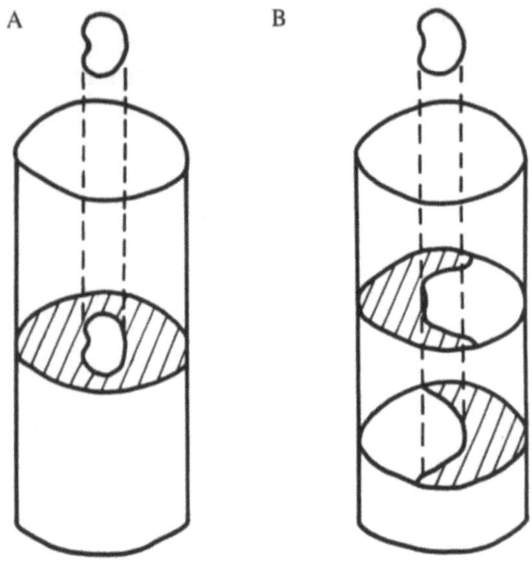

FIGURE 4.33. En face (base) images of the velopharyngeal portal of moving structures that lie on (A) one plane or (B) two different vertical planes. From en face image alone one cannot determine if moving structures lie on one or more vertical planes.

lateral view) occur at the same level as the movements of the lateral pharyngeal walls (seen on frontal view). This confirmation is most vividly demonstrated in a patient with a Passavant's ridge. On the lateral view, Passavant's ridge appears as a localized bulge projecting from the posterior pharyngeal wall. In reality, it is a crescent-shaped structure beginning at the lateral margins of the velum on either side and continuing through the lateral pharyngeal wall into the posterior pharyngeal wall (Fig. 4.34). Although Passavant's ridge appears to be thickest at the position of the posterior pharyngeal walls, considerable soft-tissue shelves project from the lateral pharyngeal walls as well. The appreciation of these lateral shelves and their relationships to the lateral margins of the velum are best delineated on the oblique projection.

Supplemental Views of the Velopharyngeal Portal

Oblique View

We have already observed that the oblique view is useful in relating movements of structures seen on the lateral view to those seen on the frontal view,

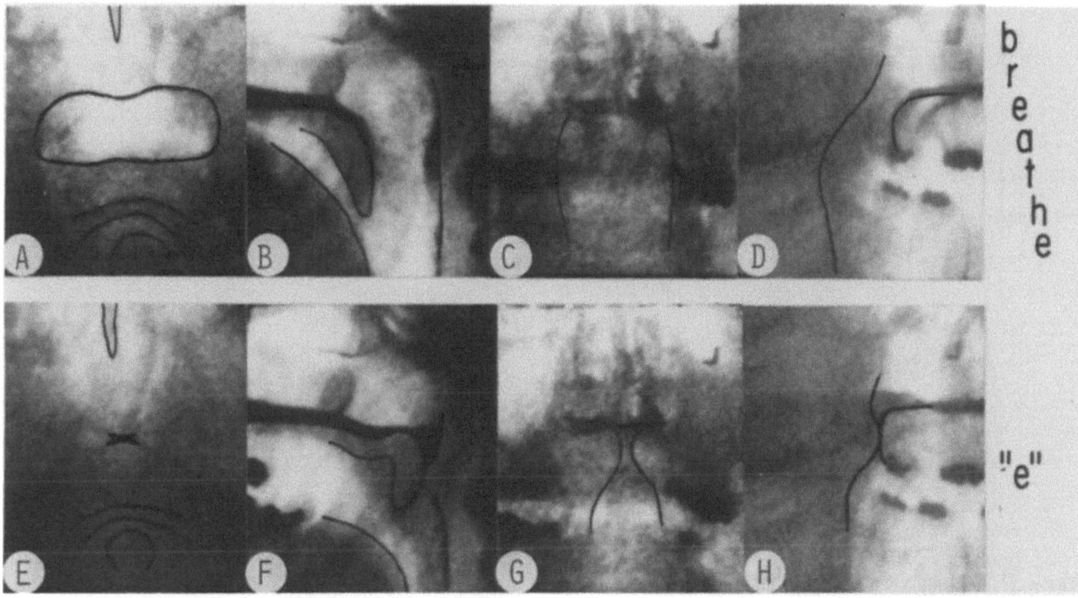

FIGURE 4.34. Patient with Passavant's ridge during breathing and phonation of "e." Use of oblique view to prove that movements of posterior pharyngeal wall seen on lateral view occur on same plane as medial movements of lateral pharyngeal walls seen on frontal view and thus correspond to pattern of sphincteric closure seen on base view. Lateral views: B, F. Frontal views: C, G. Oblique views: D, H. Base views: A, E.

especially movements of the pharyngeal walls. When it is technically impossible to get a satisfactory base view, either because the patient has large adenoids or cannot physically hyperextend his neck within the conventional fluoroscopic equipment, the oblique view may also help better relate palatal and pharyngeal wall movements to each other.

The oblique view is performed by having the patient slowly rotate as he continuously repeats a consonant-vowel combination such as "pa-pa." The patient begins in the lateral position. He repeats the phrase as he slowly rotates his entire head and body toward the frontal position and then continues to the opposite lateral by making slight incremental turns of his feet. The head and body should rotate as a single unit during this maneuver. The total time of the maneuver should take about 10 to 15 seconds.

Asymmetries of lateral pharyngeal wall movement of one versus the other side, asymmetries within one lateral wall from the more lateral to the more posterior margins of the wall, or, for that matter, asymmetries in movement of one side of the palate with reference to the opposite side can be appreciated using this rotational maneuver.

Towne's View

This view has been described by Zwitman (5) and Stringer (6) as an alternative to the base view. We do not think the Towne's view is a substitute for the base view, but we will describe it nonetheless for those who may want to try it.

The view is performed by having the patient positioned as for an erect or supine frontal view while hyperflexing his head on his neck. The x-ray beam intersects the plane of the VP portal in a perpendicular manner so as to project an en face image of the VP portal (Fig. 4.35). As with the base view, final positioning to image the VP portal optimally is done fluoroscopically. The patient slightly changes head position (either flexing or extending) until the barium-marginated VP portal is best seen. A foam wedge placed under the head may be necessary to increase the amount of flexion if the patient cannot obtain the needed flexion just by bringing his chin against his neck.

We believe that there are several problems with the Towne's view. First, the x-ray beam must tranverse through the cervical vertebrae, which increases the amount of scattering and decreases the quality of the image obtained as compared with that produced by positioning the patient prone for the base view. Second, when the head is flexed upon

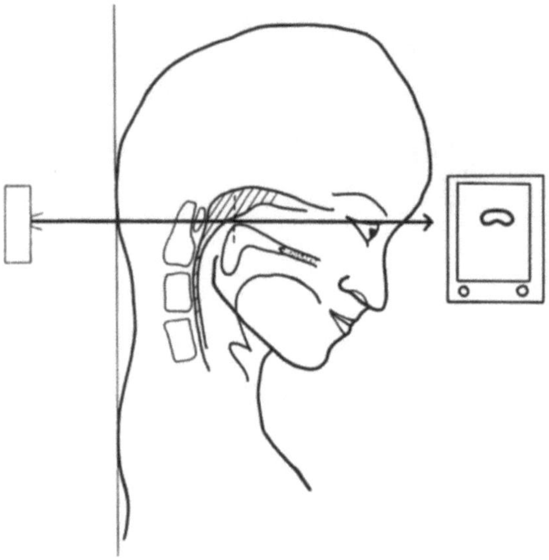

FIGURE 4.35. En face view of velopharyngeal portal with head positioned in Towne's projection (frontal view with extreme flexion of head upon neck). Plane of velopharyngeal portal (dotted line) is perpendicularly intersected by horizontal x-ray beam (solid line).

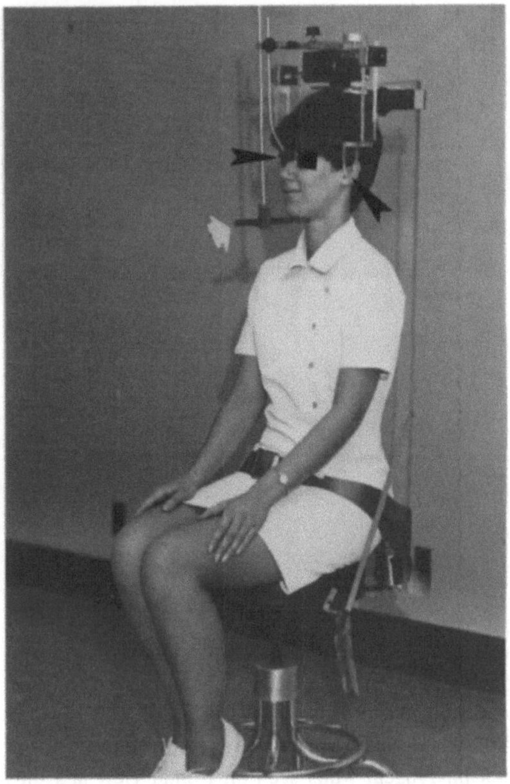

FIGURE 4.36. Specially constructed chair with forehead block (horizontal black arrowhead) and ear pins (diagonal black arrowhead) to prevent head movement. Seat belt across patient's waist prevents patient from sliding down or forward. Metal ruler with holes a centimeter apart (white arrowhead) is positioned under chin in midsagittal plane to project centimeter scale on fluoroscopic image.

the neck, the volume of the nasopharynx is reduced because the distance between the posterior surface of the palate and posterior pharyngeal wall is narrowed. Thus, patients who have marginally adequate or slightly inadequate closure mechanisms on lateral view with the head in neutral position may erroneously appear to have adequate mechanisms when the head is flexed (Fig. 4.20).

To Immobilize or Not to Immobilize

During the first four years in which we have performed multiview videofluoroscopy, patients were immobilized for all views.

While head immobilization insures reproducible positioning, it requires specially designed hardware that is not commercially available. For the lateral, oblique, and frontal views we constructed a chair with a radiolucent back (Fig. 4.36). The patient was held in this chair by a seat belt across his waist. The head was immobilized by a three-point system consisting of pins in both ears and a bar pressing upon the bridge of the nose. These points of immobilization were adjustable from patient to patient. The seat of the chair rotated in unison with the back and head immobilization device so that the entire assembly with patient in it could be turned from lateral into frontal position without disturbing the patient. Because we were obtaining views in lateral position with and without barium, the patient first had to be inserted in the immobilization device for the nonbarium lateral view, then removed to have the barium instilled in the nasopharynx, and finally reinserted for the repeat lateral and frontal views. First to remove and then to reposition the patient in the head immobilization device took several minutes and slowed down our ability to perform the procedure rapidly. More important, the presence of the immobilization device frightened a number of the smaller children and resulted in our inability to do studies on some of them.

To immobilize the head for the base view (Fig. 4.32), we modified a commerical device used to restrain shoulders during cervical myelography. The device clamped to the fluoroscopic table and locked the patient's head with a pair of ear cushions and a pressure bar on the bridge of the nose. Like the chair used for the other two views, it frightened some patients and took time to properly adjust.

Since 1975 we have used no immobilization devices, and believe that the latter way is the preferred way. We have found that the absence of immobilization devices is not a hindrance to obtaining good-quality clinical studies. Once the patient is properly positioned, he usually holds that position, especially if there is someone, or some mark upon which he can focus his eyes so as to maintain constant head orientation. For the occasional young patient who fidgets and cannot maintain a constant head position, we have been able to immobilize the head readily by having the operator place a lead-gloved hand on the patient's head and hold it in a given position. An extra long-leaded glove is used so that the operator's hand is shielded from scattered radiation. The hand is not in the primary beam.

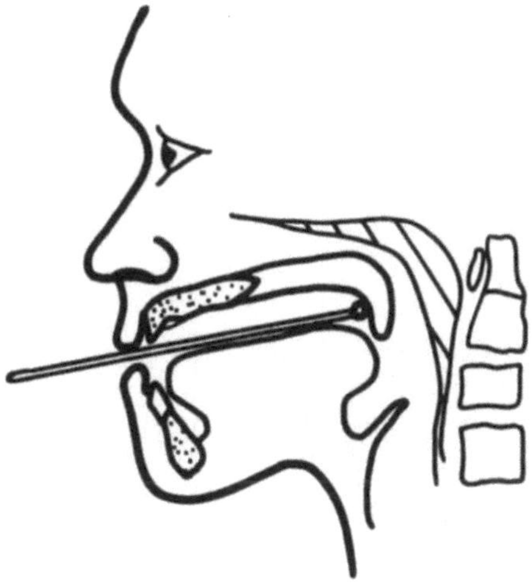

FIGURE 4.37. Lateral view of velopharyngeal portal with magnification marker in position. Five-millimeter metal ball attached to maleable metal rod is inserted into mouth like a lollipop. Ball is positioned deep in oropharynx close to or touching oral surface of velum.

Techniques of Measuring Velopharyngeal Dimensions

Because of differences in the size of TV monitors, television optical systems, and geometry of image intensifiers, the dimensions of the pharyngeal structures as displayed on the TV monitor may vary considerably depending upon the type of equipment used. In order to make accurate measurements of distances and dimensions, one must have a way of demagnifying the final TV images. The most precise way involves placing a known size radiopaque structure in the pharynx close to the plane of the VP portal and recording the image of this marker on each view before the speech tasks are performed (Fig. 4.37). We have used a 5-mm diameter metal ball welded to a maleable metal wire of approximately 15 cm in length (a common bile-duct dilator will serve this function). The subject inserts the device into his mouth with the ball end in the oropharynx very much as one would insert a lollipop. Because the ball is spherical, it projects the same dimension for all three views. By comparing the magnified dimension of the ball on the TV screen to its actual dimension, a ratio is constructed that can then be used to demagnify dimensions measured from the TV monitor or measured from any photographic recording of the TV image. For example, if the actual diameter of the ball is 5 mm, the magnified diameter of the ball is 8 mm, and the magnified distance between the palate and posterior pharyngeal wall on lateral view is 4 mm, then the demagnified distance between the palate and pharyngeal wall is 2.5 mm. A ratio is constructed such that

$$\frac{\text{actual diamater of ball}}{\text{magnified diameter of ball}}$$

$$= \frac{\text{actual distance}}{\text{magnified distance}} \text{ or } \frac{5}{8} = \frac{A}{4}.$$

A then equals $\frac{20}{8}$, or 2.5 mm (Fig. 4.38A)

Even if the tip of the magnification marker cannot be precisely positioned at the plane of the portal, no significant distortion will occur if the tip is within 1-2 cm of the plane of the portal in any direction. The amount of magnification change occurring by a shift of 1-2 cm in the tip of the magnification marker is insignificant compared with the overall magnification of the nasopharynx. For example, if we vary that 20 cm distance [(C) in Fig. 4.38B] by 2 cm, we would obtain a magnification error of 0.04 cm, or 2.5%.

A much simpler method of correcting the magnification is to use a structure of known dimension

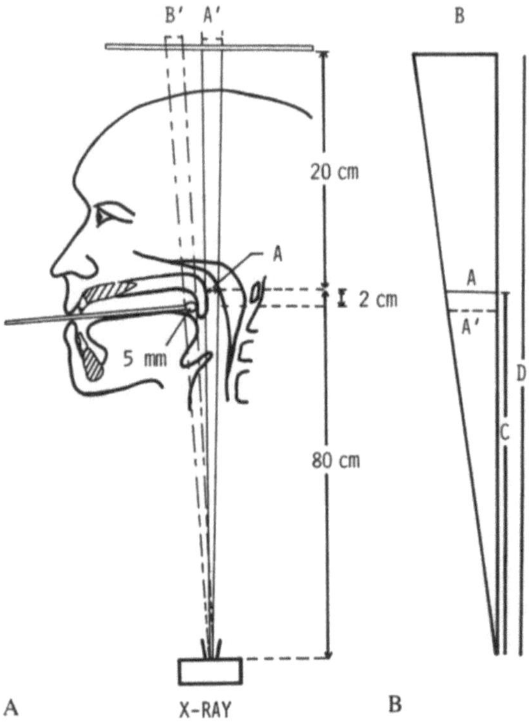

FIGURE 4.38. (A) To measure diameter of velopharyngeal portal (assuming ball is at plane of portal):

$$\frac{A'}{B'} = \frac{A}{B}$$

A = Demagnified diameter of VP portal
B = Diameter of ball
A' = Magnified distance
B' = Magnified diameter of ball

Measure A' and B' from video image on TV monitor. Ball diameter is 5 mm.

$$\frac{5(A')}{B'} = A$$

(B) To calculate the magnification error if the ball is not at the plane of the velopharyngeal portal, let us use the method of congruent triangles.

A = Actual dimension of structure
B = Magnified dimension of structure on fluoroscopic screen
C = Distance from x-ray tube to structure
D = Distance from x-ray tube to fluoroscopic screen

If the magnified dimension (B) of the VP portal equals 2 cm, and the actual plane is 20 cm from the fluoroscopic screen, the C equals 100 - 20 or 80 cm. D equals 100 cm (a given).
Thus, to find A:

$$\frac{A}{C} = \frac{B}{D}$$

$$A = \frac{CB}{D}$$

$$A = \frac{80 \times 2}{100}$$

$$A = 1.6 \text{ cm}$$

If plane A were moved 2 cm to plane A', A' would equal 1.56 cm, a change of 0.04 cm or 2.5%.

within the head that is projected on the TV monitor and compare its magnified dimension to its actual dimension as a basis for demagnifying other projected images. Such structures include the width of a molar tooth or the width of the odontoid or the arch of C1. For children of a given age, these dimensions are relatively standard. The width of the odontoid or anterior arch of C1 is a more accurate measurement as it is quite close to the plane of the VP portal. Because the amount of magnification distortion is minimal between the lateral and the base view, even if one can only accurately see the odontoid in the lateral view, it can be used as a basis for demagnifying structures seen in other views as well.

References

1. Skolnick ML, Glaser (Cohn) ER, McWilliams BJ: The use and limitations of the barium pharyngogram in the detection of velopharyngeal insufficiency. Radiology 135:301, 1980
2. Glaser (Cohn) ER, Skolnick ML, McWilliams BJ, Shprintzen RJ: The dynamics of Passavant's ridge in subjects with and without velopharyngeal insufficiency — A multiview videofluoroscopic study. Cleft Palate J 16:24, 1979
3. McWilliams BJ, Musgrave RH, Crozier PA: The influence of head position upon velopharyngeal closure. Cleft Palate J 5:117, 1968
4. Skolnick ML: Videofluoroscopic examination of the velopharyngeal portal during phonation in lateral and base projections — A new technique for studying the mechanics of closure. Cleft Palate J 7(4):803, 1970
5. Zwitman DH, Gyepes MT, Sample F: The submentovertical projection in the radiographic analysis of velopharyngeal dynamics. J Speech Hear Dis 38:473, 1974
6. Stringer DA, Witzel MA: Velopharyngeal insufficiency on videofluoroscopy: Comparison of projections. Am J Roentgenol 146:15, 1986

5
Anatomy of Velopharyngeal Portal

The purpose of this chapter is to describe the structure and function of the underlying musculature so that forces responsible for velopharyngeal movement can be more completely understood. However, it is not our intent to advance a theory of velopharyngeal physiology in this forum, but merely to summarize current thinking and controversies. A secondary purpose is to describe briefly the impact of growth, adenoidal changes, and sex on velopharyngeal structure and function.

Above all, the reader must recognize that videofluoroscopy does not demonstrate muscle type, insertion, or origin. These limitations are important to recognize when interpretating velopharyngeal structure and function.

Muscles of Velopharyngeal Function

The physiology of velopharyngeal closure has long been a subject of disagreement and heated debate. As noted by Dickson et al. (1): "Opinions regarding the specific muscles responsible for these patterns of movement are almost as numerous as authors who discuss them." Not only do sources vary as to the function of specific muscles, but they also disagree on the location of many of the muscular fibers.

Velar Activity

The velum forms the anterior margin of the velopharyngeal sphincter. The velar movements are characterized by elevation, posterior elongation, and depression.

Velar Elevation

It is generally agreed that the paired levator palatini muscles are the main muscles responsible for velar elevation (Fig. 5.1).

According to Perkins and Kent (2), this action is assisted by the tensor palatini muscles, which stretch the velum. However, others, such as Dickson and Dickson (3) and Rood (4), dispute this function of the tensor veli palatini by virtue of this muscle's attachment to the cranial base, the Eustachian tube membrane, and the medial pterygoid plate.

The musculus uvulae (Figs. 5.1 and 5.2) is contained within, but is not to be confused with the uvula of the soft palate. Based upon evidence from the histologic section of early human fetal specimens, Langdon and Klueber (5) speculate that the contraction of the lateral branches of musculus uvulae may raise the center of the muscle on the velum's dorsal midline. Dickson (6) and Azzam and Kuehn (7) are also in agreement that the musculus uvulae may influence the bulk and shape of the velar eminence.

Velar Depression

Velar depression is thought to be accomplished by two muscles, the palatopharyngeus and palatoglossus (1) (Figs. 5.1 and 5.2).

Lateral Pharyngeal Wall Movement

The source of medial movement of the lateral pharyngeal walls has been a continuing point of controversy. Although lateral pharyngeal wall movement has been attributed to the salpingopharyngeal muscle, evidence for this is weak, as the muscle is

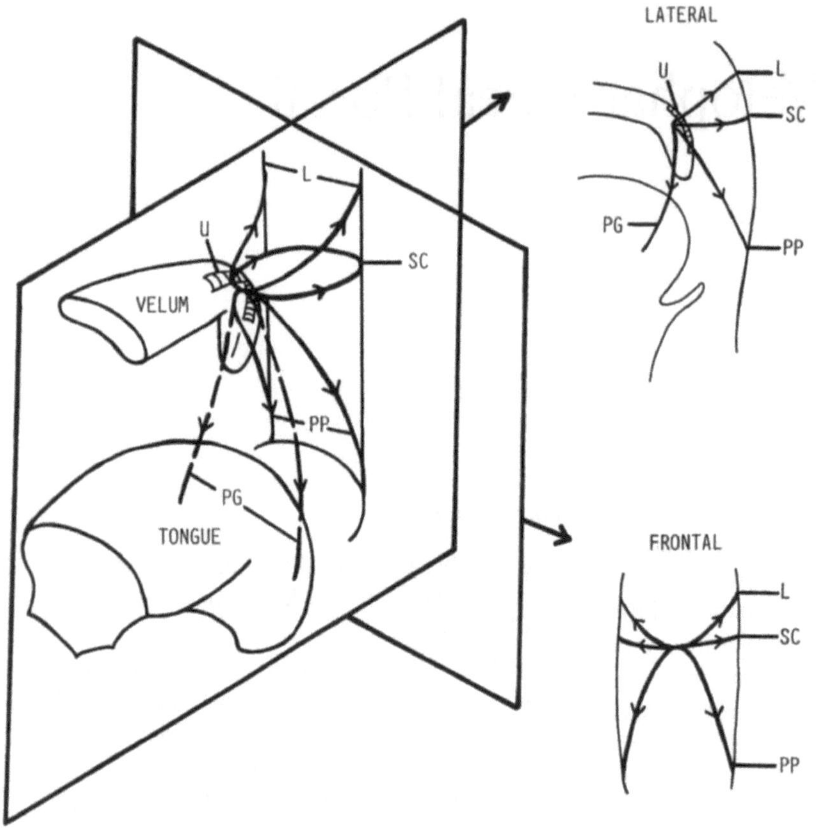

FIGURE 5.1. Schematic representation of the major muscles controlling closure of the velopharyngeal portal. Here they are shown at rest. (Arrows denote direction of motion.)

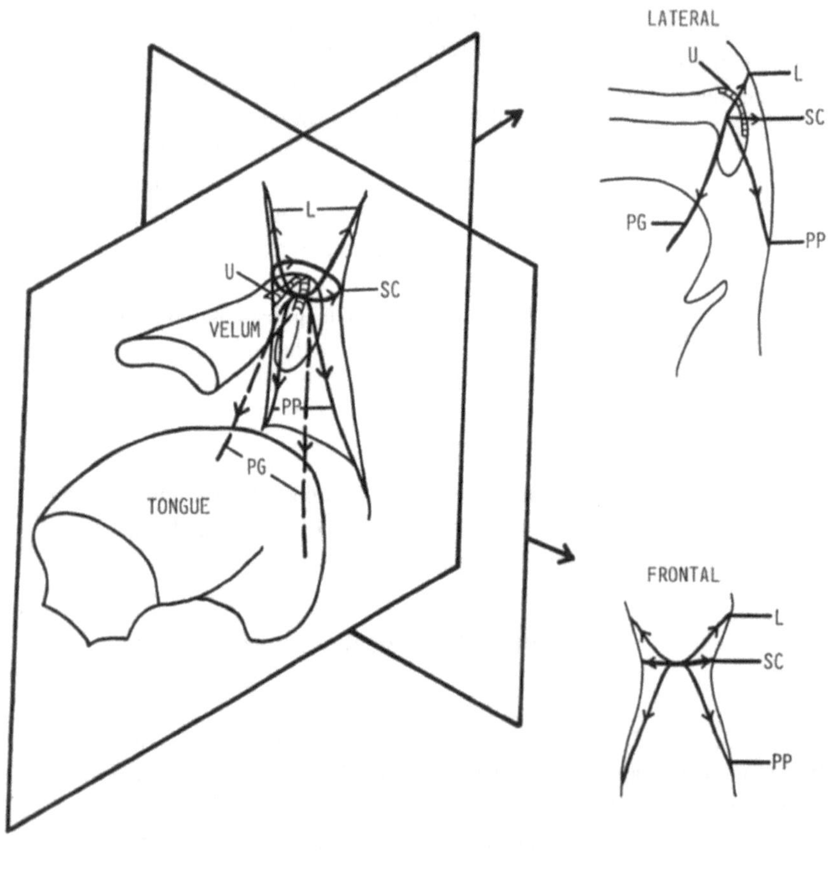

FIGURE 5.2 Schematic representation of the major muscles controllin closure of the velopharyngeal portal. They are shown here during partial closure (partial rather than complete closure was chosen for better graphic representation of the muscles). (Arrows denote direction of motion.) Muscles: *L*, levator palatini; *SC*, superior constrictor; *PP*, palatopharyngeus; *PG*, palatoglossus; *U*, uvula.

often small and inconsistently present in man (1, 3, 8). Instead, research by Dickson and Dickson (3) and Dickson et al. (1) offers support for the levator as the sole muscle of velopharyngeal function.

The work of others provides evidence that the muscle responsible for lateral pharyngeal wall movement is separate from that which creates velar movement. Thus, the superior constrictor muscles, and not the levator veli palatini muscles would probably account for lateral pharyngeal wall motion (Figs. 5.1 and 5.2) (9-12).

Passavant's Ridge

It is generally agreed that Passavant's ridge is formed by the contraction of muscle fibers which course around the pharynx in a horizontal direction, inserting into the palate. However, controversy remains as to whether these fibers should be assigned to the superior constrictor muscle, the palatopharyngeal muscle, or both. In addition, there are possible intersubject variations in the muscle constituency of Passavant's ridge (13).

A Sphincteric Model of Muscle Function

The division of the components of the velopharyngeal valving mechanism has been presented to facilitate description of the musclar components of the valve. It must be remembered, however, that the velum, the lateral pharyngeal walls, and, when present, Passavant's ridge generally function together to create the velopharyngeal sphincter (Figs. 5.1 and 5.2).

Observations of incongruous movements between velum and lateral aspects of the pharyngeal walls by Shprintzen et al. (12) point to possible variability in the physiology of velopharyngeal valving. Further evidence for this variability is offered by Skolnick et al. (14) in descriptions of multiple patterns of velopharyngeal valving. The presence of these multiple movement patterns, described in Chapter 8, suggest that "variability in the physiology of valving may be the rule," according to Shprintzen et al. (12). Whereas some individuals valve primarily with the levator muscle, others valve with the levator and the superiormost fibers of the superior constrictor. Shprintzen et al. (12) thus suggest that "the muscles contributing to valving probably vary among individuals in the percentage of valving effort they perform."

Because of reports of the emergence of Passavant's ridges in patients who were previously observed not to have any, it is conceivable that the relative contributions of various muscles may change over time. Such changes may be attributable to growth, surgical advancement of the maxilla, or altered patterns of function.

Normal Variations in Anatomy

Growth

The anatomical relationships of the oral and pharyngeal structures, and thus, of the velopharyngeal valve, are altered by normal growth and development. The nature of these anatomical changes are easily remembered because of their seeming purposefulness.

The structure of the infant's vocal tract similar to that of lower species allows for simultaneous breathing and eating, and provides for protection against choking. The infant epiglottis is situated so high in the pharynx that it can touch the soft palate; the tongue, the hyoid bone and the pharynx are also positioned higher in the infant than in the adult pharynx. This arrangement provides for more airway protection than in the adult (2).

The position and movements of the infant velum also differ from those of the adult. In the infant, the velum is parallel to the roof of the pharynx and moves in a superior-inferior direction to close against the adenoidal region. In the adult, the velum develops a 90° bend while moving in an anterior-posterior direction. Closure occurs between the vertical portion of the velum and posterior pharyngeal wall (15) (Fig. 5.3).

By the end of the first six months of life, significant remodeling of the vocal tract has occurred, so the anatomy no longer resembles that of a monkey or chimpanzee. The infant gains the ability to close off the nasal tract, but loses the capacity to eat and breathe at the same time. The development of a right-angle craniovertebral relationship creates a right angle bend to the vocal tract. By the six-month mark, the tongue, hyoid bone, epiglottis, and larynx have migrated in an inferior direction, thus enlarging the pharynx (2).

As maturation progresses, growth of the mandible from the cranial base proceeds downward and forward. The nasopharynx increases in vertical

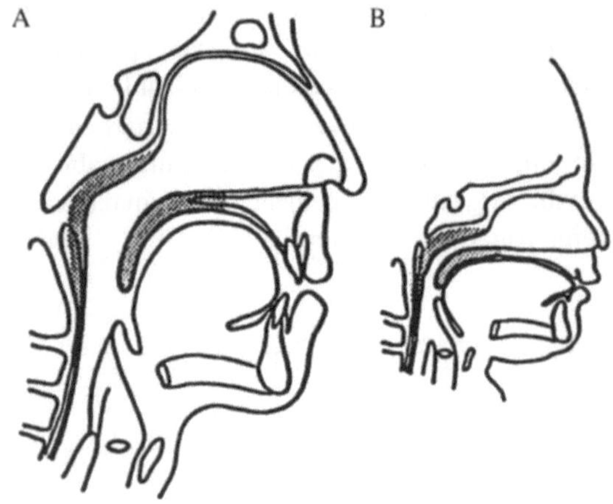

FIGURE 5.3. Lateral view comparing relative positions of velum and cranial base in adult (A) and infant (B). Reprinted by permission from Bzoch KR, Williams WN: Introduction, rationale, principles and related basic embryology and anatomy. In Bzoch KR (Ed): Communicative Disorders Related to Cleft Lip and Palate. Boston: Little, Brown and Company, 1979, p 17.

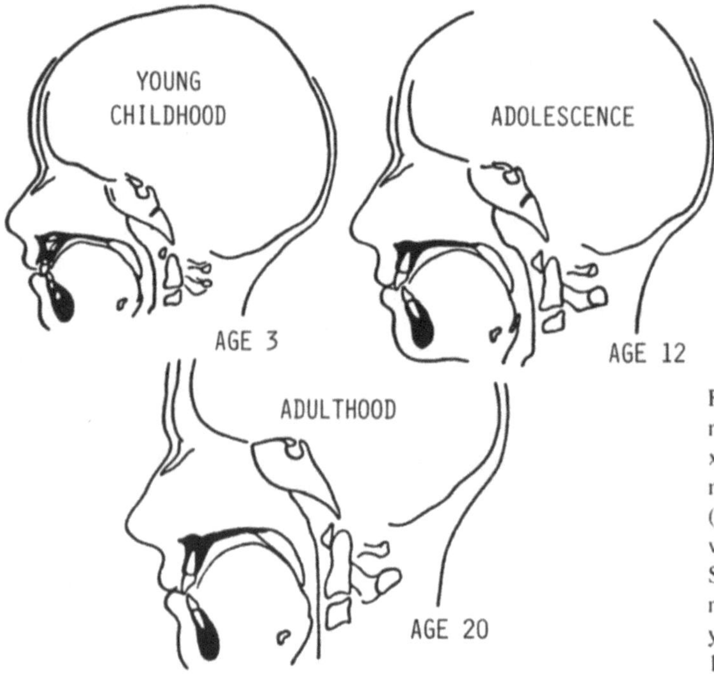

FIGURE 5.4. Age-related changes in adenoidal size derived from cephalometric x-ray tracings of the same patient. Also note increase of nasopharyngeal space (velum-nasopharyngeal wall distance) with age. Reprinted by permission from Subtelny JD, Koepp-Baker H: The significance of adenoid tissue in velopharyngeal function. Plast Reconstr Surg 17:235, 1956.

dimension. There is greater space between the velum at rest and posterior nasopharyngeal wall, and the velum increases in length (Fig. 5.4).

Sex Differences

Sex differences in cleft lip and palate are well documented. Isolated clefts of the palate occur more frequently in females and clefts of the lip, with or without cleft palate, occur more frequently in males, with males having more serious deformities than females (16, 17).

Evidence exists that males and females may also show differences in velopharyngeal structure and function. These encompass gender-related differences in velar orientation, elevation, length, contact with posterior pharyngeal wall, and in nasal airflow during speech. In the female the velum may have more of a right-angle orientation to the posterior pharyngeal wall during velopharyngeal closure and a greater extent of contact between velum and posterior pharyngeal wall than is seen in the male (Fig. 5.5). The nature of these differences are summarized in Table 5.1. However, most of these sex-

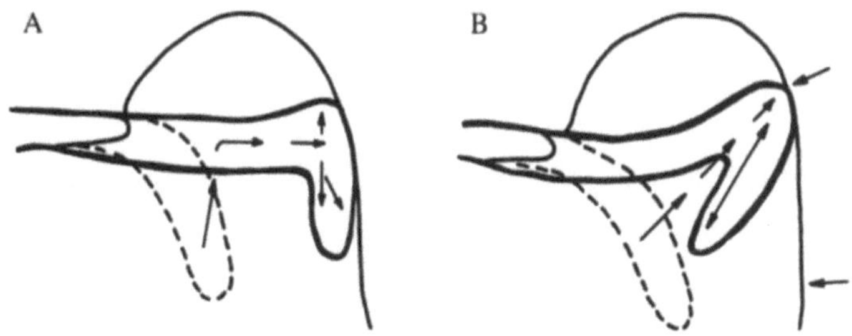

FIGURE 5.5. (A) "Squared off" female and (B) "acute" male velar orientation. The velum at rest is depicted by broken lines, arrows designate the direction of movement. Reprinted by permission from McKerns D, Bzoch KR: Variations in velopharyngeal valving: The factor of sex. Cleft Palate J 7:652, 1970.

TABLE 5.1. Summary of sex differences in velopharyngeal structure and function.

Parameter	Trend	Reference(s)
Orientation of velum to posterior pharyngeal wall[a]	Males: acute angle Females: close to a right angle	McKerns and Bzoch (19)
Velar elevation	Greater in males	McKerns and Bzoch (19)
Velar length	Greater in males	McKerns and Bzoch (19)
Height of velar elevation	Greater in males	McKerns and Bzoch (19)
Changes in velar elevation	Greater in females	Seaver and Kuehn (20)
Amount of velar-posterior pharyngeal wall contact	Greater in females	McKerns and Bzoch (19)
Nasal airflow during speech (on initial vowel in:/inil)	Greater in females	Thompson and Hixon (21)

[a]No differences found by Iglesias et al. (18); differences described by McKerns and Bzoch (19) diagrammed in Figure 5.5.

related variations are so subtle as not to be readily evident during routine clinical videofluoroscopic evaluation.

The reasons for sex differences in velopharyngeal valving are undetermined. Although McKerns and Bzoch (19) did not find the structural differences to be related to the ratio of velar length and pharyngeal depth, they suggested that males and females may differ in the areas of muscle insertion and attachment. Thompson and Hixon (21) attributed the increased nasal airflow in females either to biochemical differences between males and females or to a greater tolerance in females for nasalization. McWilliams et al. (17) conclude that the apparent sex differences in velopharyngeal physiology are not signficant for clinical management.

Adenoidal Changes

It is very important that readers of the videofluoroscopic x-ray appreciate the age-related changes in the adenoidal tissue (Fig. 5.4), and how these may affect velopharyngeal function. Normal age-related changes in adenoidal tissue will not result in velopharyngeal insufficiency in the vast majority of patients. An increase in adenoidal tissue usually occurs until puberty, when the tissue gradually begins to atrophy. At the same time, as seen in Figure 5.3, downward growth of the mandible causes lowering of the palate relative to the adenoidal tissue, and increased airway space. However, the normal velopharyngeal mechanism is able to respond to this loss in pharyngeal tissue. Neiman and Simpson (22) point out that patients who successfully adapt to surgical expiration of the adenoids evidence: "(1) increased velar mobility; (2) increased height of velopharyngeal closure; (3) increased percent of velar stretch; and (4) anterior movement of the posterior pharyngeal wall when amount of velopharyngeal contact becomes tenuous."

The adenoids may or may not be in a position to participate in velopharyngeal closure (Fig. 5.6). In

A

B

C

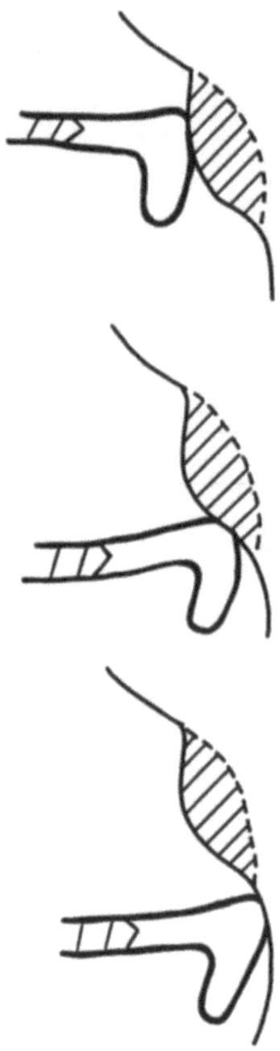

FIGURE 5.6. Variations in velar-adenoidal relationships during velopharyngeal closure. Patients (A) and (B) show different patterns of velar-adenoidal contact. In patient (C) the adenoidal tissue does not participate in velopharyngeal closure.

some patients, the palate contacts the posterior pharyngeal wall below the area of this tissue, and thus velopharyngeal competence will not be affected by adenoidal size. In patients with distinct velar-adenoidal contact, surgical removal of the adenoids will only infrequently result in transient or permanent velopharyngeal insufficiency (VPI) (22). This may occur because insufficient velar tissue, often a result of submucous cleft palate or repaired cleft palate, renders the velum incapable of occluding the surgically created space. Functional or neurological problems of velar movement may also compromise adenoid-velar contact. Finally, growth-related diminution of the adenoid pads can also create VPI in a minority of patients.

An adenoidal pad that is excessively large may create the opposite problem, that of nasopharyngeal obstruction. Such occlusion may result in mouth breathing, hyponasal or denasal speech, dysphagia, snoring, and in severe cases, sleep apnea and cor pulmonale (22). The seriousness of these symptoms and their effects upon health will depend upon the extent of the obstruction.

References

1. Dickson DR, Grant JCB, Sicher H, Dubrul EL, Paltan J: Status of research in cleft palate anatomy and physiology. Cleft Palate J 11:471, 1974
2. Perkins WH, Kent RD: Functional Anatomy of Speech, Language and Hearing: A Primer. Boston: Little, Brown and Company, 1986
3. Dickson DR, Dickson WM: Velopharyngeal anatomy. J Speech Hear Res 15:272, 1972
4. Rood SR: A functional anatomic study of the muscles related to the Eustachian tube. Unpublished PhD thesis, University of Pittsburgh, 1972
5. Langdon HL, Klueber K: The longitudinal fibromuscular component of the soft palate in the fifteen-week human fetus: Musculus uvulae and palatine raphe. Cleft Palate J 15:337, 1978
6. Dickson DR: Anatomy of the normal velopharyngeal mechanism. Clin Plast Surg 2:234, 1975
7. Azzam NA, Kuehn DP: The morphology of musculus uvulae. Cleft Palate J 14:78, 1977
8. McMyn JK: Anatomy of the salpingo-pharyngeus muscle. J Laryngol Otol 5:1, 1940
9. Fritzell B: The velopharyngeal muscles in speech. Acta Otolaryngol (suppl) 250, 1969
10. Shprintzen RJ, Lencione RM, McCall GN, Skolnick ML: A three diminsional cinefluoroscopic analysis of velopharyngeal closure during speech and nonspeech activities in normals. Cleft Palate J 11:412, 1974
11. Shprintzen RJ, McCall GN, Skolnick ML, Lencione RM: Selective movement of the lateral aspects of the pharyngeal walls during velopharyngeal closure for speech, blowing, and whistling in normals. Cleft Palate J 12:51, 1975
12. Shprintzen RJ, Rakoff SJ, Skolnick ML, Lavorato AS: Incongruous movements of the velum and lateral pharyngeal walls. Cleft Palate J 14:148, 1977
13. Glaser (Cohn) ER: A multiview videofluoroscopic study of Passavant's ridge in patients with and without velopharyngeal insufficiency. Unpublished PhD thesis, University of Pittsburgh, 1980
14. Skolnick ML, McCall GN, Barnes B: The sphincteric mechanism of velopharyngeal closure. Cleft Palate J 10:286, 1973
15. Fletcher SG: Cleft palate: A broader view. J Speech Hear Dis 31:3, 1966

16. Warkney J: Congenital Malformations. Chicago: Year Book Medical Publishers, 1971
17. McWilliams BJ, Morris HL, Shelton RL: Cleft Palate Speech. Philadelphia: BC Decker, 1984
18. Iglesias A, Kuehn DP, Morris HL: Simultaneous assessment of pharyngeal wall and velar displacement for selected speech sounds. J Speech Hear Res 23:429, 1980
19. McKerns D, Bzoch KR: Variations in velopharyngeal valving: The factor of sex. Cleft Palate J 7:652, 1970
20. Seaver E, Kuehn, DP: Cineradiographic and electromyographic investigation of velar positioning in non-nasal speech. Cleft Palate J 16:216, 1980
21. Thompson AE, Hixon TJ: Nasal air flow during normal speech production. Cleft Palate J 16:412, 1979
22. Neiman GS, Simpson RK: A roentgencephalometric investigation of the effect of adenoid removal upon selected measures of velopharyngeal function. Cleft Palate J 12:377, 1975
23. Paradise JL: Tonsillectomy and adenoidectomy. In Bluestone CD, Stool SE (Eds): Pediatric Otolaryngology, Vol. 2. Philadelphia: WB Saunders, 1983, p 992

6 Interpretation of the Videofluoroscopic Study

Years ago there was a coffee and doughnut shop in New York City called "Mayflower," which emphasized as part of its advertising slogan that one should keep an eye on the doughnut and not on the hole. In analyzing videofluoroscopic studies, we, on the other hand, should pay equal attention to the hole (velopharyngeal portal) and the doughnut (palate and pharyngeal walls ringing the VP portal). When the sphincter does not close, there is a hole. Unlike the doughnut, however, this hole is not symmetrically round but has different contours along different margins. In order to understand the manner in which the velopharyngeal portal closes, we must examine the contributions that the palate (velum) and pharyngeal walls each make. This understandably is best appreciated by examining the component movements of the palate and pharyngeal walls in the lateral, frontal, and base views.

Lateral View

Nonbarium Study

The initial recording is made without barium in order to insure that some information about velopharyngeal dynamics is obtained should the patient refuse to permit nasal installation of barium. As a secondary benefit, a pharyngeal flap may be more readily detected in the lateral examination without barium because the barium may obscure the flap (for more about flaps, see page 64).

However, it is important to emphasize that the lateral view with barium (except in a patient with a pharyngeal flap—see Chapter 10) provides more complete information than this view without barium. For example, in the absence of nasal barium it is difficult to differentiate a small velopharyngeal gap from velopharyngeal closure because the amount of air present between the velum and pharyngeal wall in a small gap does not produce enough negative contrast to define the margins of velum and pharyngeal wall. Likewise, without nasal barium a central defect in the posterior surface of the palate as a result of a submucous cleft palate may not be appreciated because the amount of air in the defect is not enough to contrast against the sides of the velum (Fig. 6.1). Therefore, most of our interpretations of the lateral view are based upon the observations made when the mucosal surfaces are coated with barium.

Velum

We begin the interpretation of the lateral view by examining the velum. Its length and contour are best appreciated during quiet breathing. The normal velum has a tapered and arcuate configuration that parallels the curvature of the dorsum of the tongue and extends into the oropharynx. A localized region of slightly increased thickness bulges from the mid superior surface.

When the patient breathes through the mouth, the inferior velar surface is separated from the tongue by an air gap, whereas when the patient breathes through the nose, the velum rests on the dorsum of the tongue (Fig. 6.2). In the nonbarium lateral view, the length and contours of the velum are only well delineated during oral breathing. However, in the barium-coated view, there is usually enough of a barium coating on the dorsum of the tongue, either

FIGURE 6.1. Repaired-cleft-palate patient. Dorsal midline groove formed at site of palatal anastomosis. (A) Lateral x-ray without nasal barium coating. Groove not seen because air within groove does not provide enough x-ray contrast to distinguish groove from adjacent surfaces of velum. (B) Lateral x-ray with barium coating. Dense barium within groove provides enough x-ray contrast to silhouette groove from adjacent velum.

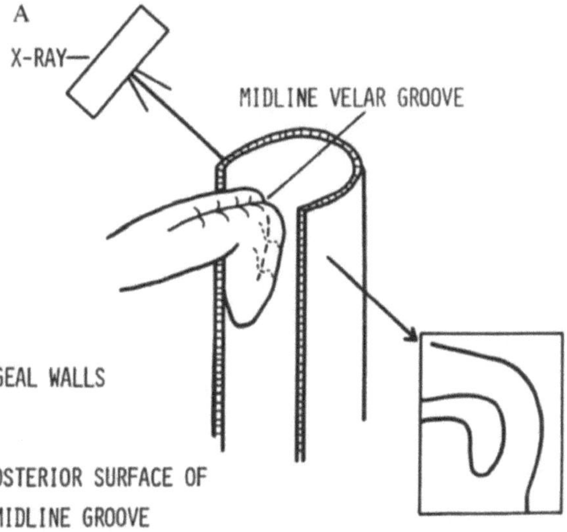

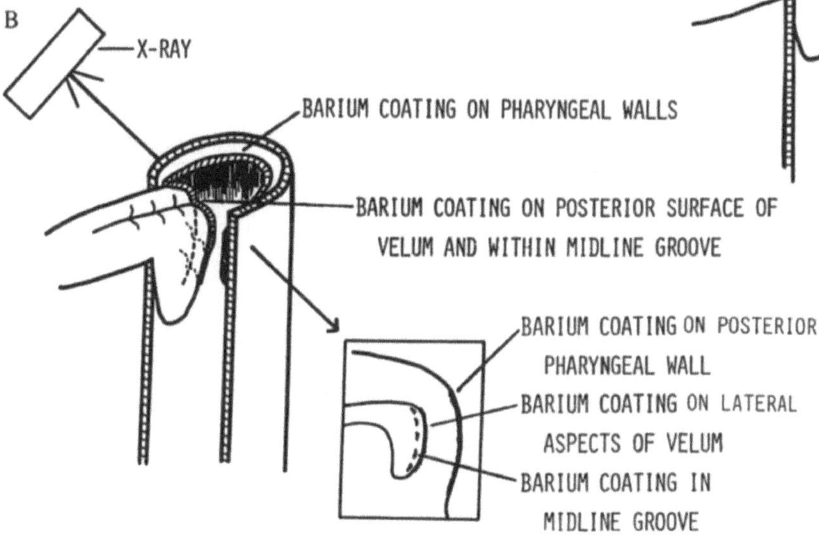

from the nasally inserted barium or from the barium swallow (that is performed before nasal installation of barium), to define the interface between the tongue and the velum during nasal breathing.

When the normal subject produces non-nasal speech, the palate elevates to approach or contact the posterior pharyngeal wall. Elongation of the palate, or "velar stretch," also occurs (1). A localized knee or eminence forms at the junction of the more anterior and approximately horizontal portion of the velum and the more posterior and vertical region. This junction represents the region of insertion of the levator muscles into the velum. The upper surface of the eminence protrudes slightly above the level of the horizontal portion of the palate, and is usually located approximately two-thirds of the distance from the posterior margin of the hard palate to the tip of the uvula. The region of velopharyngeal contact is usually between the vertical portion of the palate and the posterior pharyngeal wall, below the level of the eminence (Fig. 6.3).

Abnormal eminence position or configuration may be seen during phonation in patients with velopharyngeal insufficiency. In some noncleft palate patients with a normal appearance to the velum during breathing, we have observed a normally shaped but more anteriorly positioned eminence during phonation. It appears as if the site of levator muscle insertion is more anterior than normal (Fig. 6.4). In many patients with a repaired cleft palate, the velum does not elevate to the hard-palate level, and a distinct 90° eminence is not formed. Instead the posterior velar surface has a more gently sloping shape (Fig. 6.5).

Abnormalities of velar contour, length, and thickness may be seen in cleft palate patients following primary closure of the cleft. Often these palates are shorter than normal, and there may not be a well formed uvula (Fig. 6.6). The palates may also be thinner or thicker than normal. The increased thickness may be attributable to the surgical repair and the thinness to tissue insufficiency. Furthermore, a groove or indentation centrally located in the dorsum of the palate that represents the site of surgical closure of the cleft may be present. Nasally instilled barium may enter this defect and appear as a second line within the palate adjacent to the dorsal surface (Fig. 6.7). This defect may permit nasal escape of air during speech.

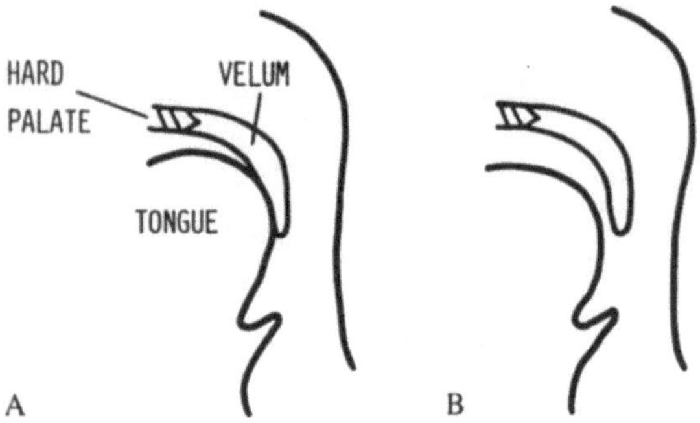

FIGURE 6.2. Lateral view: nasal versus oral breathing. (A) Nasal breathing: Velum rests on tongue. (B) Oral breathing: Velum is elevated above tongue.

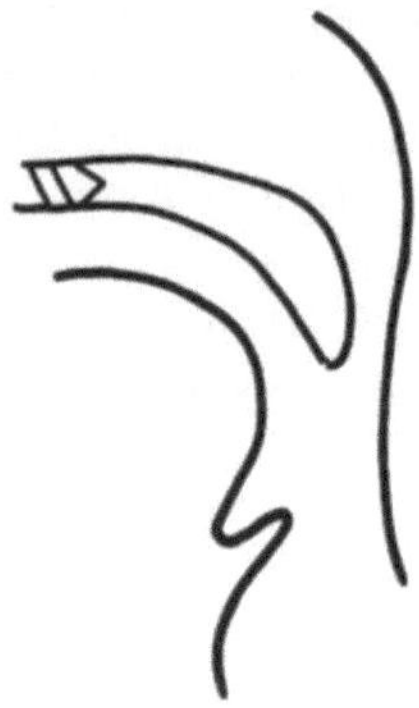

FIGURE 6.3. Lateral view, normal subject with velopharyngeal closure. Vertical portion of velum makes contact with posterior pharyngeal wall. Eminence rises above horizontal portion of velum.

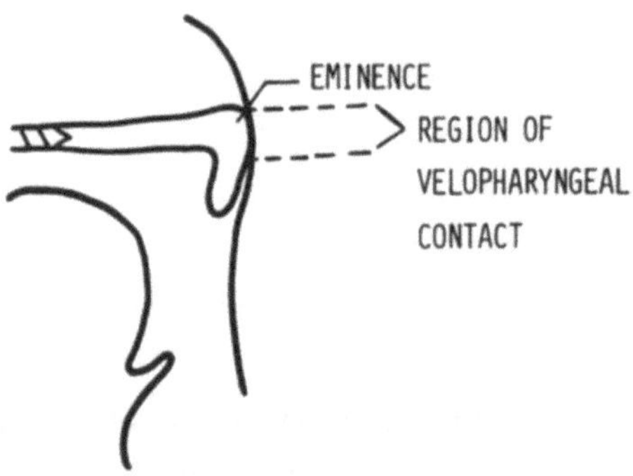

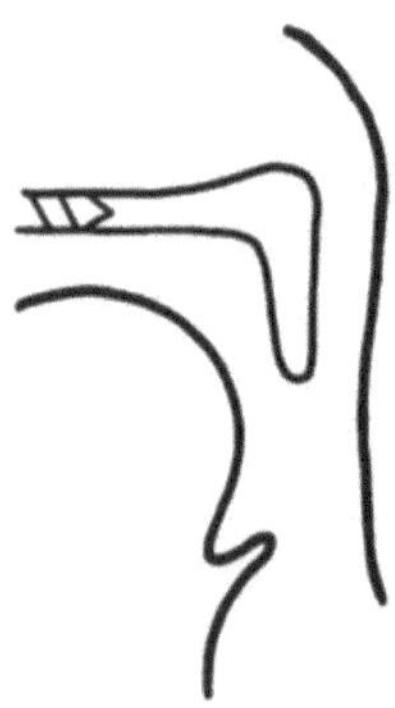

FIGURE 6.4. Lateral view during phonation in non-cleft-palate patient with velopharyngeal insufficiency. Because velar eminence is more anterior than normal (compare with Fig. 6.3), horizontal portion of velum is too short to extend to posterior pharyngeal wall.

FIGURE 6.5. Lateral view: phonation. Repaired cleft-palate patient with velopharyngeal insufficiency. Absence of distinct velar eminence because velum does not elevate enough.

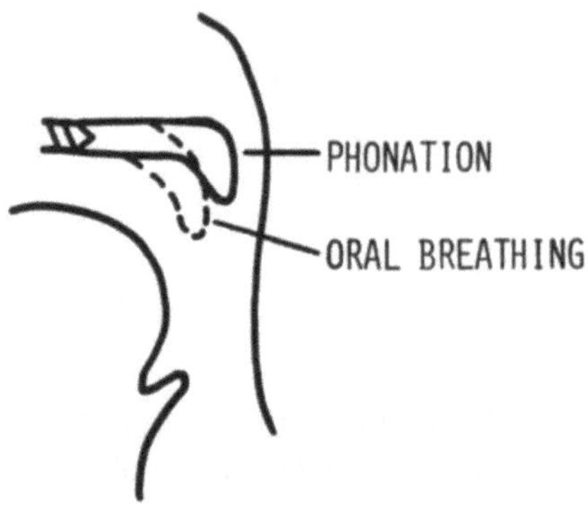

FIGURE 6.6. Lateral view. Repaired cleft palate with velopharyngeal insufficiency. Palate is shorter than normal (compare with Fig. 6.3).

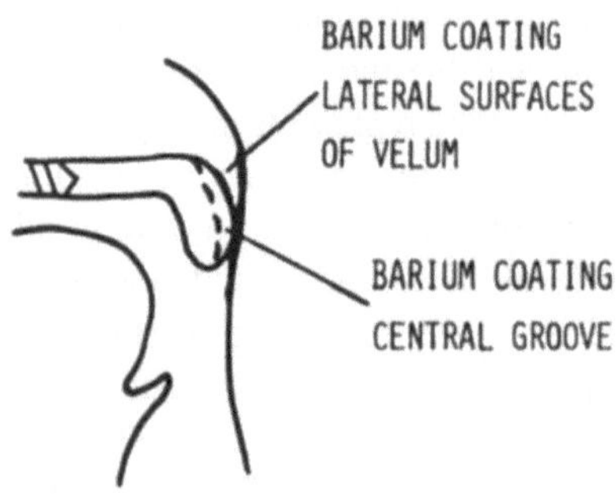

FIGURE 6.7. Lateral view. Repaired cleft palate with mild velopharyngeal insufficiency secondary to dorsal groove in palate. While lateral surfaces of palate touch posterior pharyngeal wall, air escapes through central groove.

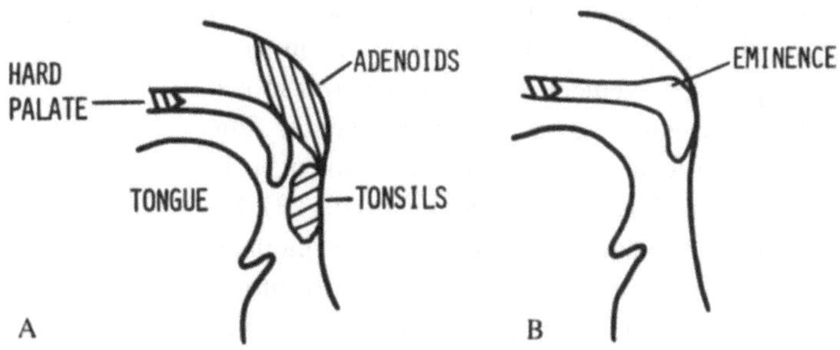

FIGURE 6.8. Lateral view. Effects of enlarged adenoids on velar contours and elevation during phonation. (A) Enlarged adenoids. Velar elevation limited and no eminence forms. (B) No adenoids. Velum fully elevates and distinct eminence forms.

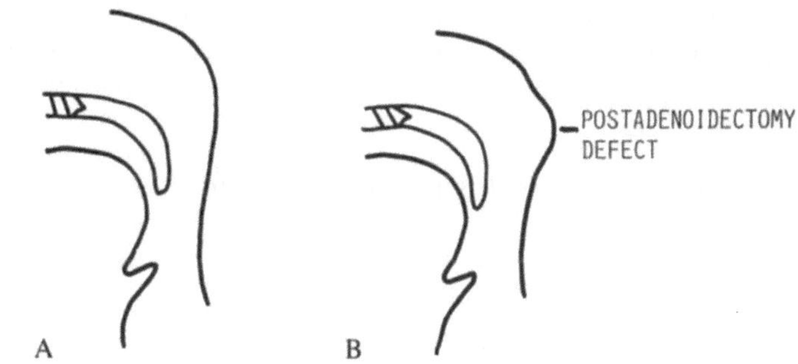

FIGURE 6.9. Lateral view. Post adenoidectomy. (A) Smooth contour of posterior pharyngeal wall. (B) Localized postadenoidectomy defect.

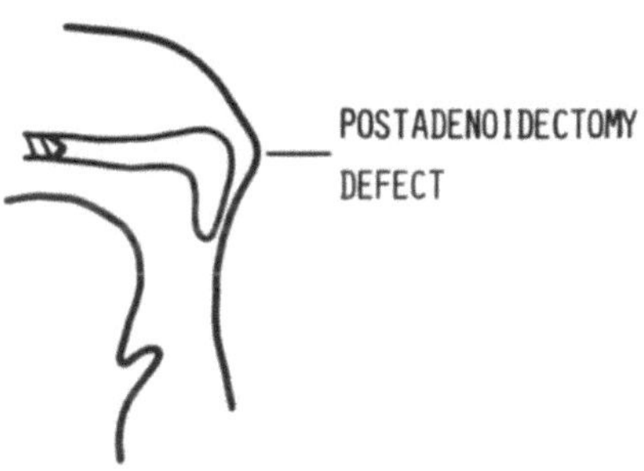

FIGURE 6.10. Lateral view, phonation. Velopharyngeal insufficiency is present because concave defect in posterior pharyngeal wall following adenoidectomy prevents velum from sealing against pharyngeal wall.

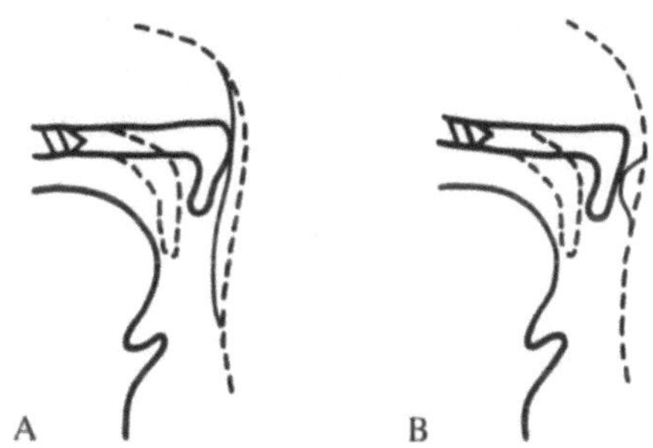

FIGURE 6.11. Lateral views. Variations in posterior pharyngeal movements during phonation. (A) Broad anterior movement of pharyngeal wall along nasopharynx. (B) Localized anterior movement: Passavant's ridge. Broken line designates positons of tissues at rest. Solid line designates positions of tissues during phonation.

Posterior Pharyngeal Wall

Quiet breathing is a good time to examine the posterior pharyngeal wall. The presence and size of both the adenoids and the tonsils should be noted. When barium coats these structures, one may see flecks or arcs of barium within the barium-coated margins of the adenoids or tonsils. These flecks represent barium within tissue crypts. It is important to assess the extent to which the adenoids and tonsils respectively impinge into the naso- and oropharyngeal airway, and to correlate these observations to any upper-airway respiratory problems the patient may have. In patients with enlarged adenoids, the extent of velar elevation during phonation may be limited and a distinct velar eminence may not appear (Fig. 6.8)

When no adenoid mass is seen, the posterior pharyngeal wall contour should be scrutinized to see if it is a smoothly continuous margin or if there is a localized posterior indentation high in the nasopharynx (Fig. 6.9) that may be a result of adenoid removal. If the patient has clinical signs of VP inadequacy, this postoperative tissue defect should be carefully scrutinized to determine if it is causing or contributing to the inadequacy (Fig. 6.10).

The posterior pharyngeal wall is not always the static structure many assume it to be. It can show two types of movement: a uniform anterior movement of the entire nasal pharyngeal length, (Fig. 6.11A) or a localized anterior movement of a selected region (Passavant's ridge) (Fig. 6.11B). Either type of movement may contribute to velopharyngeal portal closure. The selected anterior movement represents the bulge produced by Passavant's ridge, which will be described in Chapter 9. It is worth reemphasizing that this ridge is often not appreciated in a lateral view without a barium coating on the pharyngeal wall.

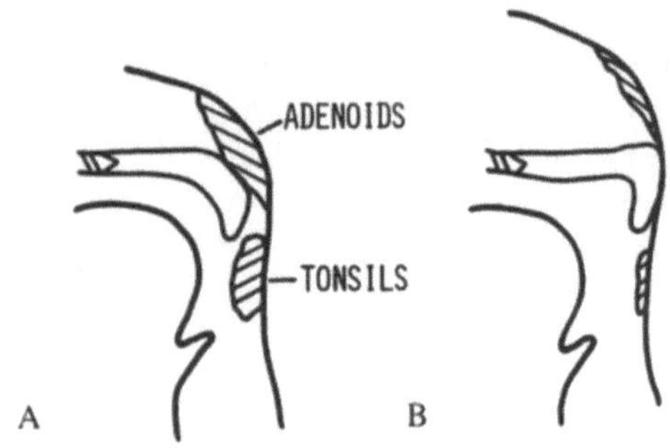

FIGURE 6.12. Lateral views. Variations in patterns of velopharyngeal closure with age. (A) Young child. Velum closes against prominent adenoids. Vertical dimension of nasopharynx is small. (B) Older child. Velum closes against posterior pharyngeal wall below adenoids. Adenoids have diminished and vertical dimension of nasopharynx has increased. Note changes in tonsillar tissue with age.

Velopharyngeal Closure/Insufficiency (VPI)

In this section we will discuss the videofluoroscopic findings that suggest VP competence or insufficiency.

The region of contact or greatest narrowing between the palate and posterior pharyngeal wall is variable, and depends upon age, the size of the adenoid mass, and the length and degree of velar elevation.

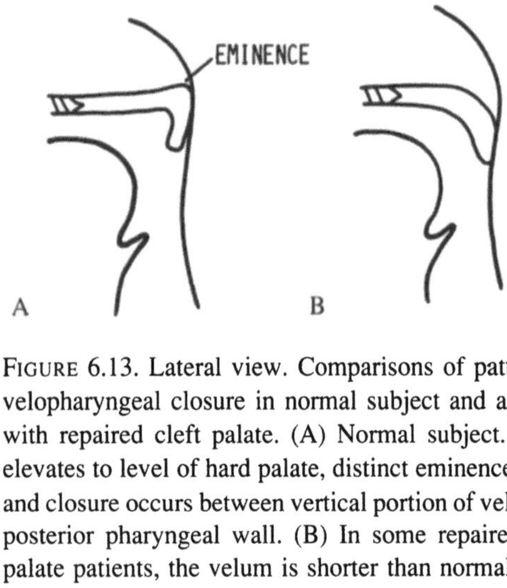

FIGURE 6.13. Lateral view. Comparisons of patterns of velopharyngeal closure in normal subject and a patient with repaired cleft palate. (A) Normal subject. Velum elevates to level of hard palate, distinct eminence forms, and closure occurs between vertical portion of velum and posterior pharyngeal wall. (B) In some repaired cleft-palate patients, the velum is shorter than normal, and it does not elevate to the level of the hard palate. No distinct eminence forms. The region of the velum contacting the posterior pharyngeal wall may be at a more distal position than in the normal subject.

Age Factors

In the younger child, the vertical dimension of the nasopharynx is less than in the older child. The younger child also has a larger adenoid mass than the older one. When the adenoids are enlarged, palatal elevation may be limited by the protruding adenoid bulk and eminence formation is less distinct. The region of VP contact is then usually between the region of the eminence and the inferior surface of the adenoids because the adenoids prevent the palate from reaching the posterior pharyngeal wall (Fig. 6.12A).

As the child grows, the vertical dimension of the nasopharynx increases and the size of the adenoid mass usually decreases. When there is enough room for full velar elevation, the region of approximation or contact occurs between the vertical portion of the velum (distal to the eminence) and the posterior pharyngeal wall (Fig. 6.12B). The increase in length of the palate during phonation [termed "velar stretch" by Graber (1)] is greater in adults than children (2) and greater in older than younger children (3). In the repaired-cleft-palate patient, regardless of age, the velum may not elevate to the same height as in the normal subject. Instead, the region of contact or greatest approximation between the palate and posterior pharyngeal wall may occur at a lower level in the nasopharynx. The velum in the repaired cleft palate may also lack the distinct knee or eminence seen in normal subjects (Fig. 6.13).

These age-related differences in velopharyngeal structure and function are described in further detail in Chapter 5.

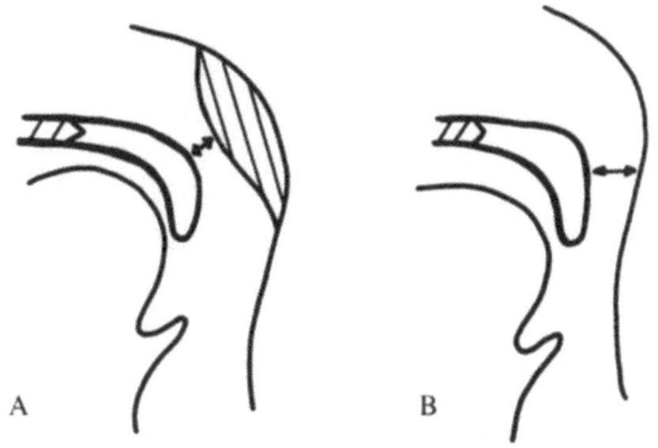

FIGURE 6.14. Lateral views, phonation. Technique for measuring degree of velopharyngeal insufficiency. The site of measurement is the region at which the velar surface is closest to the surface of the pharyngeal wall (<——>). (A) Minimum distance is between velum and enlarged adenoids. (B) Minimum distance is between velum and posterior pharyngeal wall.

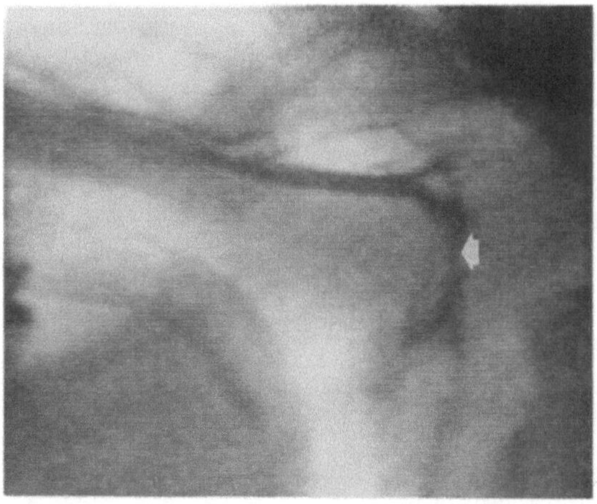

FIGURE 6.15. Lateral view during phonation. Thin line of barium (arrowhead) between velum and posterior pharyngeal wall indicates that the opposing velar and pharyngeal surfaces are in contact producing effective closure.

Definition of Velopharyngeal Gap

In patients who do not show velopharyngeal closure, the velopharyngeal gap is defined as the distance between the superior (nasal) surface of the velum and posterior pharyngeal wall, and is measured at the level of greatest narrowing between these two opposing surfaces. This level may be between the nasal surface of the velum and the inferior surface of the adenoid mass in the younger child, or between the vertical portion of the velum and the posterior VP wall in the older child (Fig. 6.14).

It is important to remember that a significant velopharyngeal gap is always present with nasal phonemes and of course the widest gap is seen in nasal breathing. The assessment of the presence or extent of a velopharyngeal gap should be made while the patient is producing non-nasal consonants. (See Chapters 11 and 14.)

Signs of Velopharyngeal Closure

Effective velopharyngeal closure is shown by the presence of a thin line of barium between the velum and posterior pharyngeal wall (Fig. 6.15). In some patients, the nasal surface of the velum may actually indent the posterior pharyngeal wall at the site of closure (Fig. 6.16). It is our impression that closure with velar indentation of the pharyngeal wall indicates a very forceful or effective closure.

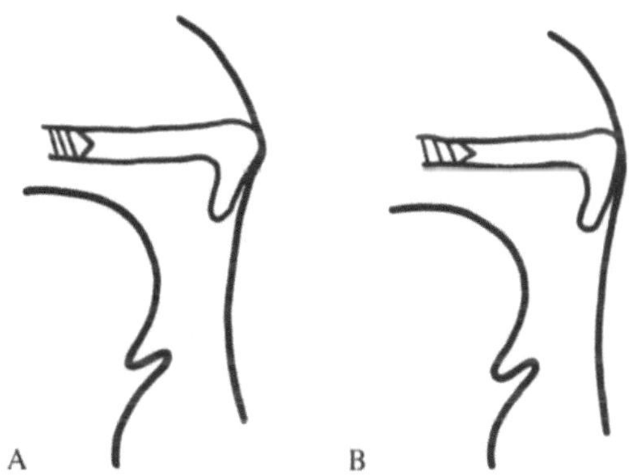

FIGURE 6.16. Lateral view during phonation. Variations in velopharyngeal closure. (A) Velum indenting surface of posterior pharyngeal wall. (B) Velum touching surface of posterior pharyngeal wall.

Borderline Velopharyngeal Insufficiency

When the line of barium interposed between the velum and posterior pharyngeal wall widens, it is indicative of a small velopharyngeal gap because the velum and pharyngeal wall are no longer in contact (Fig. 6.17). As the velopharyngeal gap increases, the barium-coated velum and pharyngeal wall appear as two separate lines of barium with a space inbetween representing air (Fig. 6.18).

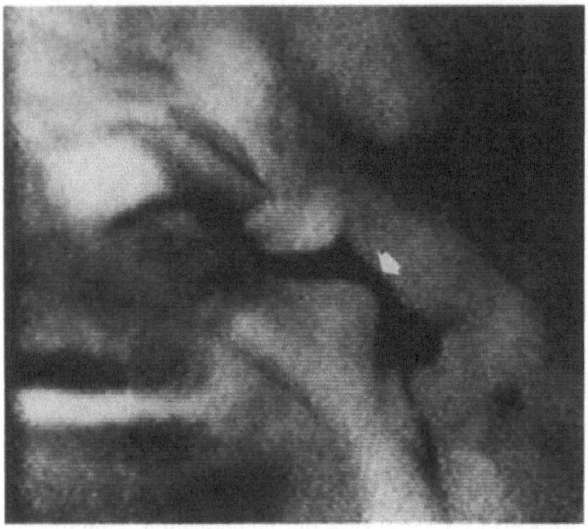

FIGURE 6.17. Lateral view during phonation in patients with velopharyngeal insufficiency. Barium (arrowhead) between velum and pharyngeal wall has increased in thickness (compare to Fig. 6.15) because opposing surfaces do not touch. Barium mixture is thick enough and velopharyngeal gap small enough that barium can fill space between velum and pharyngeal wall.

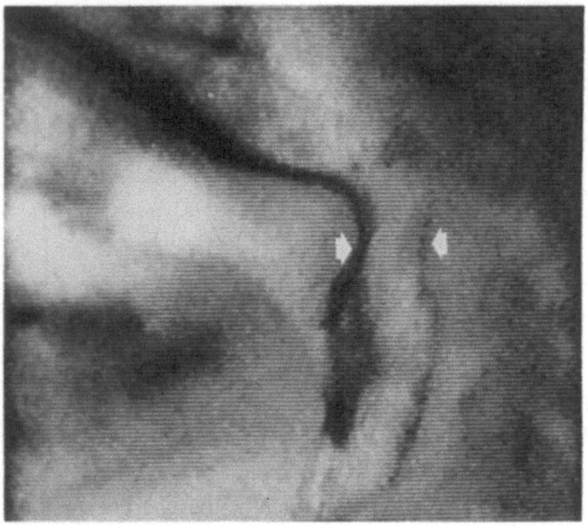

FIGURE 6.18. Lateral view during phonation in patient with velopharyngeal insufficiency and greater distance between velum and pharyngeal wall than in Figure 6.17. Barium now individually coats surfaces of velum and pharyngeal wall (arrowheads). Space between coated surfaces contains air.

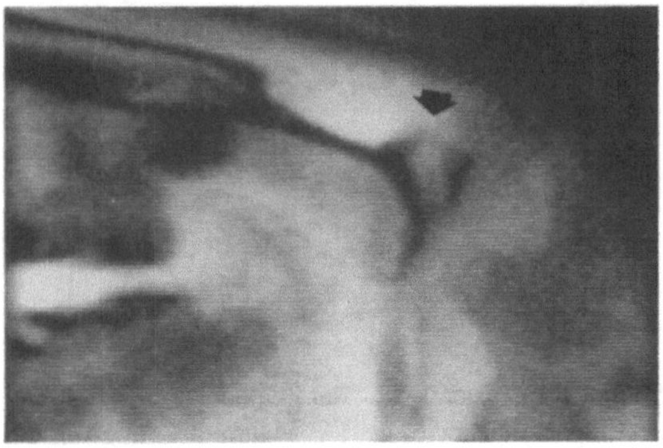

FIGURE 6.19. Lateral view during phonation in patient with minimal velopharyngeal insufficiency. The presence of barium bubbling (arrowhead) above the velopharyngeal portal indicates that there is a small gap between the velum and pharyngeal walls through which air is blowing.

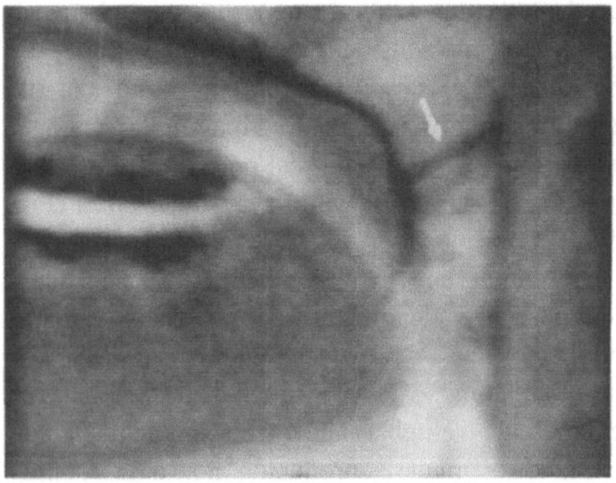

FIGURE 6.20. Lateral view during phonation in patient with velopharyngeal insufficiency. Diagonal line of barium (arrow) is a strand of mucus coated with barium.

Another sign of a minimal velopharyngeal gap is the appearance of barium-coated bubbles arising from the superior surface of the velopharyngeal portal during phonation (Fig. 6.19). Bubbling occurs when the velopharyngeal gap is narrow and the air rushing through mixes and carries the barium with the stream. The presence of bubbling does not, however, indicate where within the velopharyngeal portal the gap exists. The frontal and base views are required to define the actual locus of the gap. As the gap widens bubbles are less likely to appear.

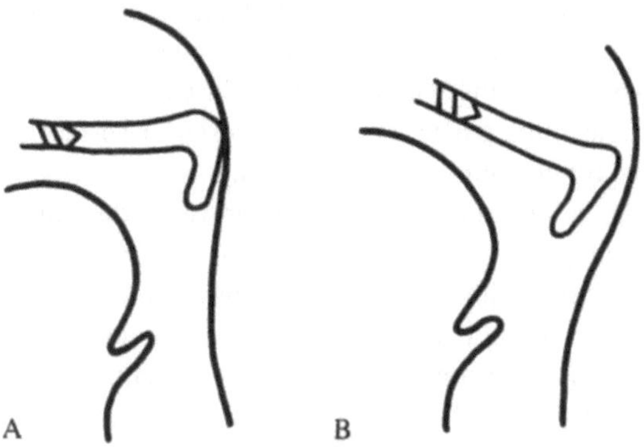

FIGURE 6.21. Lateral view during phonation comparing effects of neutral versus hyperextension on velopharyngeal closure. (A) Neutral position: marginal closure. Gentle touch between velum and posterior pharygneal wall. (B) Hyperextension position: definite velopharyngeal gap appears.

Barium Artifacts

Occasionally, a horizontally or diagonally oriented line of barium is seen between the velum and posterior pharyngeal wall, which changes in shape or position during phonation (Fig. 6.20). This line represents the barium coating of a strand of mucus connecting the velum and posterior pharyngeal wall, and should not be confused with barium coating the superior surface of a pharyngeal flap. Besides being inconsistent in appearance, this line will vanish when the patient blows his nose.

Timing of Palatal Movement with Speech

The timing of palatal movement in relationship to speech is another important area of observation.

In the normal subject, the palatal movement should be smooth and approximately coincident with speech production. The term *approximately coincident* is used because actual palatal movement is initiated slightly before speech is heard (4). When the speech task consists of a series of intermixed non-nasal, nasal, and non-nasal phonemes, one should observe palatal elevation with the non-nasal phoneme, a momentary palatal depression for the nasal phoneme (although velar height remains above rest position), and a rapid reelevation for the non-nasal phoneme (5). Closure is not always achieved during vowel production, especially when vowels are next to or near a nasal consonant (6). However, these subtle changes in VP function may not be detectable unless the videotape is reviewed in slow motion. In general, when viewing videofluoroscopy of connected speech with no nasal phonemes, the palate should remain elevated throughout the entire speech task.

We have seen several types of abnormal timing. Some palates move slowly and sluggishly whereas others move in a very staccato and discontinuous manner. The resultant speech parallels the abnormalities seen in palatal motion. When the motion is slow, the speech has a sluggish quality and the individual phonemes are often not discretely enunciated. With the staccato type of palatal motion, the palate partially descends after each consonant sound is uttered, rather than remaining elevated in a closed position until the next sound. When abnormal timing of palatal motion is present, the lateral pharyngeal walls and tongue should be examined because their movements may be similarly abnormal.

Velopharyngeal Relationships and Head Position

A change in head position can change the relationships between the velum and posterior pharyngeal wall. The usual lateral view is obtained with the head in neutral position. In practice this means that the plane of the hard palate is parallel to the plane of the floor when the patient is erect. When the head is hyperextended upon the neck, the anterior–posterior dimension of the nasopharynx increases, and when the head is flexed, the converse occurs. In patients who demonstrate a marginal mechanism of closure on the lateral view in neutral position (i.e., there is gentle touch between the posterior pharyngeal wall and the opposing velar surface), a definite VP gap may be seen in the hyperextension position because of the increase in anterior–posterior dimension of the nasopharynx (Fig. 6.21). This maneuver, first described by McWilliams and Musgrave (7), has been very helpful in discriminating between a minimally inadequate VP mechanism and a normal mechanism. However, when there is an obvious midsagittal VP gap with the head in neutral position, there is no need to repeat the examination in hyperextension.

Another effect of the hyperextension position is that it increases the size of the Passavant's ridge, or at times demonstrates a Passavant's ridge that is not seen in the neutral position (Fig. 6.22) (8). Whatever the mechanism is that produces Passavant's ridge in the neutral position, it is placed under greater stress because of the increased anterior–posterior dimension of the nasopharynx in the hyperextension position. Therefore, the ridge becomes more obvious in the latter position (see Chapter 9 for further discussion).

Pharyngeal Flap

The presence of a pharyngeal flap is appreciated on the basis of two observations. First is the visualization of the flap as a soft-tissue bridge connecting the posterior pharyngeal wall to the posterior surface of the velar eminence. However, if the flap is thin, one may have difficulty in seeing it. When barium is introduced into the nasopharynx, it may mix with the nasal mucus and obscure this soft-tissue bridge. The second observation relates to the inability of the palate connected to a pharyngeal flap to rest against the dorsal surface of the tongue during quiet nasal breathing with the mouth closed. In a normal subject, the palate touches the posterior surface of the tongue during nasal breathing. If a pharyngeal flap is present, then the flap prevents the palate from going so far forward. This observation of nasal breathing, although valid even after the nasal instillation of barium (Fig. 6.23), is best done without nasal barium so as to enhance visualization of the flap as well.

If the patient breathes instead through his mouth, the palate elevates because the air flows via the oral pharynx into the trachea, and the effect of the flap restricting anterior movement of the palate cannot be appreciated.

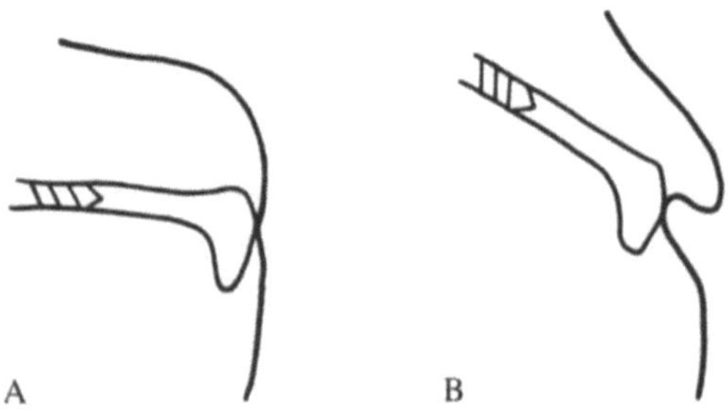

FIGURE 6.22. Lateral view during phonation demonstrating appearance of Passavant's ridge in hyperextension position (B) as compared with its absence in neutral position (A).

Frontal View

The main purpose of the frontal view is to visualize the lateral pharyngeal walls and evaluate their movement during speech. Normally, both lateral nasopharyngeal walls exhibit the same degree of medial movement with speech, and the vertical extent of the medially moving pharyngeal walls are bilaterally symmetrical (Figs. 6.24, 6.25A). A secondary purpose is to identify the level of the velar eminence so as to relate the region of maximum medial movement of the lateral pharyngeal walls during speech to the level to the velum. In normal subjects the plane of the velopharyngeal sphincter lies just below the eminence so on frontal view the maximum region of lateral pharyngeal movement is slightly below the level of the eminence (Fig. 6.25A). In certain patients, especially those post

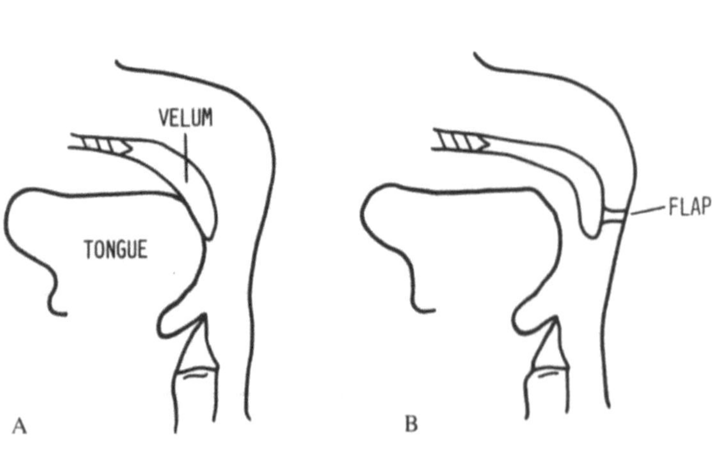

FIGURE 6.23. Lateral views during nasal breathing comparing velar position in patients without and with a pharyngeal flap. (A) No flap: Velum rests on dorsum of tongue. (B) Flap present: There is a gap between velum and tongue because flap restrains velar descent.

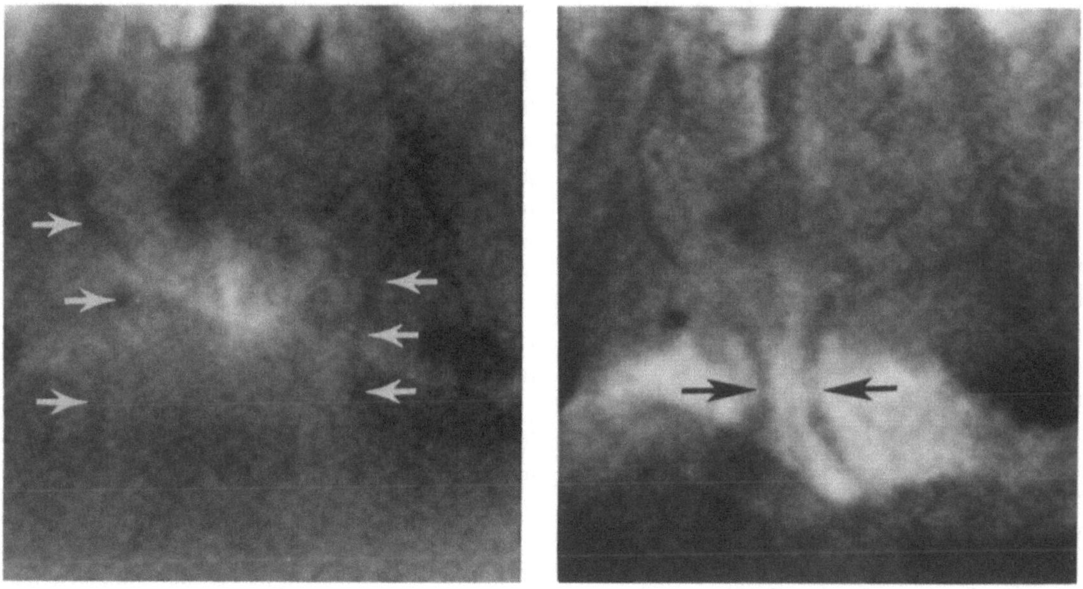

FIGURE 6.24. Frontal views, normal subject. (A) Quiet breathing. Barium-coated lateral pharyngeal walls (white arrows) are relaxed. (B) Phonation. Localized medial movement of lateral pharyngeal walls (black arrows).

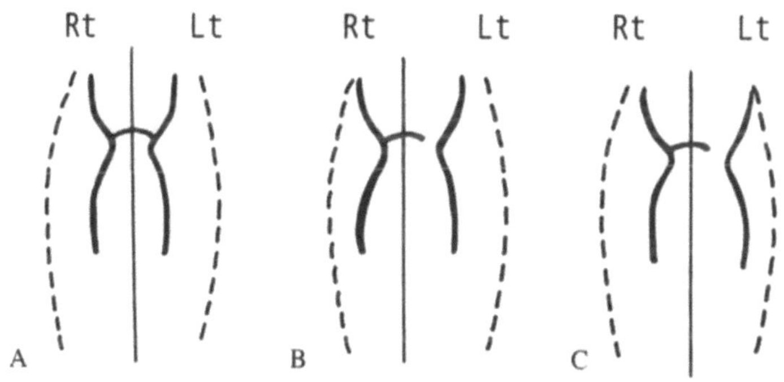

FIGURE 6.25. Frontal views during rest (dotted lines) and phonation (solid lines). (A) Lateral pharyngeal walls symmetrically positioned on either side of midline during rest and phonation. During phonation both lateral walls touch sides of velum. (B) Asymmetrically positioned lateral pharyngeal walls at both rest and phonation. Both pharyngeal walls have the same amount of movement. Because the right wall is closer to the velum, closure occurs on the right but not left side. (C) At rest both walls are equidistant from the midline. During phonation the right wall has more medial movement than the left so closure occurs only on the right side.

cleft palate repair, the velar eminence will not be seen on frontal view because the velum does not elevate enough to form a distinct eminence.

Asymmetric Lateral Pharyngeal Walls (LPW)

Occasionally a patient may show an asymmetric position of the lateral pharyngeal walls at rest. One wall may be more laterally positioned from the midline than the opposite one. Such asymmetries can be the cause of VP insufficiency if the pharyngeal wall that is further from the midline cannot move as close to the midline as the opposite lateral pharyngeal wall (Fig. 6.25B). Thus, one should carefully search for asymmetries in lateral pharyngeal wall movement at rest or during quiet breathing. One may also occasionally see symmetrically positioned lateral pharyngeal walls at rest that during phonation exhibit asymmetric medial movement with one wall having greater medial movement than the other (Fig. 6.25C).

Quantifying Lateral Pharyngeal Wall Motion

Because evaluation of pharyngeal wall movement during speech is the main concern of the frontal view, we have devised a method that we have found to be clinically practical and easy to learn for quantifying LPW motion. A vertical line running through the midplane of the nasopharynx is used as a reference plane. In patients in whom the lateral pharyngeal walls are equidistant from this midplane at rest, we visually divide the distance between the midplane and the position of the lateral pharyngeal wall at rest into four equal spaces. The rest position to the first division comprises 25% of the distance, to the second division 50%, to the third division 75%, and to the midplane the 100% position (Fig. 26A). We visually determine which imaginary percentile line the lateral pharyngeal walls reached during phonation. Then we define the medial movement as that percent of the total distance that it can potentially move toward the midline. For example, if a lateral pharyngeal wall moved approximately halfway between its rest position and the midline, we say that it has moved approximately 50% during phonation (Fig. 6.26B). Each lateral pharyngeal wall is separately rated because occasionally there are asymmetries in the degree of medial movement and it is important to appreciate these asymmetries. Occasionally the nasopharyngeal walls move laterally instead of medially during phonation. The same percentage system can be used, but now we will refer to a 25% or 50% lateral movement from its rest position. Such lateral movement is rarely seen, however.

When the lateral pharyngeal walls are asymmetric at rest, that is, one is more distant from the midplane than the other, the percentage system of evaluating their degree of movement will be misleading. If both lateral pharyngeal walls move medially by 50%, the wall that was further from the midline to begin with will still be further from the midline at its 50% movement position. To handle this infrequent situation, we recommend that actual measurements be made of distances of each of the LPWs from the midline at rest and at maximum medial motion during speech, using the techniques for demagnifying and measuring dimension of the velopharyngeal portal described in Chapter 4.

Lateral Pharyngeal Wall Movement as a Guide to Pharyngeal Flap Surgery

A major purpose for evaluating lateral pharyngeal wall movement is to help predict how effective a

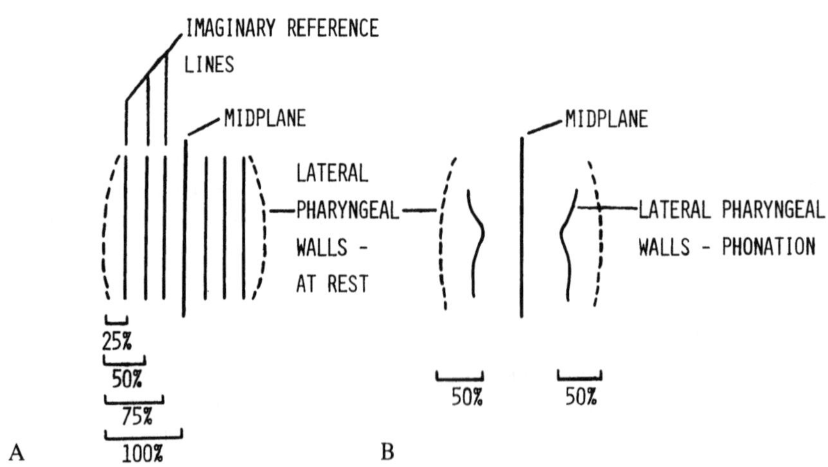

FIGURE 6.26. Method for measuring degree of lateral pharyngeal wall movement during phonation from frontal view. (A) At rest. A series of imaginary vertical reference lines divides each half of the pharynx. Zero movement is defined as the position of pharyngeal walls at rest and 100% movement occurs where the pharyngeal walls meet in the midplane. Intermediate positions of the pharyngeal walls are defined by the vertical lines at 25%-50%-75% positions from rest position. (B) During phonation. Each pharyngeal wall in this example moved halfway between rest and midplane so that each wall reached the 50% imaginary line or achieved 50% medial movement.

pharyngeal-flap operation might be for improving speech in a given patient. In patients with active and symmetric medial movement of the lateral pharyngeal walls, the pharyngeal flap operation is usually more successful because the width of the pharyngeal flap does not have to be as wide as when there is reduced pharyngeal wall activity (9) (see Chapter 10 for further details). Reduced lateral pharyngeal wall motion may require a wider flap, which may increase the chance of hyponasal speech postoperatively. The narrower flap means that the portals on either side of the flap can be wider so as to facilitate nasal breathing. Asymmetric lateral pharyngeal motion may pose a difficulty to the surgeon.

Although the surgeon can roughly design the width of the pharyngeal flap to be wide, average, and perhaps narrow to compensate for different degrees of lateral pharyngeal wall movement, these measurements are only approximate. Therefore, we feel that providing millimeter precision information concerning the distance between each lateral pharyngeal wall and the midline would not give the surgeon additional information because he cannot fashion the width of the velum with the same precision.

For those readers who do want to make precise measurements for research purposes, please refer to Chapter 4, which describes the use of the magnification marker to relate dimensions on the TV monitor to actual anatomic dimensions.

Relationship Between Naso- and Oropharyngeal Walls

The pharyngeal walls as seen on the frontal projection extend from the nasopharynx into the oropharynx. During phonation the nasopharyngeal walls usually move medially while the oropharyngeal walls may not move or may even move in the opposite direction.

During swallowing the movements of the naso- and oropharyngeal walls are completely different. Shprintzen (10) observed that in the oropharynx the pharyngeal walls moved medially to meet in the midline, whereas at the nasopharyngeal level there was no medial movement. At maximal effort during spontaneous swallowing, a Y-shaped pattern was seen. The two diagonal limbs of the Y represented the convergence of the pharyngeal walls to the junction between the naso- and oropharyngeal levels. The oropharyngeal walls opposed each other in the midline to form the central vertical portion of the Y. Palatal elevation was less in reflexive swallowing than speech and the region of velopharyngeal contact was at a lower level for swallowing than speech (Fig. 6.27). Shprintzen designated this juncture between the three lines as the junctural region between the naso- and oropharyngeal walls. These pharyngeal wall movements during swallowing were constant in both normal subjects and subjects with abnormal movement of nasopharyngeal walls during speech.

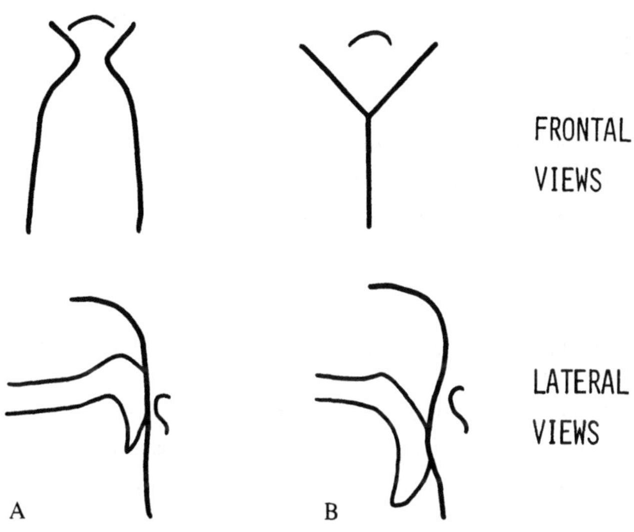

FIGURE 6.27. Comparison of palatal and lateral pharyngeal wall movements in speech and reflexive swallowing. (A) During speech the level of velopharyngeal closure occurs in the nasopharynx. The lateral pharyngeal walls show good medial movement in the nasopharynx and no movement in the oropharynx. (B) During reflexive swallowing the level of closure occurs lower at the oropharynx level. There is midline contact of the lateral pharyngeal walls within the oropharynx and no movement at the nasopharynx level. Modified from Figure 5, Shprintzen RJ, Lencione RM, McCall GN, Skolnick ML: A three dimensional cinefluoroscopic analysis of velopharyngeal closure during speech and nonspeech activities in normals. Cleft Palate J 11:412–428, 1974. Used with permission.

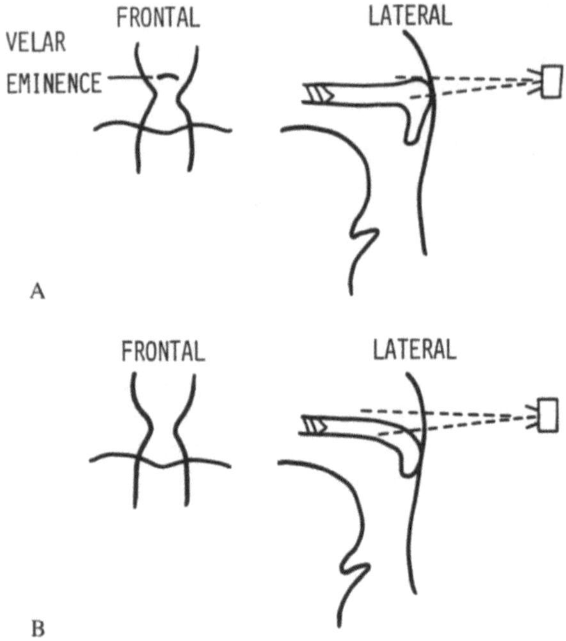

FIGURE 6.28. Frontal and lateral views during speech comparing the appearance of a velum with a distinct eminence versus one with no eminence. (A) Distinct eminence on frontal view. Arcuate line between lateral pharyngeal walls represents projection of barium-coated superior surface of eminence, which is seen because it is tangential to the x-ray beam. (B) Velum is not seen on frontal view. Because velum does not form distinct eminence, superior surface does not orient itself tangential to x-ray beam.

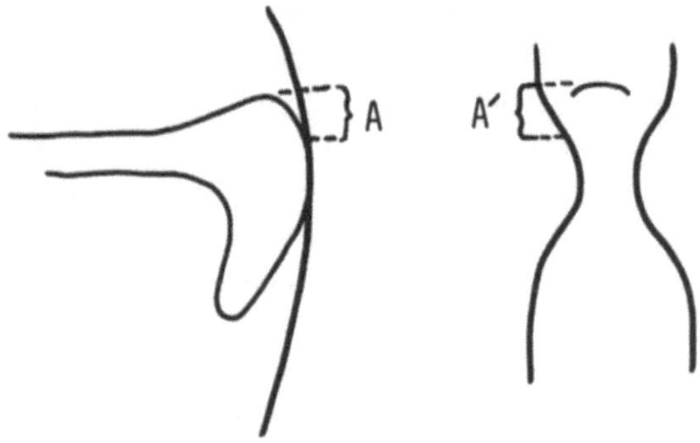

FIGURE 6.29. Using velar eminence as reference for relative level of velopharyngeal closure seen on lateral view to that on frontal view. Distance A between velar eminence and zone of midsagittal velopharyngeal contact on (A) lateral view should equal distance A' between velar eminence and maximum medial movement of lateral pharyngeal walls on (B) frontal view.

Velum in Relationship to Lateral Pharyngeal Walls

Identifying the velum on the frontal view is more difficult than identifying the margins of the lateral pharyngeal walls because the velum can only be seen when its superior surface is tangential to the x-ray beam during phonation. The only region of the velum that potentially can be visualized on the frontal view is the velar eminence. In order for a well-defined velar eminence to be visible during speech, the barium coating the superior surface of the eminence must be projected tangential to the x-ray beam. For similar reasons, the palate cannot be seen in quiet breathing.

When the palatal eminence is identified on frontal view during phonation, it is important to observe how far below the level of the eminence the maximal medial movement of the lateral pharyngeal walls occurs, and whether this maximal medial movement reaches the lateral edges of the eminence. (Usually the eminence is above the level of maximum lateral pharyngeal wall motion.)

After having determined the level at which maximum lateral pharyngeal wall motion occurs on the frontal view, and the vertical distance of that movement in relationship to the velar eminence, the lateral view should be reexamined at the same distance below the eminence, to determine what the midsagittal velopharyngeal relationships are at this level (Fig. 6.29). Because the VP portal closes in a sphincteric manner, the maximum medial movement of the lateral walls seen on frontal projection should occur at the same level as VP contact or maximum narrowing on the lateral view. If these relationships are not consistent, the patient may be improperly positioned or there may be inadequate barium coating of the velum or pharyngeal walls.

These velopharyngeal relationships are best appreciated in subjects with a Passavant's ridge. This ridge is a crescent-shaped arc of soft tissue projecting centripetally from the lateral and posterior pharyngeal walls. It thus forms an excellent marker correlating the region of maximum constriction of the pharyngeal wall movement seen on the frontal view, with the region of greatest VP narrowing seen on the lateral view. Passavant's ridge is described in greater detail in Chapter 9.

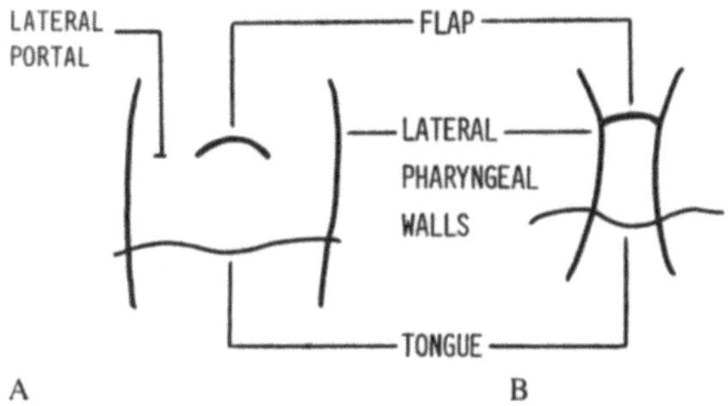

FIGURE 6.30. Frontal view: patient with velopharyngeal closure following pharyngeal flap surgery. Flap and pharyngeal walls are coated with barium. (A) At rest. Superior surface of barium-coated flap appears as a horizontal or convex superior line. Lateral portals, spaces between edges of flap and lateral pharyngeal walls, are open permitting patient to breathe nasally. (B) During phonation flap elevates and lateral walls touch edges of flap to close lateral portals.

Vertical Level of Lateral Pharyngeal Wall Movement

One should not assume that both lateral pharyngeal walls exhibit their maximum medial movement at the same vertical level, although such is usually the case. In our experience, even when both pharyngeal walls move at the same level, the degree of medial movement, and the vertical extent of this medial movement, may vary with different sounds in a given patient.

It is also important to ascertain that both lateral pharyngeal walls are well coated with barium in their rest position as well as during phonation. One should not assume that both walls move symmetrically if one wall is well seen and one wall is not well seen because of inadequate coating. Because of the possibility of asymmetric movement, observations made on movements of one wall should not be applied to the other wall. Instead, when one wall is poorly coated, the examination should be interrupted and additional barium introduced into the nasal opening on the side of the poor coating. As the barium is being dipped into the nose on the affected side, it may be helpful to have the patient tilt the head to that side as the head is hyperextended. After a 2- or 3-minute pause, the frontal examination can be repeated. It may be even more advantageous to recoat both sides if the patient can tolerate this procedure as the other side may lose its coating

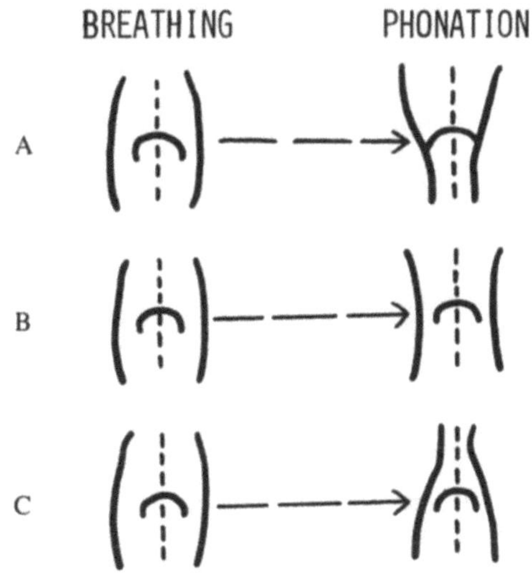

FIGURE 6.31. Frontal view: with pharyngeal flap. Patterns of movement of lateral pharyngeal walls. (A) Both lateral walls are equidistant from flap at rest and close against edges of flap during phonation. (B) Both lateral walls are equidistant from flap at rest. During phonation pharyngeal walls show greatest medial movement at level of flap but degree of movement is not enough to close lateral portals. Velopharyngeal insufficiency persists. (C) Both lateral walls are equidistant from flap at rest. Maximum medial movement during phonation occurs above level of flap. No velopharyngeal closure occurs.

during the interval wait for the barium to dip down when one side is recoated.

The variations of pharyngeal wall movement in patients with and without pharyngeal flaps as they relate to different patterns of VP narrowing or closure are discussed in Chapter 10.

Pharyngeal Flap

In patients with a pharyngeal flap, the width and the level of the pharyngeal flap are well demonstrated on the frontal view provided that the superior surface of the pharyngeal flap is coated with barium. Nonbarium views in frontal projection, however, are of no value in delineating the pharyngeal flap. The pharyngeal flap is seen on frontal view because the barium-coated superior surface of the flap is roughly in a horizontal plane (Fig. 6.30A). Therefore, from the frontal projection one can determine the width of the pharyngeal flap and the level of the pharyngeal flap both during rest and phonation. The width of the flap usually does not change during

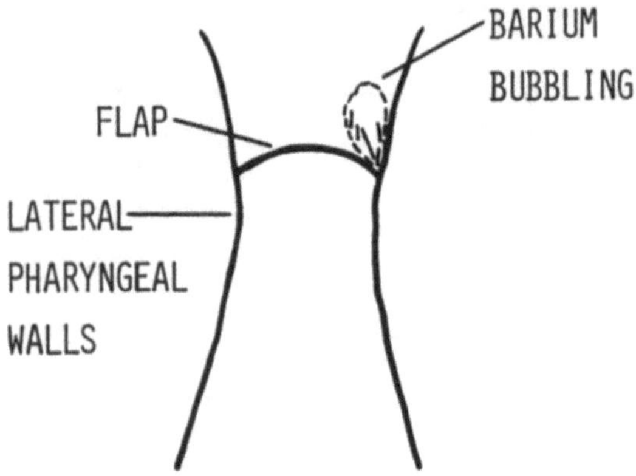

FIGURE 6.32. Frontal view: with pharyngeal flap. Lateral pharyngeal walls appear to touch edges of flap but presence of barium-coated bubbles appearing over apparently closed lateral portal indicates a small opening in the portal through which air escapes.

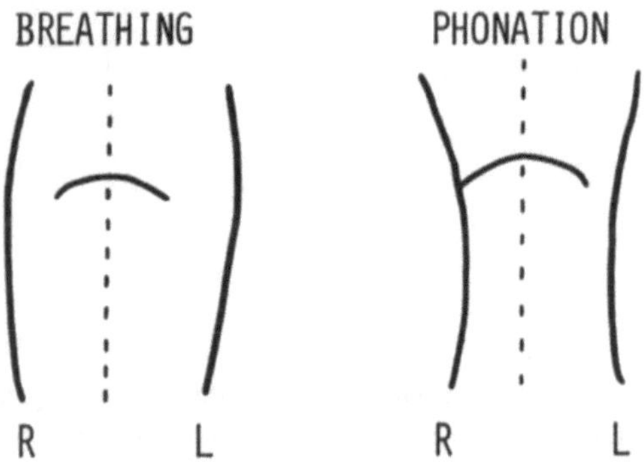

FIGURE 6.33. Frontal view: with pharyngeal flap. At rest: Flap is centered in midline (vertical dots) but one lateral pharyngeal wall (R) is closer to midline than other wall. During phonation: Both walls show equal medial movement but closure occurs only between wall closer to midline (R) and flap.

speech, but the level may rise as the palate elevates during phonation (Fig. 6.30B).

When a pharyngeal flap is present we are interested in two things that occur during phonation: (1) the vertical level of the flap in reference to the region of maximum medial movement of the lateral pharyngeal walls; and (2) the horizontal distance between each lateral pharyngeal wall and the adjacent edge of the flap at the level of the greatest medial excursion of each of the lateral pharyngeal walls. In other words, we are interested in determining whether or not the pharyngeal flap adequately fills the gap between the lateral pharyngeal walls at their maximum medial movement (Fig. 6.31) so as to effectively close the velopharyngeal portal. If the two lateral pharyngeal walls abut on the lateral edges of the flap, then VP closure is usually effective. The word "usually" is used because on a rare occasion only a portion of the lateral pharyngeal wall will touch the edge of the flap. In this situation the entire lateral pharyngeal wall does not exhibit enough medial movement to abut on the flap. As a result there is a small opening through a portion of the portal that may not be detected on frontal projection but is evident on the base view. One clue on frontal view to the presence of incomplete closure of a lateral portal is the appearance of bubbles of barium blowing through a lateral portal (Fig. 6.32).

The base view is the ideal view for detecting small areas of incompetence in a lateral portal that appears to close on the frontal projection. Although we try to obtain a base view in all patients, this view is especially important in those patients with pharyngeal flaps who appear to have VP closure on frontal projection but who still exhibit hypernasal speech.

Besides observing the level and the width of the pharyngeal flap in the nasopharynx, it is also important to observe whether the flap is positioned in the midline or off to one side. When there is an asymmetrically positioned flap, the lateral portal that has a greater width at rest may be incompetent if that pharyngeal wall cannot move medially enough to abut on the flap during phonation (Fig. 6.33).

Base View

What Is and What Is Not Revealed

The velopharyngeal portal can be likened to a tube with a section of the wall removed and replaced by the velum. The base view is the only view that examines the tube end on and projects an en face view of its lumen (Fig. 6.34). This lumen contracts during speech in a sphincteric manner.

Because x-rays superimpose structures residing or moving at different planes, it is not possible to determine from the base view alone in what planes the various barium-coated structures actually lie.

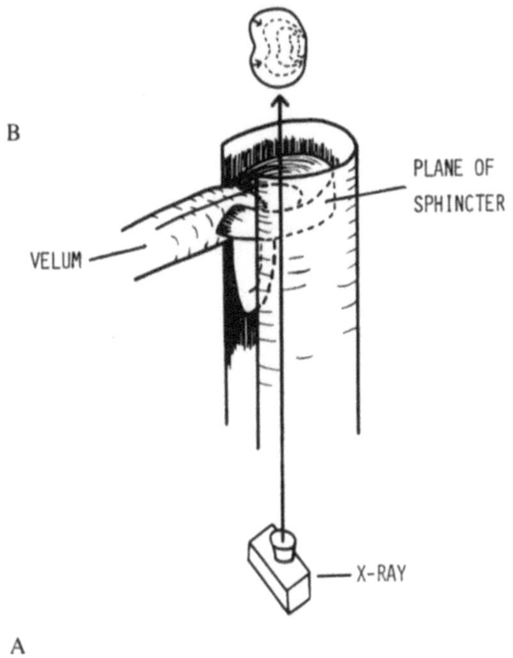

FIGURE 6.34. Base view of velopharyngeal portal. (A) Vertical x-ray beam passes through long axis of nasopharynx and perpendicularly intersects plane of velopharyngeal sphincter to produce en face or base view. (B) Projected en face image of sphincter shows all margins as they centripetally constrict during phonation.

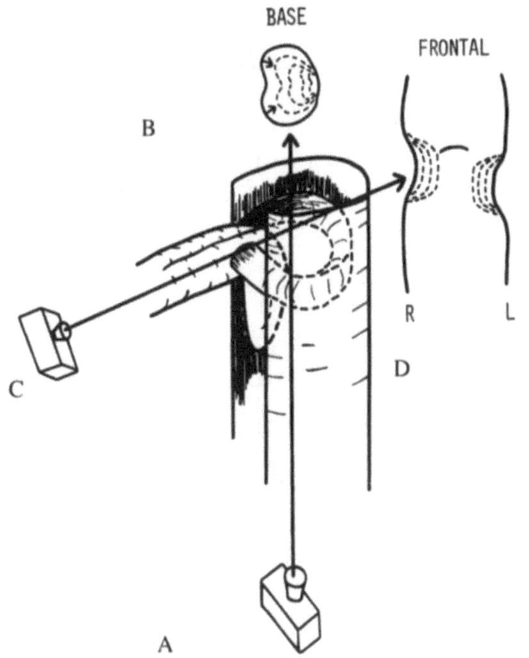

FIGURE 6.35. Limitations of base view. Lateral pharyngeal walls move asymmetrically: Right wall has maximum medial movement at more cephalad level than left wall. On base view (A-B) right and left sides of velopharyngeal portal show equal medial movement. On frontal view (C-D) level of medial movement of right pharyngeal wall is more cephalad than level of left wall.

For example, one lateral nasopharyngeal wall may show maximum movement at one vertical position in the nasopharynx while the other lateral pharyngeal wall is maximally active at a different level. On the base view one cannot appreciate these differences because the margins of these walls will be projected as parts of the same ring (Fig. 6.35). Furthermore, the oropharyngeal and nasopharyngeal margins appear superimposed upon each other on the base view. Unless one has determined from prior examination of the frontal and lateral views at what levels in the naso- and oropharynx the LPWs move, there may be confusion on the base view as to what structures the projected margins of the LPWs actually represent.

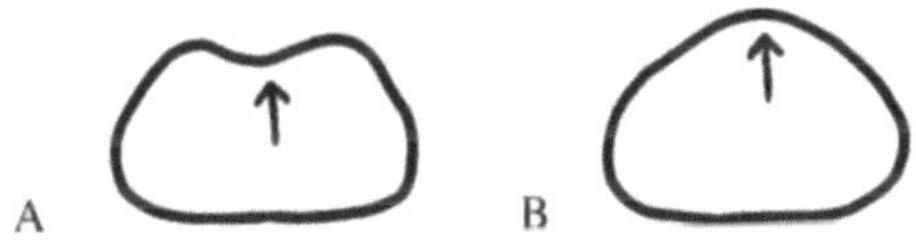

FIGURE 6.36. Base views. (A) Normal subject: Convex-posterior velar surface (↑) represents barium-coated surface of uvulae muscle. (B) Submucous cleft palate or primary repair of cleft palate. Concave posterior velar surface (arrow) represents barium in midsagittal velar groove.

Appearance During Quiet Respiration

Assuming that the patient is properly positioned, one sees the margins of the nasopharynx as a barium-coated oval. If the subject has a normal palate, the velum projects a localized posterior convexity (Fig. 6.36A), which represents the mucosal surface overlying the uvular muscle. If, on the other hand, the patient has a submucous cleft palate or has had a palate repair, one usually sees a localized concavity or notch in the posterior surface of the velum (Fig. 6.36B). In a submucous cleft palate, this concavity represents an absent uvulae muscle, and in the patient with a repaired cleft palate, it represents the site of the repair scar.

The lateral nasopharyngeal walls are arcuate structures with a convex lateral curvature that connects the lateral aspects of the velum to the posterior

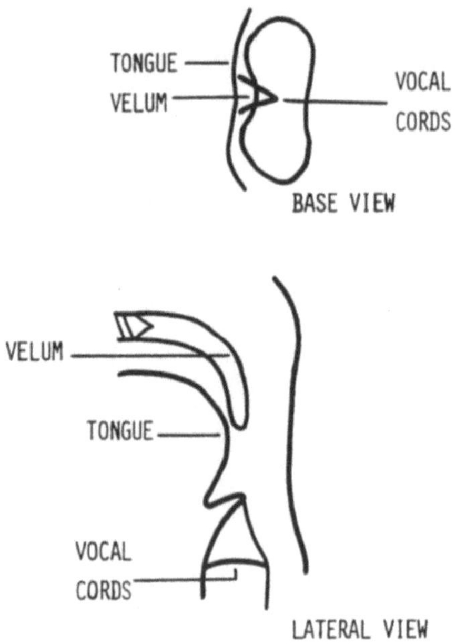

FIGURE 6.37. Tongue and vocal cords may be seen on base view. Both structures stand out because of the negative contrast from adjacent air in the oropharynx or larynx. A barium coating over the tongue may also provide positive contrast.

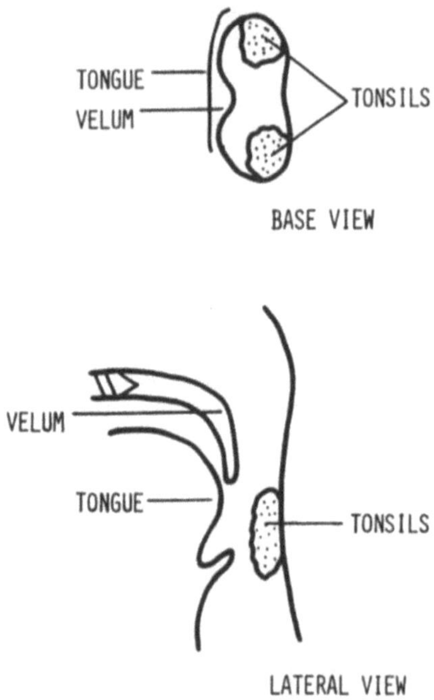

FIGURE 6.38. Enlarged tonsils projecting into base view. Patient at rest. Dots within tonsils represent barium within tonsillar crypts.

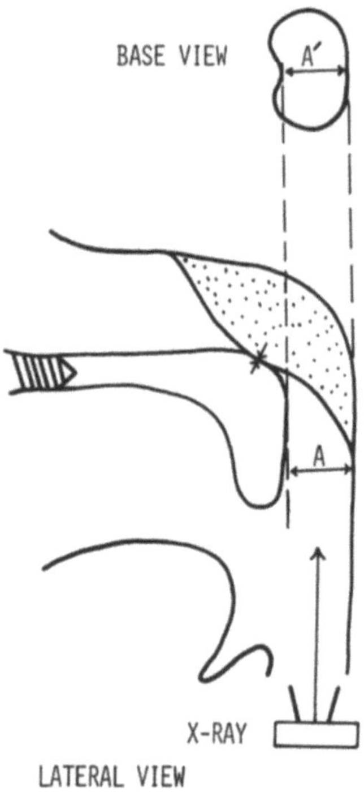

FIGURE 6.39. Enlarged adenoids are not appreciated on base view because the vertical x-ray beam is almost perpendicularly oriented to the barium coating on the adenoidal surface. Thus the barium coating does not absorb enough radiation to cast a shadow. During velar-adenoidal closure (arrows), the plane of closure is not perpendicular to the x-ray beam. The velopharyngeal gap A' projected on base view represents the distance A between the velum and posterior wall that lies below the velopharyngeal portal.

pharyngeal wall. There are no precise demarcation points between the junction of the lateral pharyngeal walls and the velum, or lateral and posterior pharyngeal walls. Rather, these structures blend into each other and project as a continuous barium-silhouetted oval.

The posterior surface of the tongue may be seen forming the anterior margin of the oropharynx. At times the epiglottis, and even the vocal cords, may project through the nasopharyngeal portal (Fig. 6.37).

Tonsillar tissue may also be seen as barium-coated convex structures projecting into the velopharyngeal portal from the lateral walls. As the tonsils contain multiple crevices, some barium may appear to project from within the tonsils because

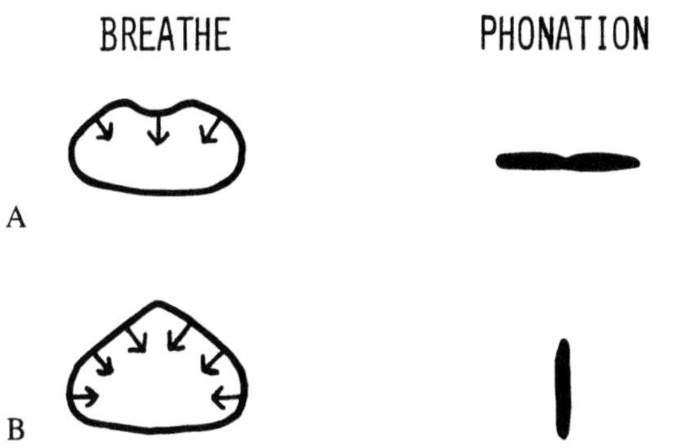

FIGURE 6.40. Base views: two patterns of velopharyngeal closure at opposite extremes. (A) Coronal pattern. (B) Sagittal pattern.

the barium was able to flow into the crevices or crypts within the tonsils (Fig. 6.38).

The adenoids are not seen on base projection because the vertically oriented x-ray beam travels almost perpendicular to the surface of the adenoids so that the barium coating is not projected as a dense line within the VP portal (Fig. 6.39). Therefore, the base view gives no information about adenoidal size or presence. In fact, when the adenoids are large, the base view is of no value in evaluating VP wall closure because the palate touches the adenoid mass. When closure occurs in this manner, the plane of the portal is approximately 45° inclined to the plane of the hard palate. Because the vertically oriented x-ray beam does not intersect the portal tangentiality, the region of contact between the palate and adenoids is not appreciated. Hyperextension alone is usually not satisfactory to bring the plane of the portal parallel to the horizontal plane because the patient's neck cannot hyperextend enough. If one has available a C-arm fluoroscopic system, then the plane of the x-ray beam can be changed from vertical to diagonal and can be directed to intersect that plane of the portal in a perpendicular manner so as to define its margins (see Chapter 3).

Appearance During Phonation

By visualizing the portal en face, the base view reveals exactly how the component movements of the velum and pharyngeal walls contribute to the closure effort. Closure during phonation results from the combined centripetal movements of the velum and pharyngeal walls. The shape of the velopharyngeal portal as it constricts may vary both among different subjects and within a given subject for different speech tasks because of the varying contribution of the palatal and pharyngeal components. In some subjects closure or attempted closure is mainly achieved by a posterior movement. At the other extreme, closure is mainly a result of a central approximation of the pharyngeal walls with little posterior midsagittal velar movement (Fig. 6.40). See Chapter 8 for a detailed discussion of patterns of VP closure.

Although the movements of the right and left portions of the velopharyngeal sphincter are usually symmetric as the portal narrows or closes, sometimes there may be regions of palatal or pharyngeal walls that exhibit locally reduced movement which can account for the presence of the velopharyngeal insufficiency.

Cautions in Interpreting the Base View During Phonation

Structures that move into the pharyngeal lumen at different levels in the naso- or oropharynx may be projected into the aperture of the VP portal on base view. It is important that the operator appreciate this situation so as not to confuse movements of structures occurring at a lower level in the pharynx with movements at the plane of the VP portal.

There are two approaches to distinguishing movements of the pharynx at the plane of the VP portal from those occurring at a lower level. First, because the lateral and frontal views have been performed prior to the base view, one knows to what extent the velum and posterior pharyngeal wall approximate to each other (lateral view) and the lateral pharyngeal walls move medially (frontal view). On the base view, one should see the same dimensional changes between the velum and posterior pharyngeal wall, and between the lateral pharyngeal walls. If, however, the margins of the VP portal on base view appear to show greater closure movements than one would expect to see from the degree of movement seen on the lateral or frontal views, one should suspect that movement is occurring at a lower level in the pharynx and that the margins of these structures are being projected into the VP portal. Lower pharyngeal structures most commonly causing such movements are the posterior surface of the tongue moving toward the oropharynx being

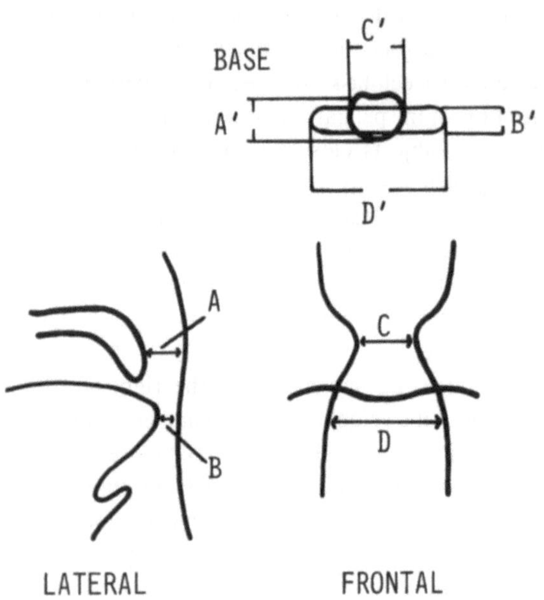

FIGURE 6.41. Projection on base view of oropharynx narrowed by tongue (B' × D') may be mistaken for projection of velopharyngeal portal (A' × C') at nasopharyngeal level. Compare dimensions of oropharyngeal narrowing (B) on lateral and (D) on frontal views to dimensions of velopharyngeal portal (A) on lateral and (C) frontal views to prevent such an error.

mistaken for the velum approaching the posterior pharyngeal wall (Fig. 6.41), and medial movement of enlarged tonsils (Fig. 6.42) or vocal cords being mistaken for movement of the lateral pharyngeal walls.

Second, one may see on the base view what appears to be margins of the VP portal moving centrally to close or narrow the portal, but on closer inspection these margins extend beyond the rim of the portal. Such an effect can only occur from the projection of a structure moving on a different plane from that of the VP portal because all the structures forming the VP portal blend into each other. The width of the oropharynx may widen more than the distance between the lateral naospharyngeal walls so that the line formed by the posterior surface of the tongue may project into the VP portal as well as extend beyond either side of the lateral pharyngal walls (see Fig. 6.41).

Appearance During Swallowing

During swallowing, the oropharynx closes in a sphincteric manner analogous to nasopharyngeal closure during speech. Although swallowing occurs at the oropharyngeal level, its appearance on the base view is as a sphincteric movement similar to velopharyngeal closure during speech at the naso-

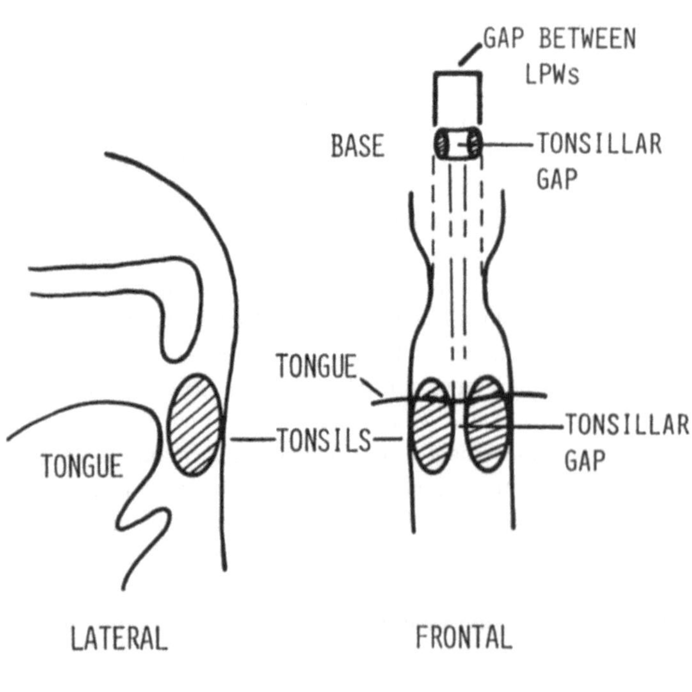

FIGURE 6.42. Projection on base view of gap between enlarged tonsils may be mistaken for medial movement of lateral pharyngeal walls. To prevent this error, evaluate size of tonsils on lateral and frontal views prior to performing base view.

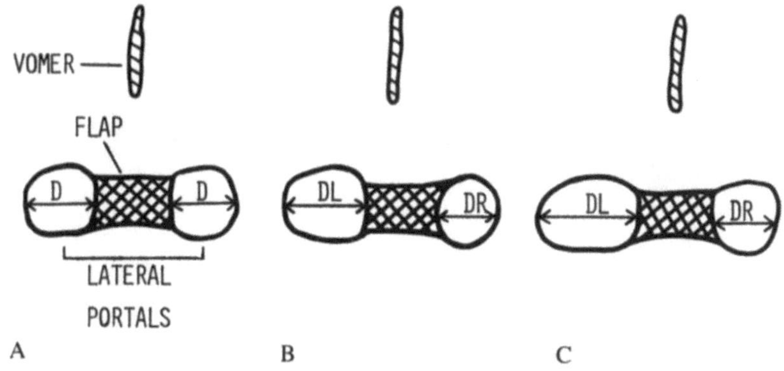

FIGURE 6.43. Base view of pharyngeal flap during breathing. (A) Flap centrally located in velopharyngeal portal. Both lateral portals are of equal diameter (D). (B) Flap asymmetrically positioned to right side of velopharyngeal portal (DL > DR). (C) Flap centrally located in velopharyngeal portal. Left lateral pharyngeal wall further from the midline than right lateral wall (DL > DR).

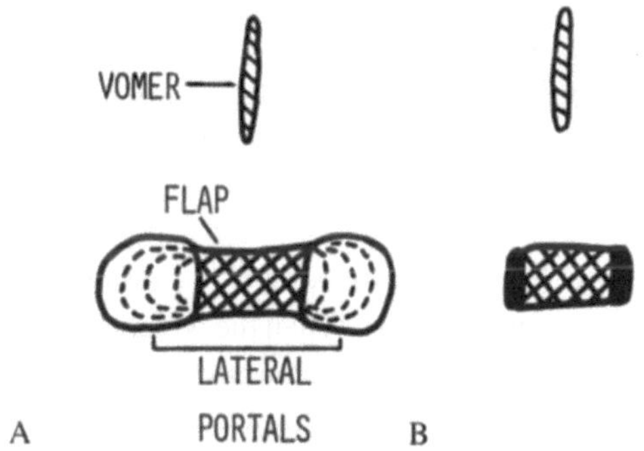

FIGURE 6.44. Base view of pharyngeal flap during phonation. Complete portal closure. (A) Pattern of movement (dotted line) as lateral portals close. (B) Final appearance during complete closure of lateral portals.

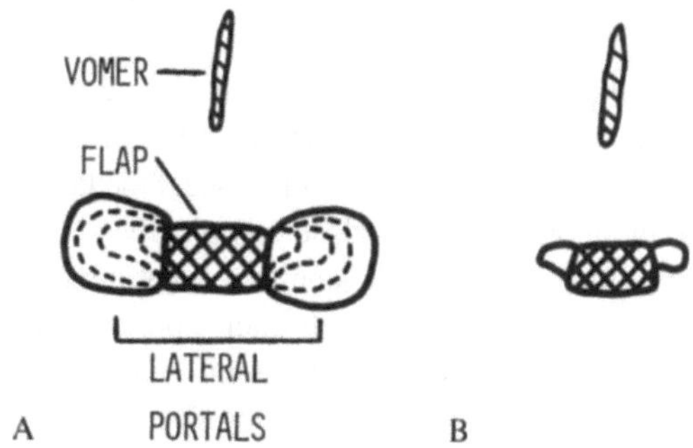

FIGURE 6.45. Base view of pharyngeal flap during phonation. Asymmetric closure of each lateral portal. (A) Pattern of movement (dotted lines) as lateral portals close. (B) Maximum movement of each lateral portal. Posterior portions of portals close. Small anterior openings remain.

pharyngeal level. Obviously the absence of associated speech allows one to differentiate swallowing from VP closure on base view.

Appearance of Pharyngeal Flap

The base view becomes especially useful in evaluating patients with pharyngeal flaps. During quiet breathing, the anterior-posterior soft tissue strip comprising the flap can be seen as a band traversing through the central portion of the velopharyngeal portal. The width of the flap, and its location within the portal, and whether it is symmetrically (Fig. 6.43A) or assymetrically (Fig. 6.43B) positioned relative to the lateral pharyngeal walls, are important points to observe.

Besides knowing whether a flap is central or asymmetric, one is interested in knowing whether the lateral margins of the two portals are equidistant from the midline (Fig. 6.43C). (To determine central position of the flap, one should relate the flap position to the position of the vomer.) During phonation, the lateral portals close as a series of concentrically shrinking arcs (Fig. 6.44). Although the margins of the portals most often move uniformly, asymmetries in movement can occur so that a portion of the arc may not abut on the lateral margin of the flap. As a result, a small region of inadequacy occurs (Fig. 6.45). If the diameter of the region of inadequacy is tiny (perhaps only 1–2 mm), one may only appreciate its presence by observing the bubbling of barium through the defect during phonation of sibilant or plosive sounds (Fig. 6.46). When

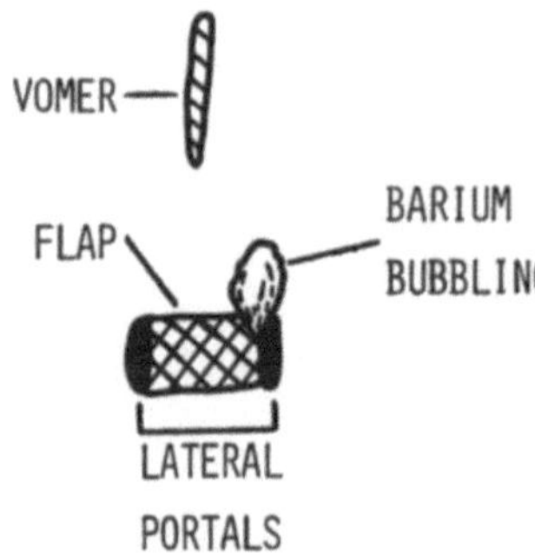

FIGURE 6.46. Base view of pharyngeal flap during phonation. Both portals appear to close completely. However, presence of barium bubbling from one lateral portal indicates that portal is not completely closed.

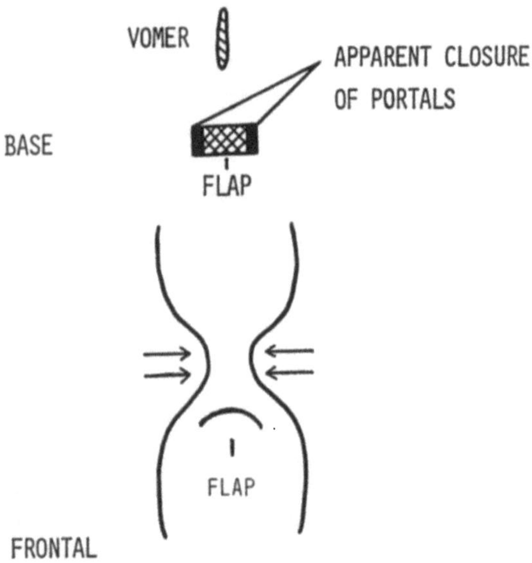

FIGURE 6.47. Base and lateral views of pharyngeal flap. Apparent closure of lateral portals during phonation. Maximum medial movement of lateral pharyngeal walls occurs above level of flap. Lateral portals are open at level of flap. However, on base view, barium-coated margins of flap and pharyngeal walls project superimposed to give appearance of closed lateral portals.

the lateral portals appear asymmetric during quiet breathing, movements of the portals must be carefully scrutinized during phonation to see if the larger of the two portals fails to close completely.

Sometimes on base view one sees what appears to be complete closure of both lateral portals (because the lateral pharyngeal walls move medially apparently to touch the edges of the flap), but the patient still remains hypernasal. In such a situation, the flap may actually be at a level below the maximum medial movement of one or both lateral pharyngeal walls (as seen on the frontal view) and the base veiw is misleading. The true anatomic relationships become obvious in the frontal view (Fig. 6.47). This situation is another example of why multiple views of the velopharyngeal portal are necessary for complete understanding of its dynamics during phonation.

Artifacts Produced by the Barium Coating

In patients both with and without pharyngeal flaps, the actual observation of contact between the mucosal surfaces of the margins of the sphincter to effect complete closure may not be appreciated because the barium may be squeezed above or below the plane of closure as the surfaces come together. The result is that a dense glob of barium mixed with mucus can obscure the actual plane of the portal as it closes. Therefore, the observer may assume the portal has a central region of inadequacy because the walls cannot be seen touching each other. In such a case, one should stop the examination and wait several minutes for the glob of barium to descend from the nasopharynx into the oropharynx and subsequently into the esophagus (as a result of swallowing). When the patient is then refluoroscoped, the margins of the nasopharynx will usually be sharply defined so that accurate information can be obtained.

Fistula Within Velum or Flap

In the occasional situation where there is a fistula in the soft palate or pharyngeal flap one may observe barium blowing through the fistula on the base view. One should be careful to distinguish this blowing of barium from barium blowing through a minimally inadequate velopharyngeal portal. Relating the position of the blowing barium on the lateral view to that of the base view will determine if the blowing is through a fistula or an inadequate portal (Fig. 6.48).

Tips for Analyzing Rapidly Changing Portal Contours

The velopharyngeal portal moves from its rest position (maximum opening) to complete closure or

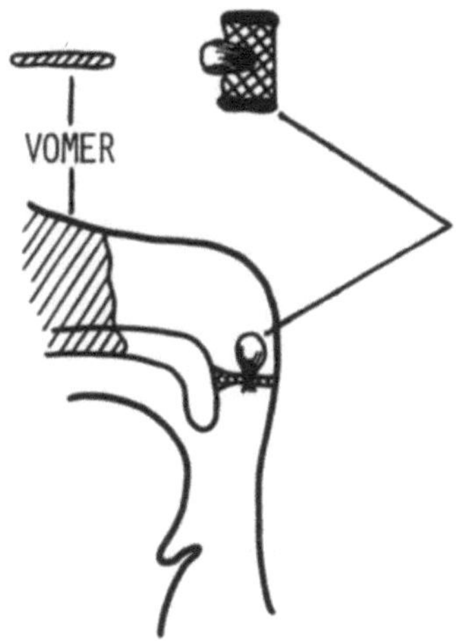

FIGURE 6.48. Base and lateral views of pharyngeal flap. Both lateral portals are closed during phonation. Barium seen blowing from region of flap on base view represents barium going through fistula in flap (better seen on lateral view).

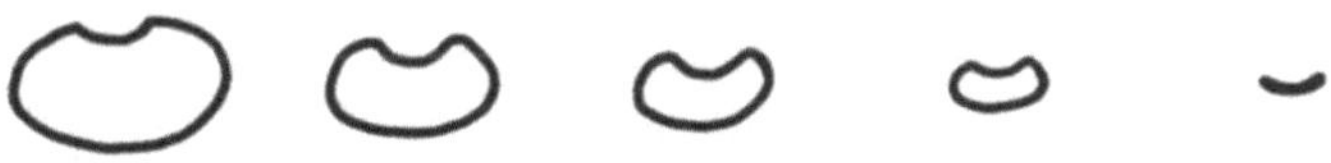

FIGURE 6.49. Base view. Series of images showing appearance of velopharyngeal portal during breathing and phonation of isolated sounds to demonstrate patterns of closure.

maximum narrowing usually in just a few TV frames. Because these movements are occurring quite rapidly, it may be difficult for the observer to appreciate the extent to which the pharyngeal and velar components contribute to the narrowing or closure of the portal, and to analyze their respective movements. When one region of the portal shows greater movement than another region, it is especially helpful for one to observe the study in slow motion. Such slow motion can be obtained by using a videotape recorder that has a frame-by-frame analysis control. Unfortunately, image degradation may occur with some recorders during slow or stop motion of the videotape recorder, and it may be difficult to appreciate the same detail that one can see when the tape recorder is running at normal speed. Another way of getting around this problem is to have the patient utter a series of prolonged phonemes ranging from nasal through low and then high vowels, ending in consonants. With this maneuver one will obtain a series of views of the portal at various stages of narrowing. With nasal sounds there is minimal narrowing. Greater narrowing occurs with low vowels and still further narrowing occurs with high vowels. Maximum narrowing or closure occurs with non-nasal consonants (Fig. 6.49).

Oblique Views

When a satisfactory base view cannot be obtained because of large adenoids or the inability to hyperextend the head fully on the neck, oblique views (as described in Chapter 4) are useful in further assessing the movements of the pharyngeal walls. These views demonstrate relationships of multiple points along the curving pharyngeal walls relative to the margins of the velum and can reveal an area of localized incompetence that may not be seen on the frontal view.

When using the oblique view, it is important to remember the cylindrical concept of the portal that we discussed in Chapter 4. The margins of the cylinder are curved. Therefore, to appreciate the movements of various margins of the curved pharyngeal walls, one must image these walls in a series of changing tangential projections. Just using a specific tangential projection such as the 45° oblique view, one obtains information only about the movement that occurs when that margin of the portal is tangential to the x-ray beam. Thus, for the oblique views to be effective, the patient should be rotated in a continuous manner from lateral to frontal view, first with the head turning in one direction

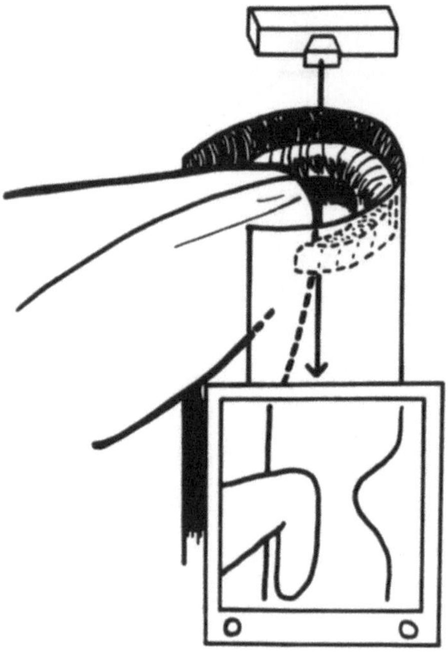

FIGURE 6.50. Patient with Passavant's ridge positioned oblique to x-ray beam so as to view tangentially one pharyngeal wall and its relationship to the adjacent margin of the velum.

and then with the head turning in the opposite direction in order to obtain symmetric views of both lateral pharyngeal walls.

Each oblique view is good for examining only one lateral pharyngeal wall and its relationship to that side of the velum. For example, when the head is turned to the right of the midline, then the left pharyngeal wall and its relation to the left lateral margin of the velum is seen tangentially so that one can appreciate whether or not the pharyngeal wall touches the surface of the velum during phonation. The space between the opposite pharyngeal wall and its adjacent velar margin are not tangentially seen. Thus, the size of this portion of the VP portal cannot be assessed during phonation (Fig. 6.50).

References

1. Graber TM, Bzoch KR, Aoba T: A functional study of palatal and pharyngeal structures. Angle Orthodont 29:30, 1959
2. Simpson RK, Austin AA: A cephalometric investigation of velar stretch. Cleft Palate J 9:341, 1972
3. Mourino AR, Weinberg B: A cephalometric study of velar stretch in 8 and 10 year old children. Cleft Palate J 12:417, 1975
4. McWilliams BJ, Morris HL, Shelton RL. Cleft Palate Speech. Philadelphia: BC Decker, 1984
5. Moll KL, Shriner TA: Preliminary investigation of a new concept of velar activity during speech. Cleft Palate J 4:58, 1967
6. Moll KL: Velopharyngeal closure on vowels. J Speech Hear Res 5:30, 1962
7. McWilliams BJ, Musgrave RH: The influence of head position upon velopharyngeal closure. Cleft Palate J 5(2):117, 1968
8. Glaser (Cohn) ER, Skolnick ML, McWilliams BJ, Shprintzen RJ: The dynamics of Passavant's ridge in subjects with and without velopharyngeal insufficiency—A multiview videofluoroscopic study. Cleft Palate J 16:24, 1979
9. Kelsey CA, Ewanowski SJ, Crummy AB, Bless DM: Lateral pharyngeal wall motion as a predictor of surgical success in velopharyngeal insufficiency. New Eng J Med 287(2): 64, 1972
10. Shprintzen RJ, Lencione RM, McCall GN, Skolnick ML: A three dimensional cinefluoroscopic analysis of velopharyngeal closure during speech and nonspeech activities in normals. Cleft Palate J 11:412, 1974

7
Congruence of Portal Dimensions Among Multiple Views

Introduction to Congruence Testing

Establishing congruence of the dimensions obtained on lateral, frontal, and base views is our method of checking the accuracy of patient positioning. This is done by relating the dimension of a given structure in one view to the same dimension of that structure in one of the other views. Because each of the views should be perpendicular to the other two, measurement of a given dimension should be the same in both views. If the measurement for a particular dimension is different, one can assume that the patient is incorrectly positioned for one of the two views.

For example, the lateral and base views demonstrate the midsagittal AP dimension of the velopharyngeal portal (1 in Fig. 7.1A, 1' in Fig. 7.1C). If these two views are perpendicular to each other, measurements of the midsagittal AP dimension of the VP portal will be the same in both views. Likewise, on the frontal and base views, the transverse distance between the lateral pharyngeal walls (2 in Fig. 7.1B, 2' in Fig. 7.1D) is seen. If these two views are perpendicular to each other, measurement of the distance between the pharyngeal walls should be the same.

In applying the test of congruence to a dimension seen on two views, one should first determine which of the two views is most likely to be properly positioned. The measurement on this view should serve as a reference for the same dimension in the second view.

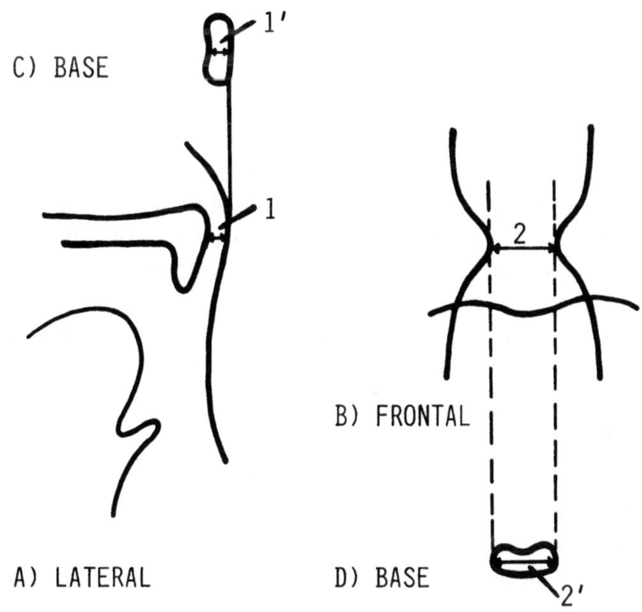

FIGURE 7.1. Demonstrating congruence among lateral, frontal, and base views of velopharyngeal portal: Midsagittal AP dimension (*1* in A) equals same dimension on base view (*1'* in C). Transverse dimension (*2* on B) equals same dimension on base view (*2'* on D).

Midsagittal Velopharyngeal Dimension

Now let us be specific. The anterior-posterior dimension of the VP portal, that is, the dimension between the posterior surface of the velum and the anterior surface of the posterior pharyngeal wall, is seen on two views—the lateral and the base (Fig. 7.1). The lateral view is used as the reference standard for this dimension because one can determine whether the lateral view is properly or improperly positioned solely on the basis of the lateral view. The lateral view is correctly positioned when the plane of the hard palate is horizontal (to eliminate flexion or extension distortions), and when the vertical rami and angles of the mandible superimpose (to indicate that the head is not rotated to one side or the other) (Fig. 4.3). This "true lateral" position is the defined position to measure the sagittal or anterior-posterior dimension of the VP portal.

In contrast to the ease of positioning of the lateral view, positioning for the base view, as described in Chapter 4, is much more difficult. Fixed osseous landmarks can neither be used to determine the plane of the VP portal on the base view nor to adjust that plane perpendicular to the vertical x-ray beam. Positioning of the base view is done on a trial-and-error basis by changing the degree of flexion or extension of the head until one perceives what appears to be a smoothly marginated en face view of the portal that moves centripetally during phonation from its rest position to its position of maximum narrowing or closure. A check on the accuracy of this position is made by comparing dimensions seen on the base view with the same dimensions seen on the lateral and frontal views. This is vital to insure that the base view was properly positioned. One of the reasons for performing the lateral view before the base view is that one observes the maximum and minimum midsagittal dimensions of the VP portal on the lateral view and looks for the same distances on the base view.

In certain situations the midsagittal distance of the VP portal on base view may not match the same dimension on lateral view. For example, VP closure may occur between the superior surface of the velum and the inferior surface of a prominent adenoid pad. When such a patient is positioned for the base view, the plane of the VP portal will be approximately 45° off the vertical. If this patient is unable to hyperextend his neck enough to bring the plane of closure perpendicular to the vertical beam (Fig. 7.2), the resultant base view image (Fig. 7.3B) will appear to show a grossly inadequate VP portal during speech. If one interprets the images on this base view without referring to the lateral view (Fig. 7.3A), one may conclude that there really is VP inadequacy. However, when the lateral view is used as the reference standard, it is obvious that there is closure in the sagittal dimension on the lateral view and, therefore, that there cannot be inadequacy on the base view. The apparent VP gap seen on the base view represents the distance between the uvula and posterior pharyngeal wall, which is below the region of sagittal closure.

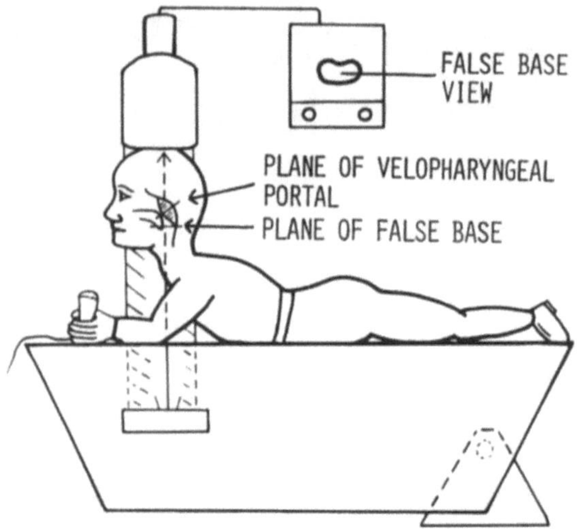

FIGURE 7.2. Base view: patient with enlarged adenoids. Closure occurs between superior surface of velum and inferior surface of adenoids resulting in plane of velopharyngeal portal that lies diagonal to vertical x-ray beam. Therefore x-ray beam does not perpendicularly intersect portal or view sphincter en face. Image projected on TV monitor is false base (below plane of VP portal).

For the sake of simplicity, we have initially defined the proper lateral view as one in which the plane of the hard palate is parallel to the horizontal plane. However, for the base view, the head may have to be hyperextended to properly position the plane of the VP portal perpendicular to the vertical x-ray beam (Fig. 7.4). This hyperextension position, as described in Chapter 4, can increase the sagittal dimension between the velum and the pos-

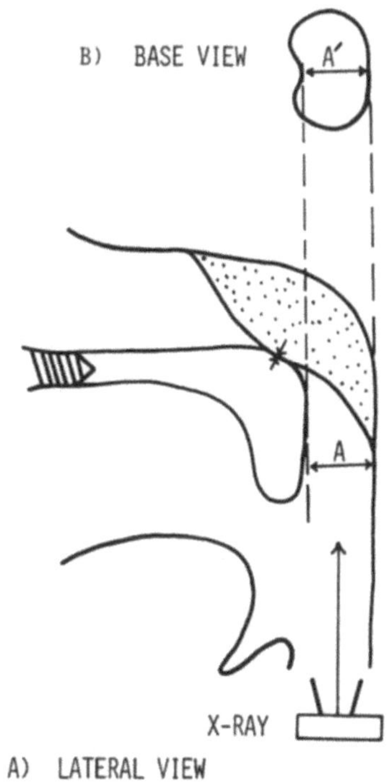

FIGURE 7.3. Patient with enlarged adenoids. During phonation of non-nasal consonants, VP closure occurs between velum and adenoids (A, double arrows on lateral view). However, on base view (B) there appears to be a gross velopharyngeal gap (distance A'), which is actually the en face projection of the gap between the uvula and posterior pharyngeal wall (distance A) that lies below the plane of the VP portal.

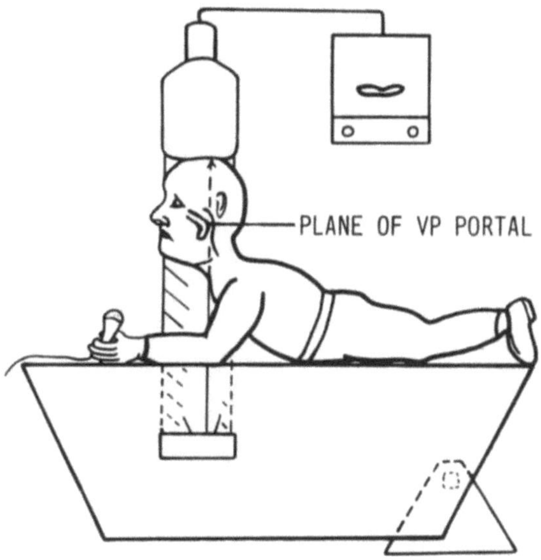

FIGURE 7.4. Because some head hyperextension is needed in this patient to achieve proper positioning for the base view (so that vertical x-ray beam perpendicularly intersects plane of portal), a small VP gap is seen on this view whereas there is none on the lateral in neutral position (also refer to Fig. 7.5).

terior pharyngeal wall. Thus, one may see fluoroscopically a slightly greater midsagittal dimension on the base view than on the lateral view in neutral position. In other words, one may possibly see a small VP gap on the base view where none exists on the lateral view in neutral position, or a greater gap on the base than seen on the lateral. If the sagittal dimension appears greater on the base than on the lateral view, that sagittal dimension measured from the base view should then be compared to the sagittal dimension seen on the lateral view with the head hyperextended (Fig. 7.5). It is difficult to determine the exact degree of hyperextension of the head for the base view because no simultaneous lateral view is obtained using a conventional fluoroscopic system. In practice, however, we have not found significant differences between the AP dimension of the VP portal in the lateral and base views that required such comparison lateral views.

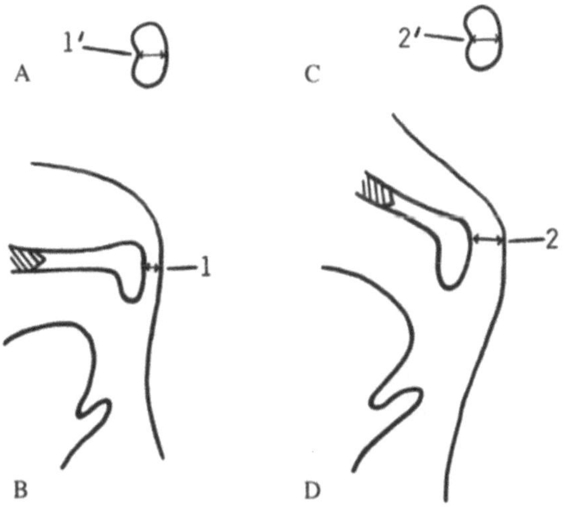

FIGURE 7.5. Head positioned in neutral for lateral view and in hyperextension (same as in Fig. 7.4) to position patient properly for base view. The VP gap appears greater on base view (*1'* on A) than on the lateral view with the head in neutral position (*1* on B). However, when head is positioned in about the same degree of hyperextension for lateral and base views and when the VP gap on lateral view (2 on D) is compared with the VP gap on base view (*2'* on C) then the gap is the same.

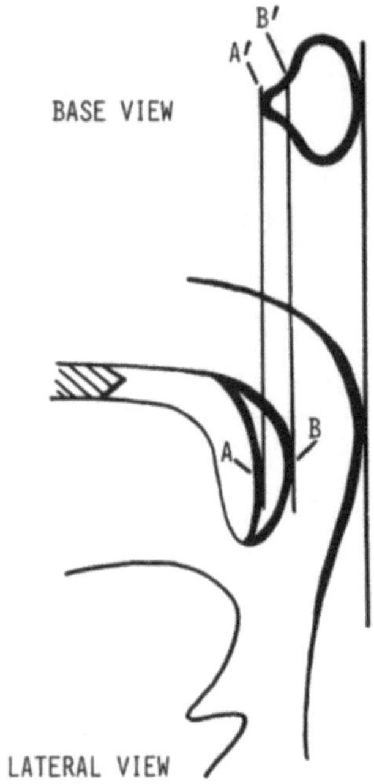

FIGURE 7.6. Patient with submucous cleft palate. The central submucous cleft and laterally positioned levator muscles on lateral view (represented by barium coatings *A* and *B*) are more clearly appreciated on base view (*A*′ and *B*′).

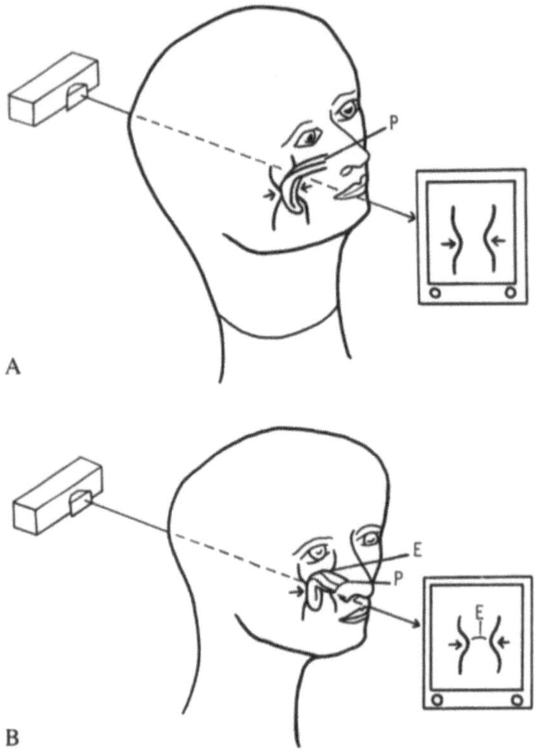

FIGURE 7.7. Patient positioning for frontal view with head in hyperextension versus neutral positions. (A) Hyperextension. The plane of the hard palate (*P*) is projected above the level of the velopharyngeal portal so that barium lying on the hard palate will not obscure the pharyngeal walls at the velopharyngeal portal (arrows). (B) Neutral position. Plane of hard palate (*P*) is horizontal. Barium resting on hard palate may obscure lateral pharyngeal walls at nasopharyngeal level. However, if there is good eminence formation, the barium-coated velar eminence (*E*) will be projected between the lateral pharyngeal walls (arrows).

Confusion in determining the sagittal velopharyngeal dimension may also occur in patients with a repaired cleft palate or with a submucous cleft palate. These patients may have a deep raphe through the central portion of the palate. On the lateral view two lines representing the palate may be seen: the more anterior one corresponding to the barium coating on the central raphe and the more posterior one corresponding to barium on the lateral margins of the velar mucosa overlying the levator muscles. Anatomic confirmation of these lines is provided by the base view, which will show the velar raphe as a localized convex or V-shaped anterior notch (Fig. 7.6).

Interpharyngeal Wall Dimensions

The second dimension for which congruence should be determined is the transverse distance between the lateral nasopharyngeal walls. This distance is seen both on the frontal and on the base view (Figs. 7.1B,D). The frontal view is the reference view because, like the lateral view, there are specific osseous landmarks for determining proper positioning. The bones on either side of the midline should be symmetrically positioned, indicating that the head is not rotated. The plane of the hard palate should remain horizontal although it is less critical for the palatal plane to be horizontal than for the face to be rotated. Since the lateral pharyngeal walls are vertically oriented structures that lie just in front of the atlanto-occipital joint (the joint controlling flexion and extension of the head upon the neck), mild flexion or extension of the head will significantly change neither the actual distance between the lat-

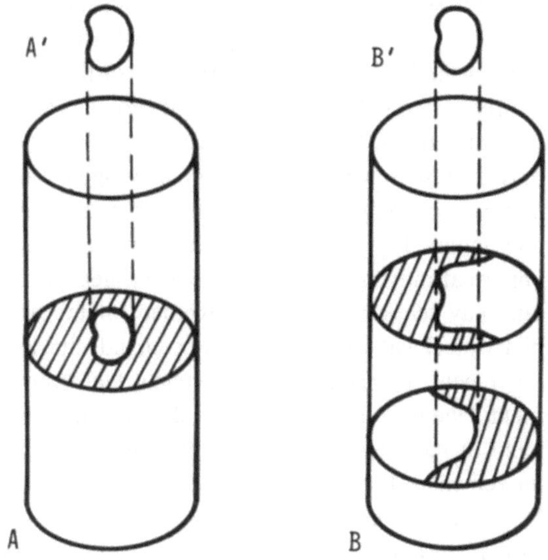

FIGURE 7.8. Tube with constriction (shaded area) at one level (*A*) and at two levels (*B*). When the tube is viewed en face (as the pharynx is viewed for base viewing), the projected image of the aperture within the constriction has the same appearance (*A*' vs. *B*') regardless of whether the constriction is at one plane (*A*) or is partial at each of two planes (*B*).

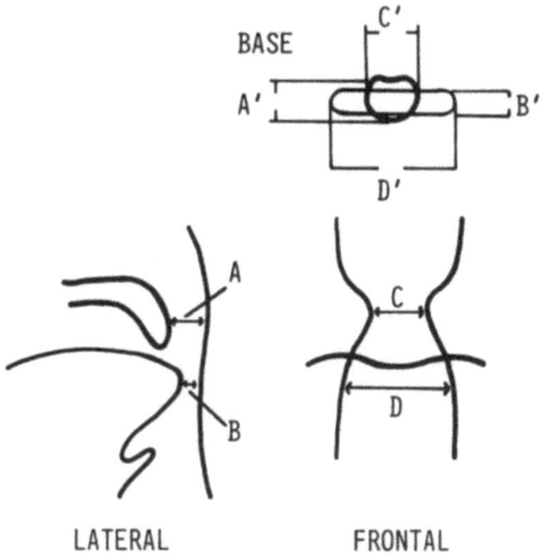

FIGURE 7.9. Projection on base view of oropharynx narrowed by tongue (*B*', *D*') may be mistaken for projection of velopharyngeal portal (*A*', *C*') at nasopharyngeal level. Compare dimensions of oropharyngeal narrowing on lateral (*B*) and on frontal (*D*) to dimensions of velopharyngeal portal on lateral (*A*) and frontal (*C*) views to prevent such an error.

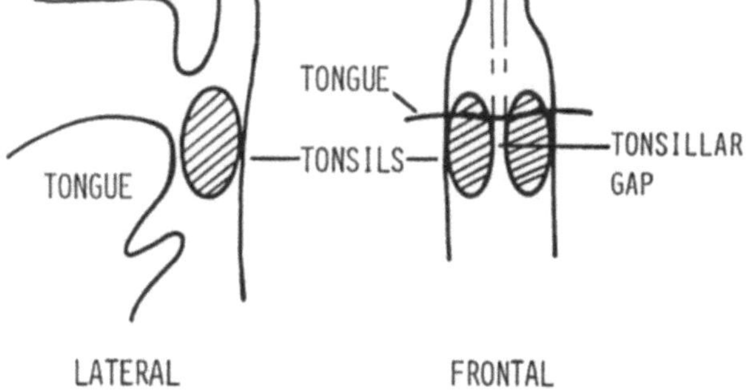

FIGURE 7.10. Projection on base view of gap between enlarged tonsils may be mistaken for gap between of lateral nasopharyngeal walls (LPWs). To prevent this error, evaluate size of tonsils on lateral and frontal views prior to performing base view.

eral pharyngeal walls nor their projected size on the TV monitor. At times, the patient is intentionally positioned with the plane of the hard palate slightly above the horizontal so as to project any residual barium that remains along the floor of the nasal cavity above the region of the nasopharynx in order not to obscure the barium-coated lateral pharyngeal walls (Fig. 7.7).

Besides the ease of accurate positioning, the frontal view is also used to determine the vertical level within the pharynx at which the nasopharyngeal walls move. This information allows one to clarify

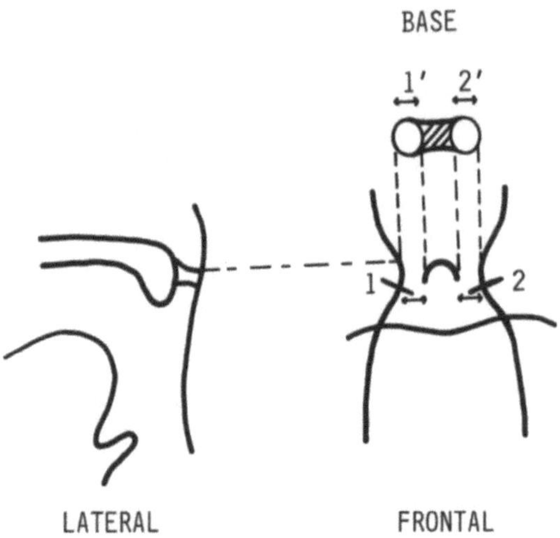

FIGURE 7.11. Patient with pharyngeal flap during phonation. Bilateral portal incompetence is present. When there is congruence between the lateral and base views, then the width of lateral portals on frontal view (*1* and *2*) equals the width on base view (*1*′ and *2*′).

movements of barium-coated surfaces seen on base view. Because the base view examines the entire pharyngeal tube on end, one cannot tell on the basis of the base view alone whether moving barium coated surfaces represent movement of structures at the nasopharyngeal level (Fig. 7.8A) (the VP portal margins) or movement of structures at multiple levels anywhere along the length of the pharyngeal tube (Fig. 7.8B). It is usual during speech for the lateral nasopharyngeal walls to move medially. At the same time the lateral oropharyngeal walls may move laterally. If the nasopharyngeal surfaces are poorly coated with barium while the oropharyngeal margins are well coated, by base view alone one may confuse movements of the oropharyngeal walls and base of tongue with the nasopharyngeal walls and velum (Fig. 7.9). Occasionally, when there is abnormal movement of the nasopharyngeal walls, one may see some medial movement of the oropharyngeal walls and erroneously conclude that this movement represents nasopharyngeal wall movement. One may also misinterpret the medial movement of large tonsils in the oropharynx as medial movement of the nasopharyngeal walls (Fig. 7.10). Such misinterpretations are avoided by using the frontal view as a reference for evaluating movement of the lateral pharyngeal walls on the base view.

Pharyngeal Flap Width and Lateral Portal Size

In patients with pharyngeal flaps one should check for congruence in flap width, in horizontal position of the pharyngeal flap between the two lateral pharyngeal walls, and in the width of the portal on either side of the flap between the frontal and the base views. Once again, the frontal view is used as the reference, and the dimensions on the base view should be congruent to those on frontal view if the base view is properly positioned (Fig. 7.11).

8
Patterns of Velopharyngeal Closure in the Nonpharyngeal Flap Patient

This chapter will describe the spectrum of patterns of velopharyngeal (VP) closure that can be seen using multiview videofluoroscopic techniques in subjects who do not possess pharyngeal flaps. A discussion of the appearance of the VP portal in patients with pharyngeal flaps will be presented in Chapter 10.

Sphincteric Concept of Velopharyngeal Closure

Before going into a detailed discussion of the various closure patterns, it is important to reiterate a basic concept; namely, that VP closure is a sphincteric movement that results from the combination of pharyngeal and palatal movements and that the contribution of the two components shows considerable variation among different subjects (1). This variation may depend upon the intrinsic anatomic variations of the individual subjects, the changes that have been produced in the VP portal following repair of the cleft palate, or a combination of both. Although a major overall pattern is usually dominant for a given subject, there is a certain amount of variation within each subject in the relative contribution of the pharyngeal and velar components, depending upon the particular speech task.

Patterns of Closure: Velopharyngeal Competence

The base projection is the one best suited to illustrate the component contributions of the palatal and pharyngeal walls to the closure mechanism and the various patterns these differing contributions produce. The lateral view alone usually gives little information about the relative contribution of the velar and pharyngeal wall components to the closure effort (except in the presence of a Passavant's ridge). The combined use of the lateral and frontal views is much more useful.

In a series of 85 patients without pharyngeal flaps (1), we observed four basic patterns of VP closure: coronal, circular, circular with Passavant's ridge, and sagittal (Fig. 8.1). One must realize, however, that the patterns we have described are not rigid categories, but are really part of a continuous spectrum produced by the interaction of palatal and pharyngeal wall movements.

Coronal Closure

At one end of the spectrum is the pattern we call coronal closure. In this pattern the palate shows a broad posterior movement. When closure occurs, the posterior surface of the palate touches the posterior wall over a wide region of contact. There is little medial movement of the lateral pharyngeal walls because the posterior surface of the velum makes such a broad contact with the posterior pharyngeal wall. When this pattern of VP closure is seen on the base projection, the region of contact

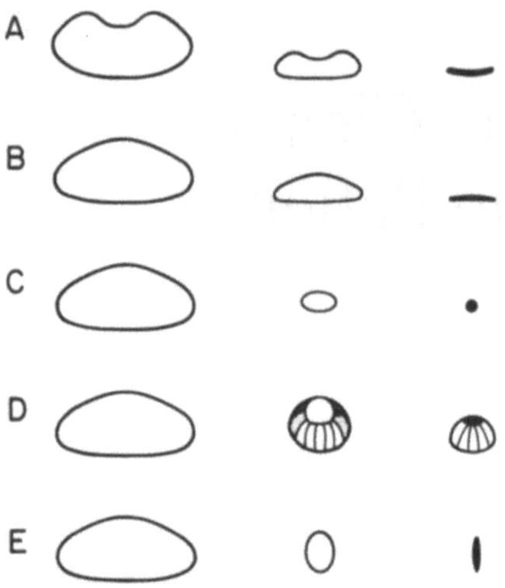

FIGURE 8.1. Schematic representation of sphincteric patterns of velopharyngeal closure, base view. Left column—contour of portal at rest; middle column—partial closure; right column—full closure. (A) Normal subject. Note convex projection of uvula portion of velum into velopharyngeal portion at rest. (B) Repaired-cleft-palate subject. Note absence of uvula muscle bulge at rest. Pattern of closure is coronal, similar to subject in A. (C) Circular closure pattern, repaired cleft palate. (D) Circular closure pattern with Passavant's ridge, repaired cleft palate. Ridge is represented by stippled and lined area in middle and right columns. (E) Sagittal closure pattern, repaired cleft palate. Reproduced with permission from Skolnick ML, McCall GN, Barnes M: The sphincteric mechanism of velopharyngeal closure. Cleft Palate J 10:286–305, 1973.

between the velum and posterior pharyngeal walls appears like a slit oriented in the coronal plane; thus, the name coronal closure pattern (Fig. 8.1A).

Circular Closure

As we progress down the spectrum from the coronal closure pattern in which velar movement contributed to almost the entire closure of the VP portal, we come to intermediate stages in which medial movement of the pharyngeal walls plays an increasingly important role in closure. The appearance on the base projection is now a coronally oriented slit that becomes progressively narrower as the amount of lateral pharyngeal wall movement increases (Fig. 8.1B, 8.2). The extreme of this closure pattern is a circle or dot (Fig. 8.1C, 8.3, 8.4).

Passavant's Ridge

The Passavant's ridge is seen on the base view as a crescent-shaped shelf of soft tissue projecting from the lateral and posterior pharyngeal walls (Fig. 8.1D, 8.4). On the lateral view the ridge appears as a mass projecting from the posterior pharyngeal wall (Fig. 8.4), and on the frontal view as localized masses projecting from the lateral pharyngeal walls at the nasopharyngeal level (Fig. 8.4). One should not confuse the closure pattern of Passavant's ridge with the circular closure pattern (Fig. 8.5). In the circular pattern one sees on base view a central circular constriction without a shelf, on the frontal view a broad vertical region of medial movement of the nasopharyngeal walls, and on the lateral view no localized projection from the posterior pharyngeal wall.

The reason the Passavant's ridge has a shelf-like appearance on the base view whereas the circular closure pattern without Passavant's ridge shows a central constricting pattern, relates to the vertical extent of the pharyngeal walls that move centrally in the two situations. In a patient with a Passavant's ridge there is a localized crescent of centrally moving tissue that projects almost at right angles to the pharyngeal walls so that one sees the pharyngeal walls below this projection as well as the Passavant's ridge projecting into the nasopharyngeal air space (Fig. 8.4B). In patients without Passavant's ridge there may be an hourglass constriction of the nasopharynx over vertical extent of several centimeters so there is no sharp demarcation between the pharyngeal walls at the level of maximum constriction and at a lower level (Fig. 8.5B). Chapter 9 describes the appearance and function of the Passavant's ridge in greater detail.

Sagittal Closure

At the opposite end of the spectrum from coronal VP closure is the sagittal closure pattern. The lateral pharyngeal walls move medially to oppose each other in the midline resulting in a sagittally oriented slit on base view at maximum portal closure. The posterior surface of the velum rests against the abutment formed by the midline apposition of the lateral pharyngeal walls (Fig. 8.1E, 8.6).

Pattern Variations with Different Phonemes

Although each subject usually has a predominant closure pattern, one may often see modifications of this closure appear during the utterance of certain sounds. For example, a given subject may show a predominantly coronal closure pattern but may exhibit more of a circular pattern during the phonation of a particular phoneme, usually a consonant (2).

Patterns of Constriction: Velopharyngeal Insufficiency

In patients with VP insufficiency one usually sees patterns of constriction occurring similar to those seen in patients with VP closure (1, 3). The only difference is that the patients with VP insufficiency do not achieve complete closure of the VP portal. Instead one sees a central opening, the size and shape of which depends upon the relative contributions of the palate and pharyngeal walls in their attempt to obtain closure.

Therefore, we can see the same spectrum of patterns ranging from the coronal to the sagittal (Fig. 8.7) as we have demonstrated in patients with complete VP closure. As in the subjects who achieve complete closure, the most common patterns seen in the inadequate group are the coronal (Fig. 8.8) and circular ones (Fig. 8.9), especially with Passavant's ridge (Fig. 8.10). The sagittal pattern of insufficiency (Fig. 8.11) occurs when there is a major defect in the posterior surface of the palate and the lateral pharyngeal walls attempt to compensate by producing broad areas of medial movement. Such an appearance may be seen in a submucous cleft palate or in a palate that has a very prominent notch in its posterior midline surface from a poor palatal repair.

There is one pattern seen only in VP inadequacy that cannot be seen with VP closure. It is one that shows no movement of the palate or pharyngeal walls with any speech task. The portal during phonation does not look any different from the portal during breathing (Fig. 8.12). What is even more unusual is that during swallowing the palate and tongue move normally to close the oropharynx. The explanation for this phenomenon is not clear, but the discrepancy in movement between the speech and swallowing suggests (1) a disturbance in motor programming and (2) that the two functions employ different neuromuscular control mechanisms. This situation should not be confused with phoneme-specific VPI, in which one or more phonemes fail to be produced because of velopharyngeal mislearning.

Posterior Velar Margin Appearance

As we have already mentioned, one should pay careful attention to the appearance of the posterior margin of the velum during the interpretation of the base projection. When the margin has a convex configuration (Fig. 8.1A, 8.2), then one can rule out a submucous cleft palate. The presence of a localized concave defect in the center of the posterior surface of the palate is typical of the repaired primary cleft palate (Fig. 8.1B, 8.3). These patients frequently demonstrate very good movement of the pharyngeal walls and lateral portions of the velum to constrict the portal in a circular manner. However, a small central defect remains, which we believe represents the absence of uvular tissue (Fig. 8.9). This absence of tissue allows air to escape during phonation to produce speech symptoms of velopharyngeal insufficiency (VPI).

Nevertheless, the presence of a concave anterior appearance to the posterior velar surface on base view during breathing cannot be used as an indicator of VPI during phonation. Most cleft palate patients who have had a primary repair of the cleft palate have a localized concavity in the center of the velum, but during phonation most of these patients achieve VP closure. Only a small number of patients following primary palatal repair in our experience have hypernasal speech from VPI because of this central palatal concavity.

Nonuniform Pharyngeal Wall Movement

There is one more point to make about the appearance of the pharyngeal walls on the base view. These walls may not move uniformly. Instead, regions of greater or lesser movement may occur

A A'

B B'

C C'

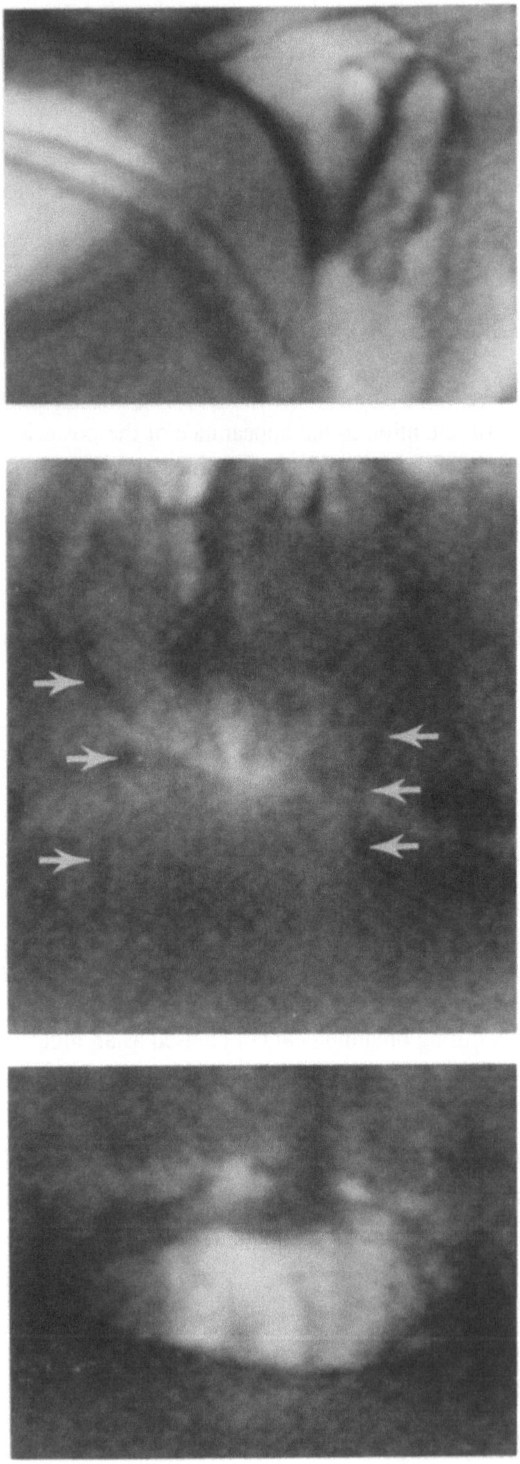

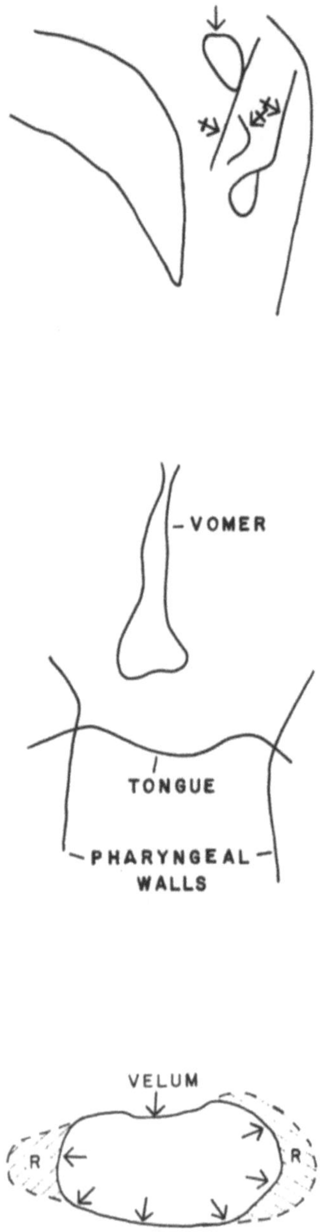

Figure 8.2. Normal subject with coronal closure pattern during breathing and phonation in lateral (A-A', D-D'), frontal (B-B', E-E'), and base (C-C', F-F') views. Closure of velopharyngeal portal results from localized medial movement of lateral pharyngeal walls (horizontal arrows) and posterior movement of velum to produce coronal closure pattern (vertical arrows) on base view. Barium beyond the margins of the velopharyngeal portal on base view represents barium lying within Rosenmuller's fossa (R) (C'). (Vertical arrow in A', D': eustachian tube orifice coated with barium; horizontal arrows: barium coated mucus.) Reproduced with permission from Skolnick ML, McCall GN, Barnes M: The sphincteric mechanism of velopharyngeal closure. Cleft Palate J 10:286–305, 1973.

D
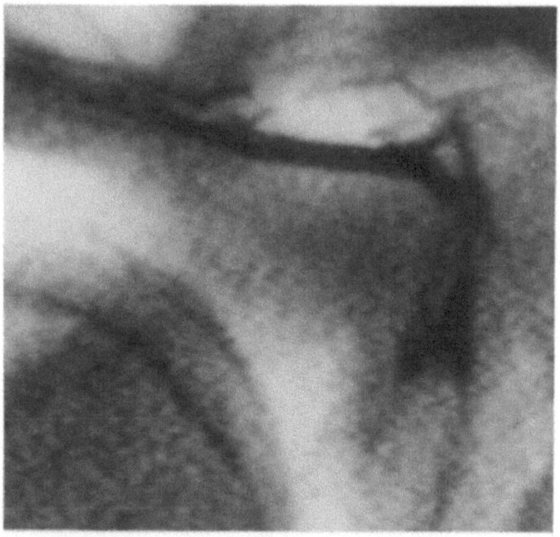

D'
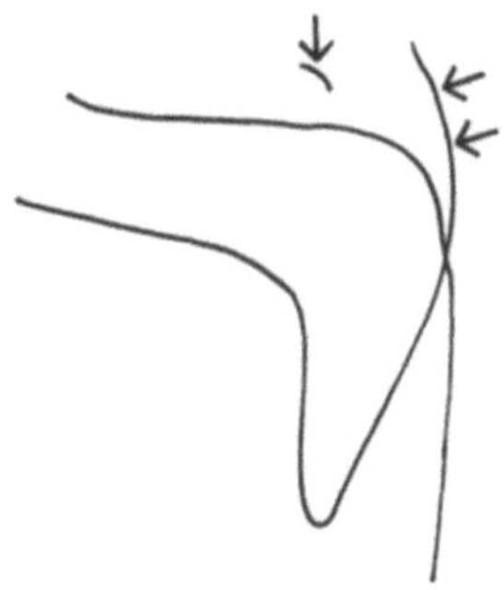

E
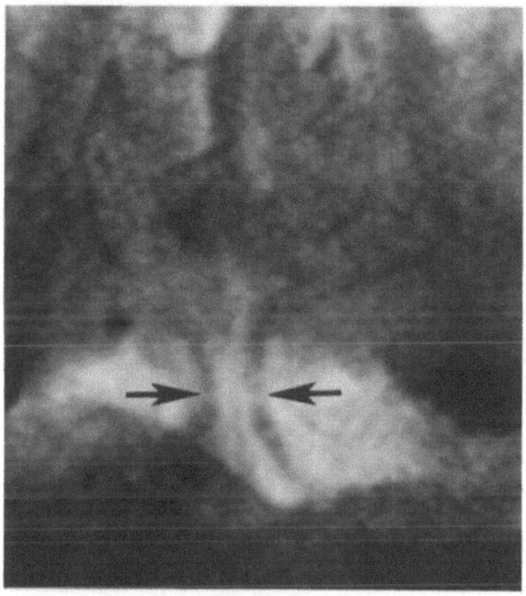

E'
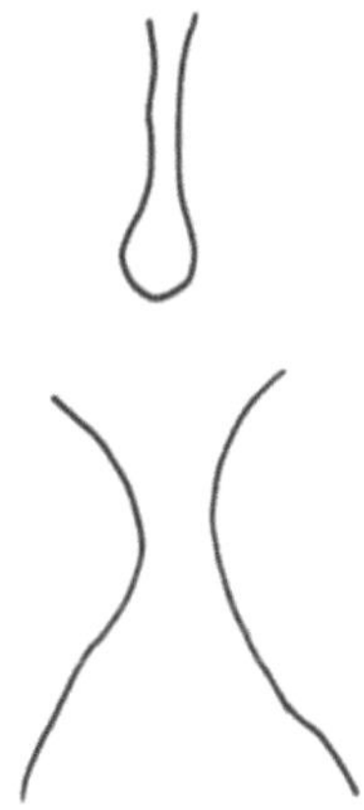

F
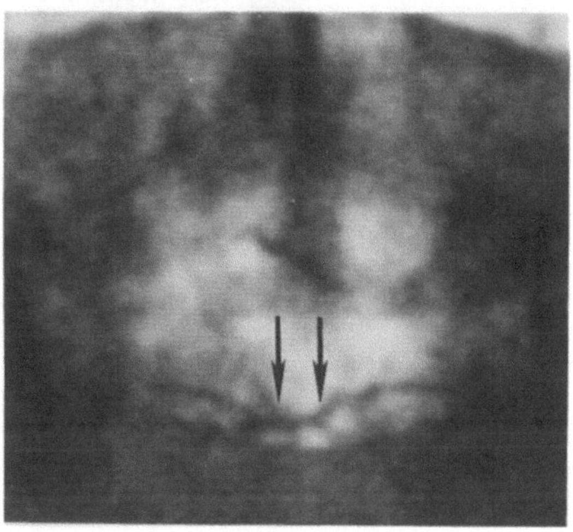

F'
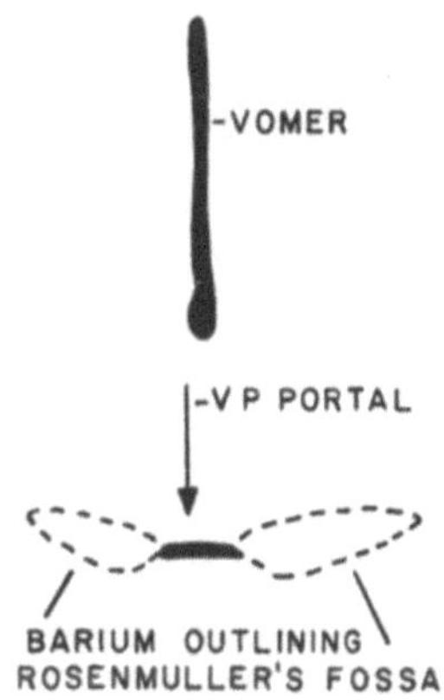

A

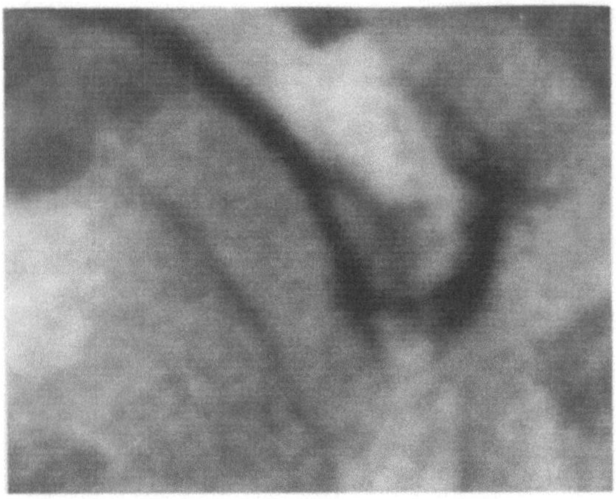

A'

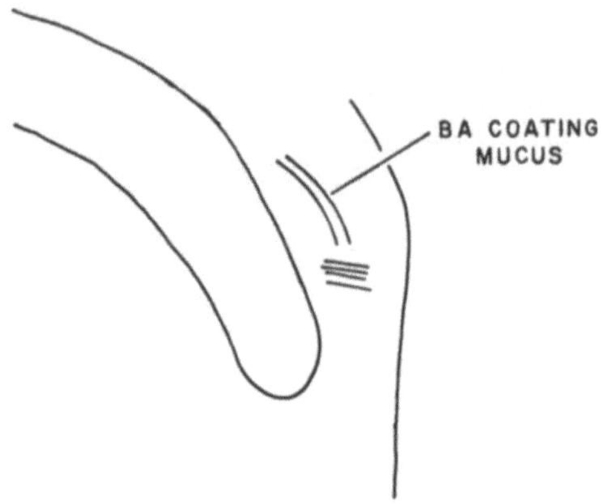

C

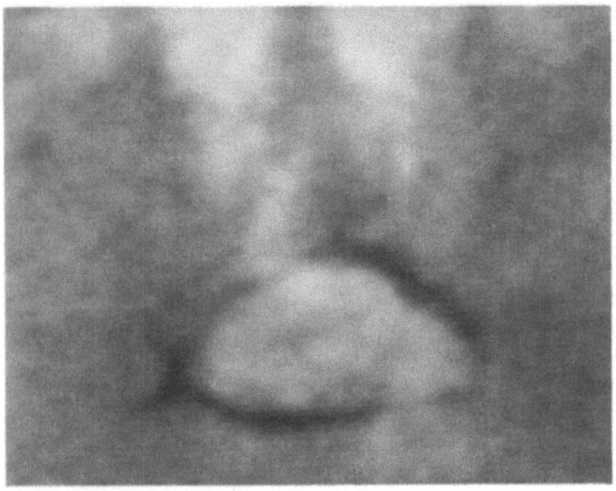

B

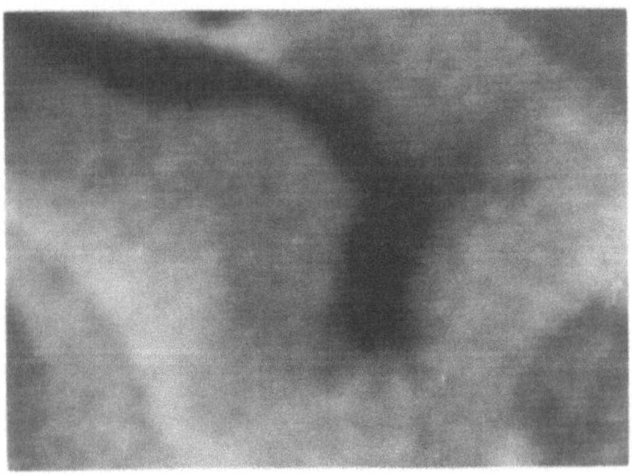

B'

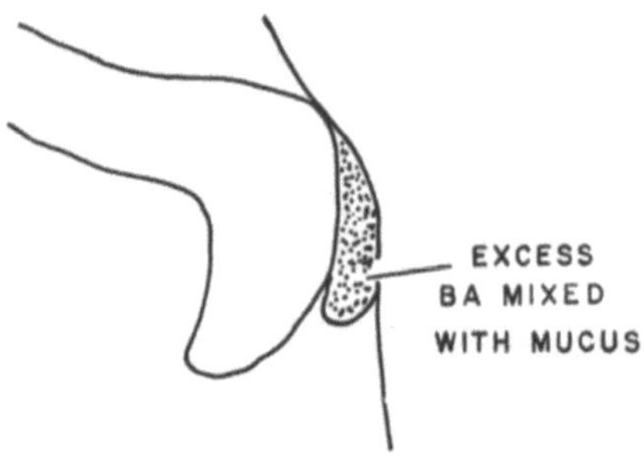

FIGURE 8.3. Repaired cleft palate patient with circular pattern of velopharyngeal closure. Lateral and base views during breathing (A-A', C) and phonation of "e" (B-B', E-E'). Base view as velopharyngeal portal is in process of closing (D-D') better demonstrates circular constriction of velopharyngeal sphincter. At stage of complete closure on base view (E-E'), appearance of velopharyngeal sphincter is seen less clearly because barium mixed with mucus lying below plane of velopharyngeal portal is superimposed onto plane of portal, obscuring its margins. Reproduced with permission from Skolnick ML, McCall GN, Barnes M: The sphinteric mechanism of velopharyngeal closure. Cleft Palate J 10:286–305, 1973.

D

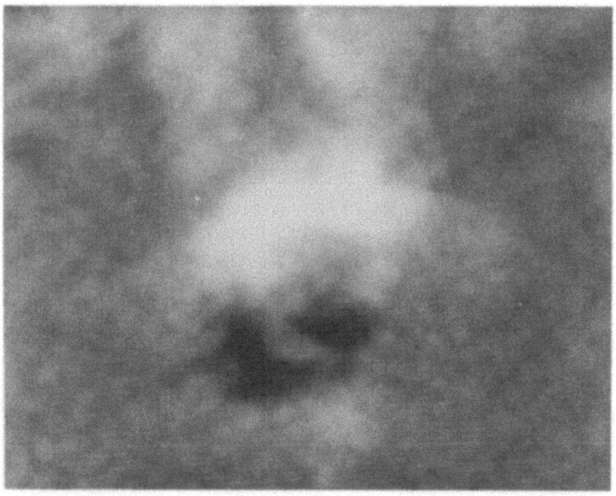

D'

E

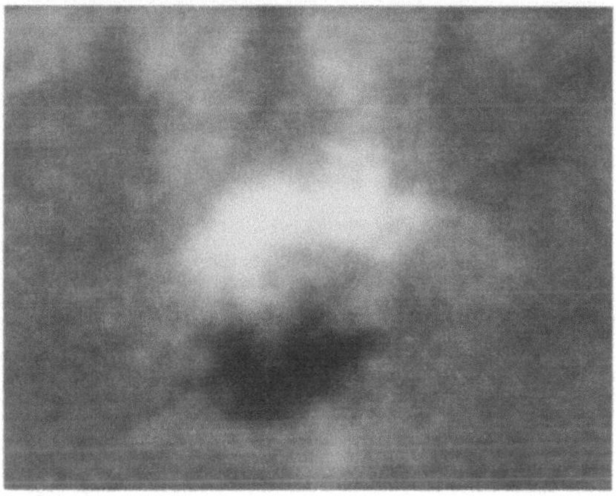

E'

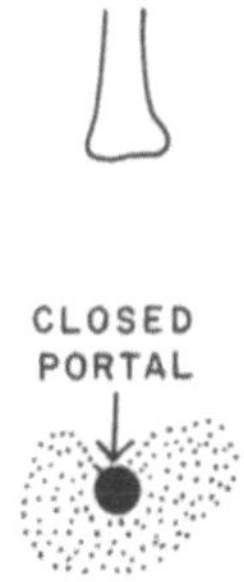

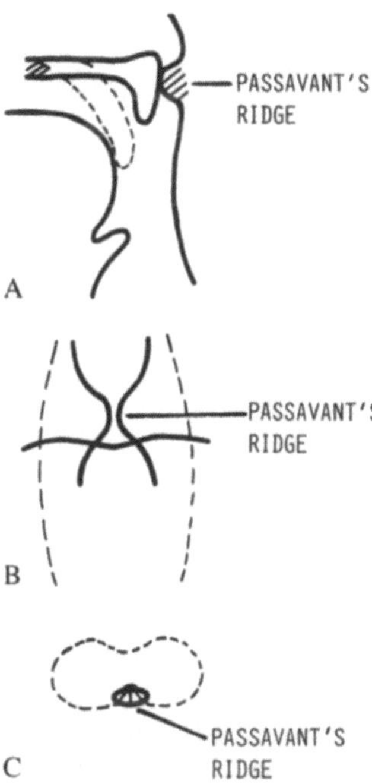

A

B

C

FIGURE 8.4

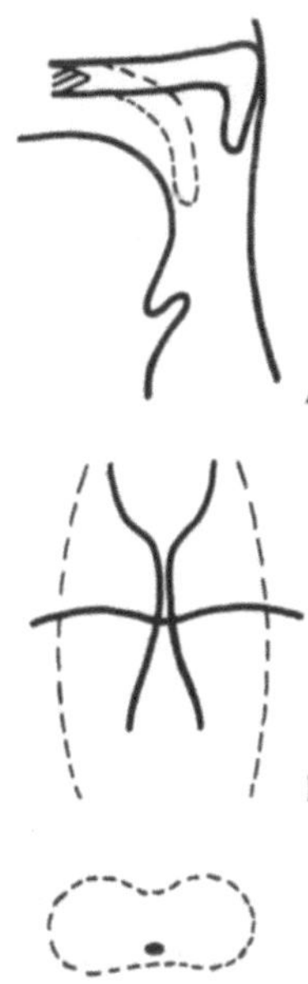

A

B

C

FIGURE 8.5

Figures 8.4 and 8.5. Comparison of circular closure patterns in patients with (Fig. 8.4) and without (Fig. 8.5) Passavant's ridge in (A) lateral, (B) frontal, and (C) base views.

A C

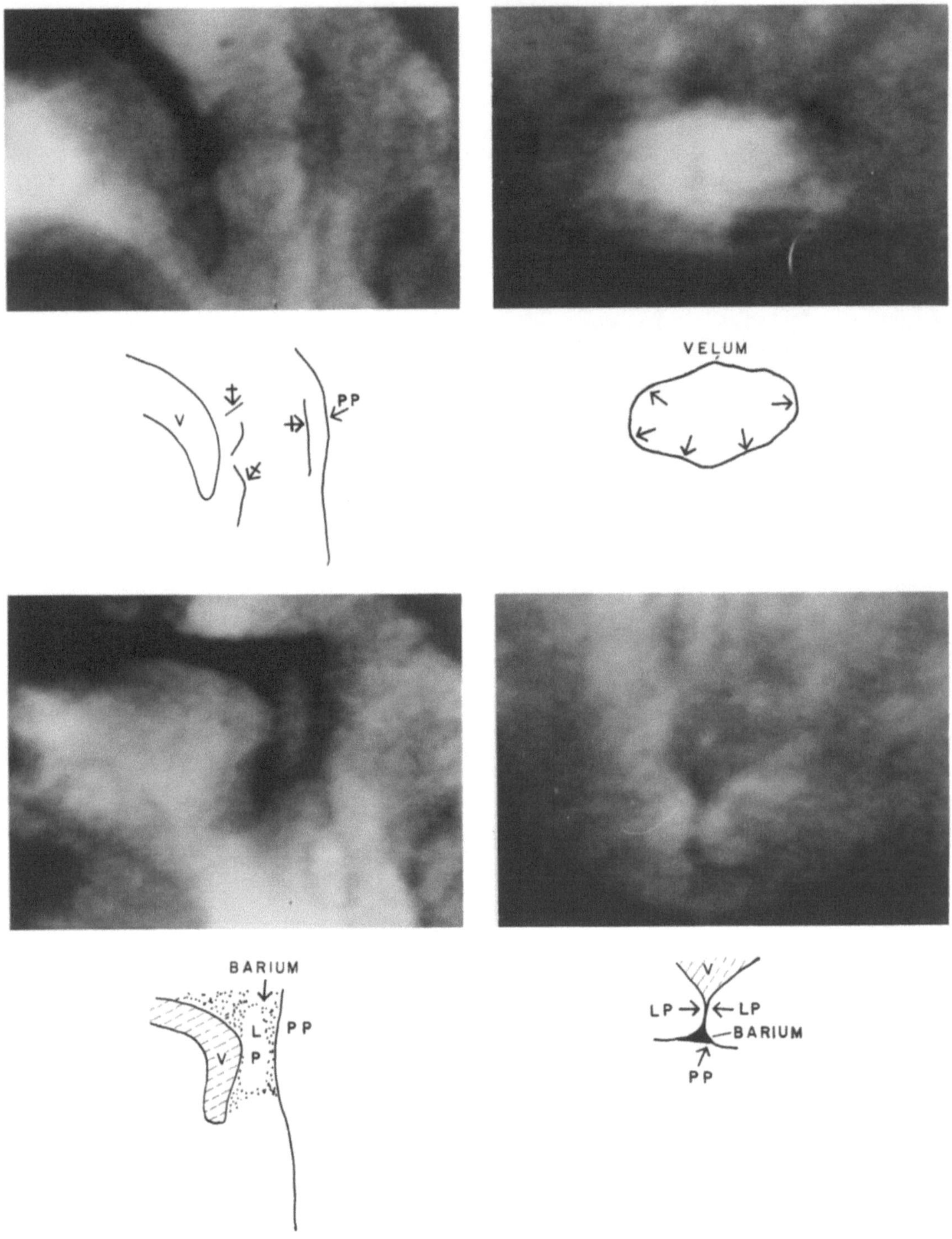

A' C'

B D

B' D'

FIGURE 8.6. Repaired cleft-palate patient with sagittal closure pattern. Lateral and base views during breathing (A-A', C-C') and phonation (B-B', D-D'). In the lateral view during phonation (B-B'), a soft-tissue mass (*LP*; the midline meeting of the pharyngeal walls) prevents the velum (V) from reaching the posterior pharyngeal wall. On the base view during phonation, one sees the lateral walls (*LP*) meeting in the midline. *PP*, posterior pharyngeal wall; (↓), barium-coated mucus. (→←), pharyngeal walls on base view. Reproduced with permission from Skolnick ML, McCall GN, Barnes M: The sphincteric mechanism of velopharyngeal closure. Cleft Palate J 10:286-305, 1973.

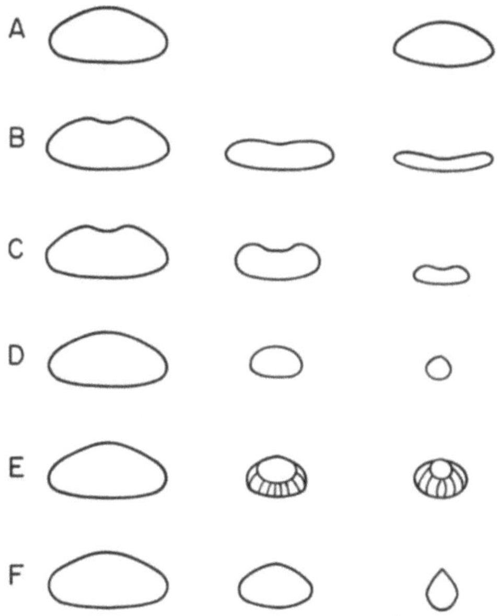

FIGURE 8.7. Schematic representation of sphincteric patterns of velopharyngeal insufficiency—base view. Extreme left column—portal at rest; middle column—partial closure; extreme right column—maximum closure. (A) Essentially no velar or pharyngeal wall movement. (B) Postadenoidectomy with good velar and almost no pharyngeal wall movement. (C) Average coronal narrowing pattern. (D) Circular narrowing pattern. (E) Circular narrowing pattern with Passavant's ridge (lined areas). (F) Sagittal narrowing pattern. Reproduced with permission from Skolnick ML, McCall GN, Barnes M: The sphincteric mechanism of velopharyngeal closure. Cleft Palate J 10:286-305, 1973.

resulting in incomplete concentric closure, especially in patients with VP inadequacy. This incomplete central movement may result in localized areas of portal inadequacy that can most readily be appreciated on the base view, although with careful examination of oblique views one may also either see or suspect these areas of inadequacy. We can only speculate that in those patients in whom regions of the pharyngeal walls show different degrees of movement, there are muscular differences reponsible for such movement.

Limitations of Lateral View for Evaluating Patterns of Closure

We have emphasized the value of the base view because this is the view that best demonstrates the relative contributions of palatal and pharyngeal wall movement to the closure mechanism. One cannot assume because there is good velar movement with significant narrowing or even closure of the velopharyngeal portal on lateral view that there is actually complete closure of the velopharyngeal portal. The lateral view alone, except in cases of Passavant's ridge, gives very little information about movement of the pharyngeal walls. The velum may significantly narrow the velopharyngeal portal on lateral view but there may be no evidence of lateral pharyngeal wall movement on frontal (Fig. 8.13) or base views, or there may be no velar motion with extremely active lateral pharyngeal motion (Fig. 8.14) (4). There can even be midsagittal velopharyngeal closure on lateral view with hypernasal speech because the lateral portals remain open even though the central part of the portal closes (Fig. 8.15). To further emphasize the lack of parallelism between velar and pharyngeal wall movements, examine Figures 8.16 and 8.17. Although both figures show similar degrees of velar movement on lateral view, in Figure 8.16 there is almost no lateral wall motion, whereas in Figure 8.17 the lateral pharyngeal walls almost meet in the midline.

Thus, to reiterate, the lateral view alone is usually inadequate to evaluate the effectiveness of velopharyngeal closure because it does not provide adequate information about lateral pharyngeal wall movement.

There are, however, two circumstances in which the lateral view can provide significant information about pharyngeal wall movement. When a localized posterior bulge (the Passavant's ridge) is seen, one can assume that there is good medial movement of the pharyngeal walls as well, because the Passavant's ridge is a crescent-shaped structure extending from the posterior into the lateral pharyngeal walls, not just a localized bulge projecting solely from the posterior wall. When we see the Passavant's ridge on lateral view (Fig. 8.10), we can also assume there is also a circular closure pattern on the base view. Thus, even if a base view cannot be obtained, we know that there is good lateral pharyngeal wall movement, a fact important for the surgeon to know if he is planning a pharyngeal flap. The lateral view also may provide information about lateral pharyngeal wall movement in the case of a sagittal closure pattern. Here one sees a soft-tissue mass interposing itself between the posterior surface of the palate and the posterior pharyn-

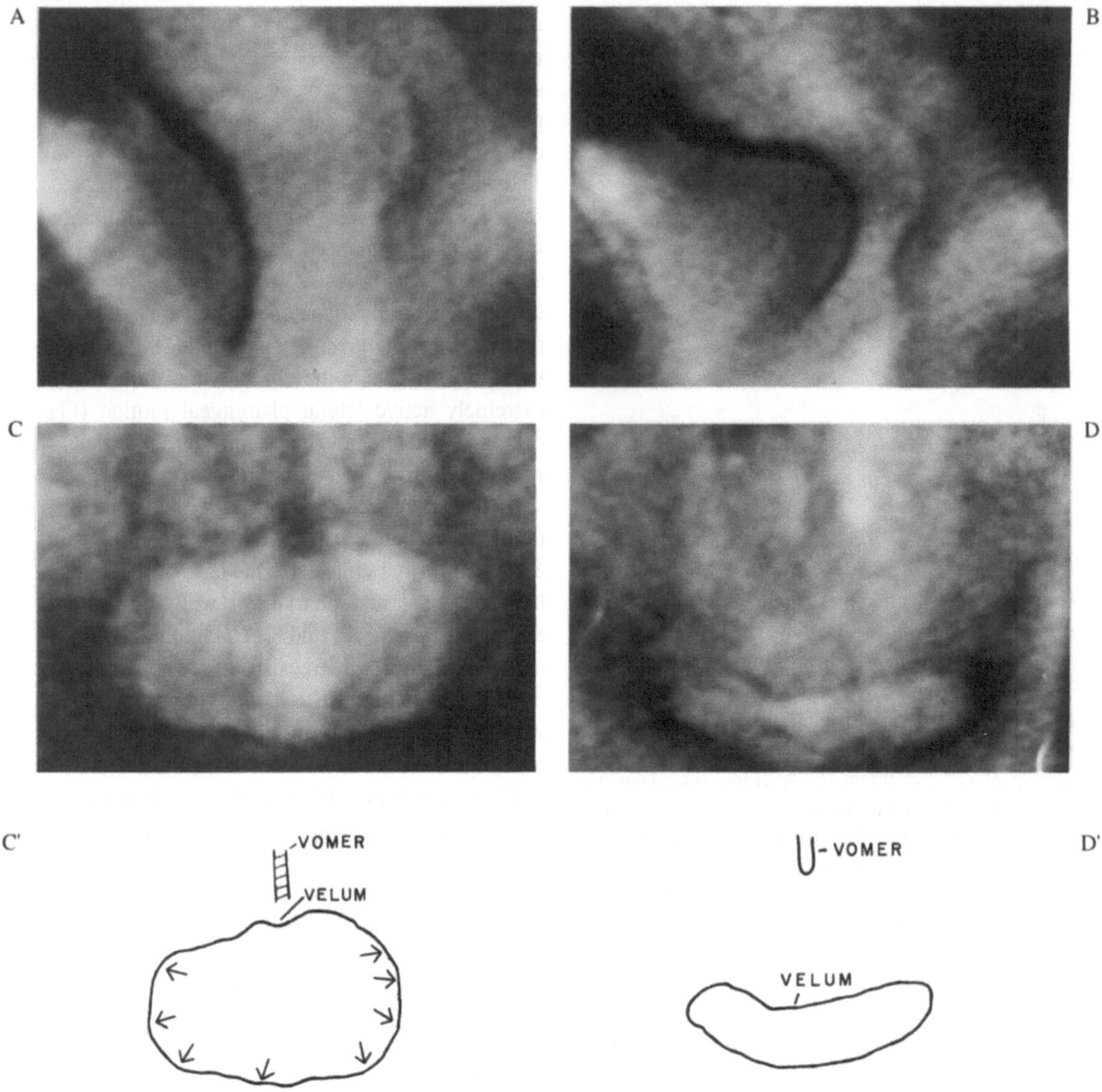

FIGURE 8.8. Velopharyngeal insufficiency postadenoidectomy. Lateral and base views during breathing (A, C-C') and phonation (B, D-D'). Moderate velar but almost no lateral pharyngeal wall motion during phonation results in a coronal pattern of narrowing of the VP portal. Reproduced with permssion from Skolnick ML, McCall GN, Barnes M: The sphincteric mechanism of velopharyngeal closure. Cleft Palate J 10:286–305, 1973.

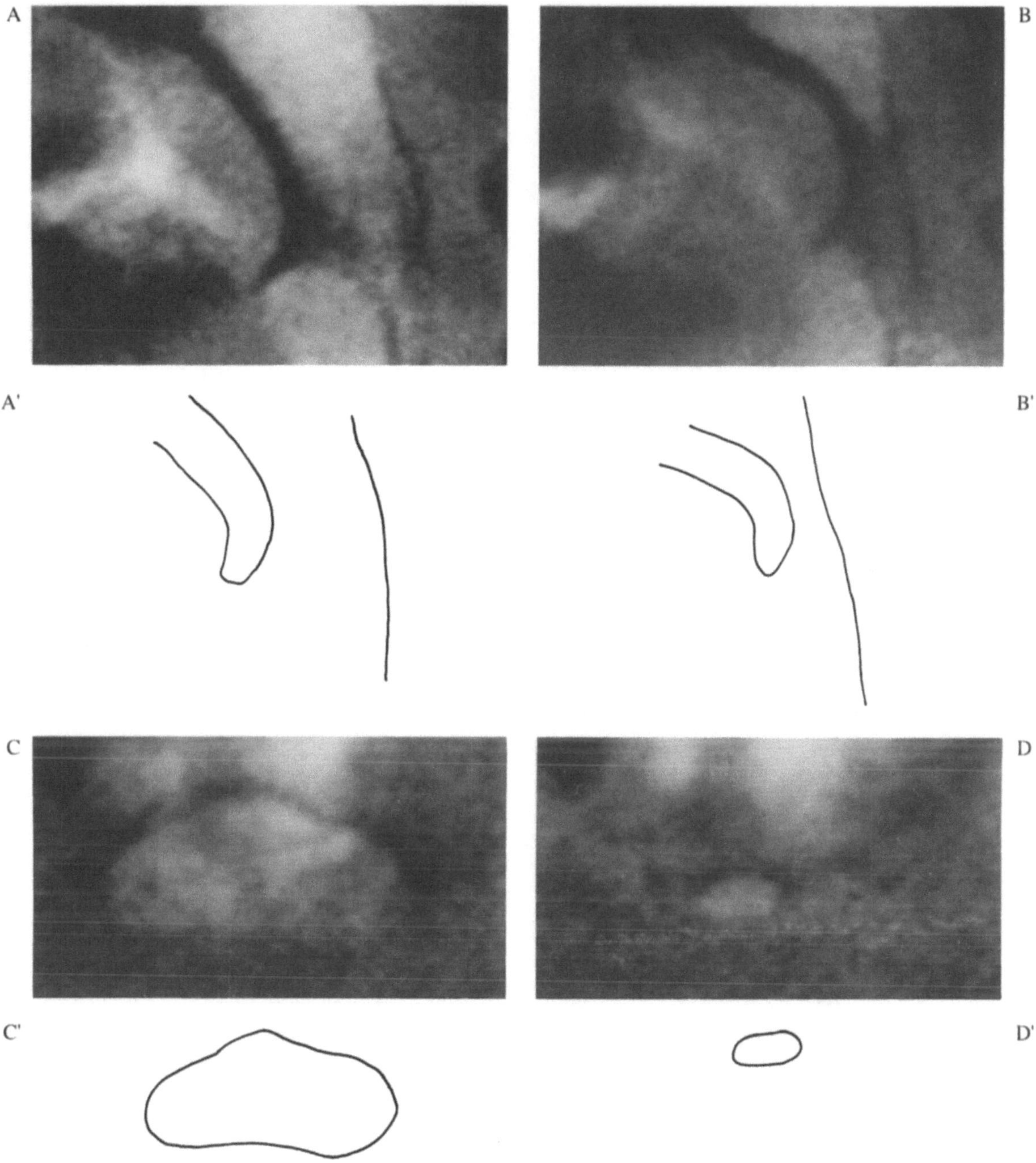

FIGURE 8.9. Repaired cleft palate patient with circular narrowing pattern. Lateral and base views during breathing (A-A', C-C') and phonation (B-B', D-D'). Reproduced with permission from Skolnick ML, McCall GN, Barnes M: The sphincteric mechanism of velopharyngeal closure. Cleft Palate J 10:286–305, 1973.

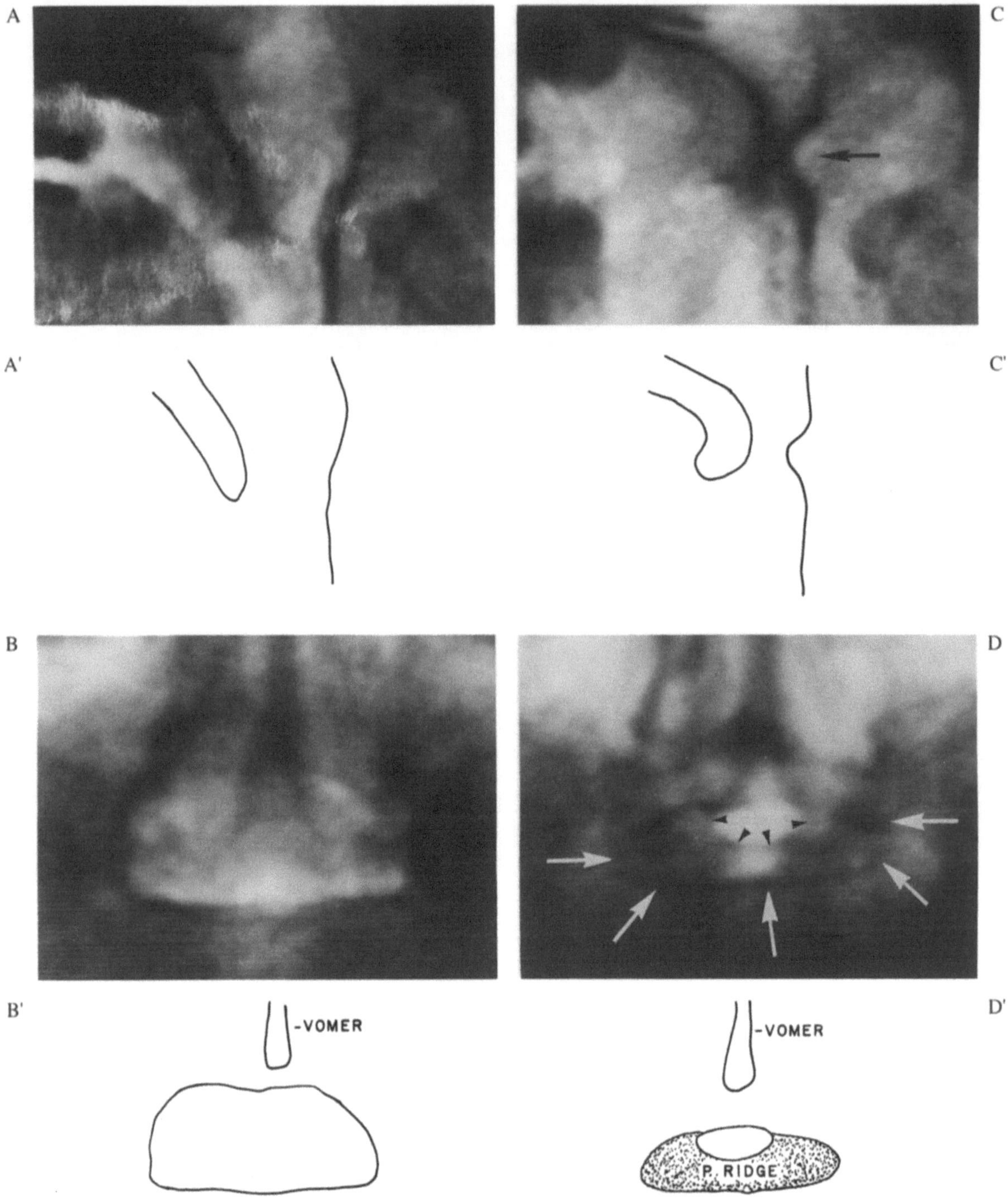

FIGURE 8.10. Repaired cleft palate patient with velopharyngeal insufficiency. Circular narrowing pattern with Passavant's ridge. Lateral and base views during breathing (A-A', B-B') and phonation (C-C', D-D'). Passavant's ridge on lateral view (arrow). Passavant's ridge on base view during phonation (arrows pointing to arrowheads). Reproduced with permission from Skolnick ML, McCall GN, Barnes M.: The sphincteric mechanism of velopharyngeal closure. Cleft Palate J 10:286–305, 1973.

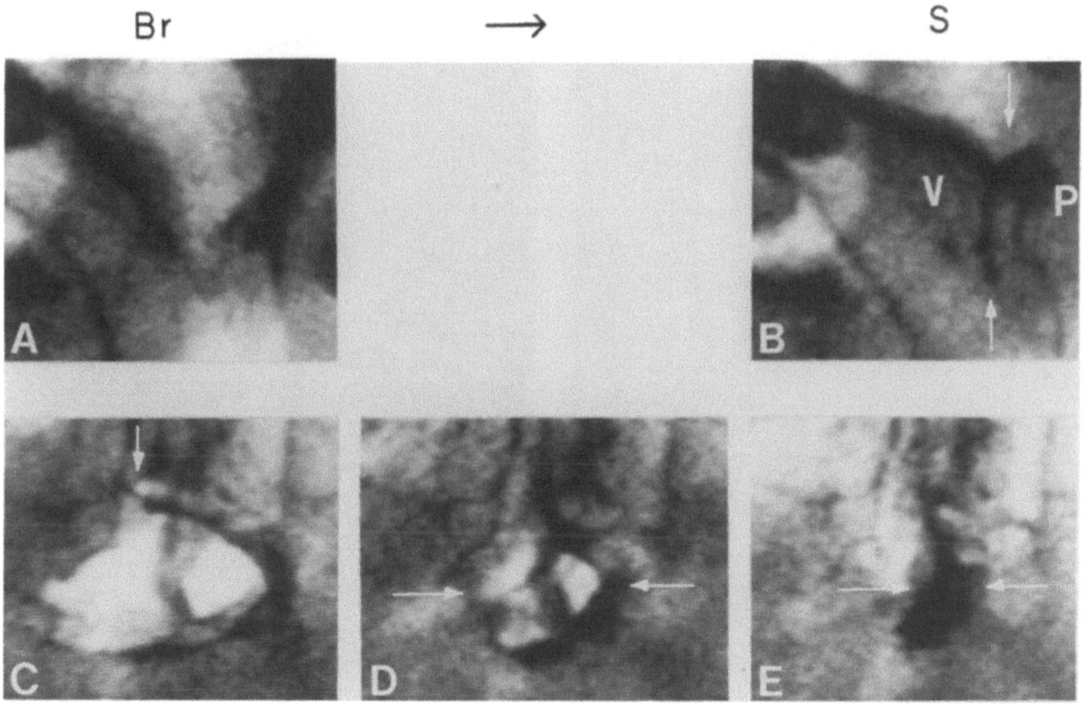

FIGURE 8.11. Repaired cleft palate patient with velopharyngeal insufficiency and sagittal narrowing pattern. Lateral and base views during breathing (Br) (A,C) and maximum portal narrowing during phonation of "S" (B,E). Intermediate stage of narrowing on base view (D). A significant amount of tissue is absent from central portion of velum. (Notch on posterior velum [arrow] on base view [C]). Narrowing of portal mainly results from medial movement of lateral pharyngeal walls (horizontal arrows) (D,E). Barium within portal on base view (E) obscures central gap between pharyngeal walls. On lateral view (B) bulk of lateral walls (vertical arrows) is interposed between velum (*V*) and posterior pharyngeal wall (*P*).

geal wall. This interposed soft tissue represents the opposition of the two lateral pharyngeal walls as they approach each other or touch in the midline. The posterior surface of the palate rests against the anterior edge of these abutting lateral pharyngeal walls (Fig. 8.6). However, the base view is still needed to determine whether there is complete portal closure or whether an opening between the lateral walls exists.

The lateral view can also suggest a submucous cleft palate or repaired cleft palate by demonstrating a double barium line along the posterior surface of the palate. The more anterior of these two lines represents the central defect from the submucous cleft palate or from the absent uvular muscle seen in the repaired cleft palate, whereas the more posterior of the lines represents the more lateral margins of the palate (Fig. 8.18).

Combined Frontal and Lateral Views for Evaluating Patterns of Closure

The frontal view, in combination with the lateral view, although not as informative as the three-view study of lateral, frontal, and base, can offer a good idea of the effectiveness of velopharyngeal closure because it demonstrates the degree of medial movement of the lateral pharyngeal walls. When frontal-view information is combined with the amount of velar movement seen on the lateral view, one can predict the closure pattern. However, small foci of portal inadequacy, either centrally or laterally, can be missed if the study consists solely of the lateral and frontal views. When patients with mildly hypernasal speech appear to have an effective closure

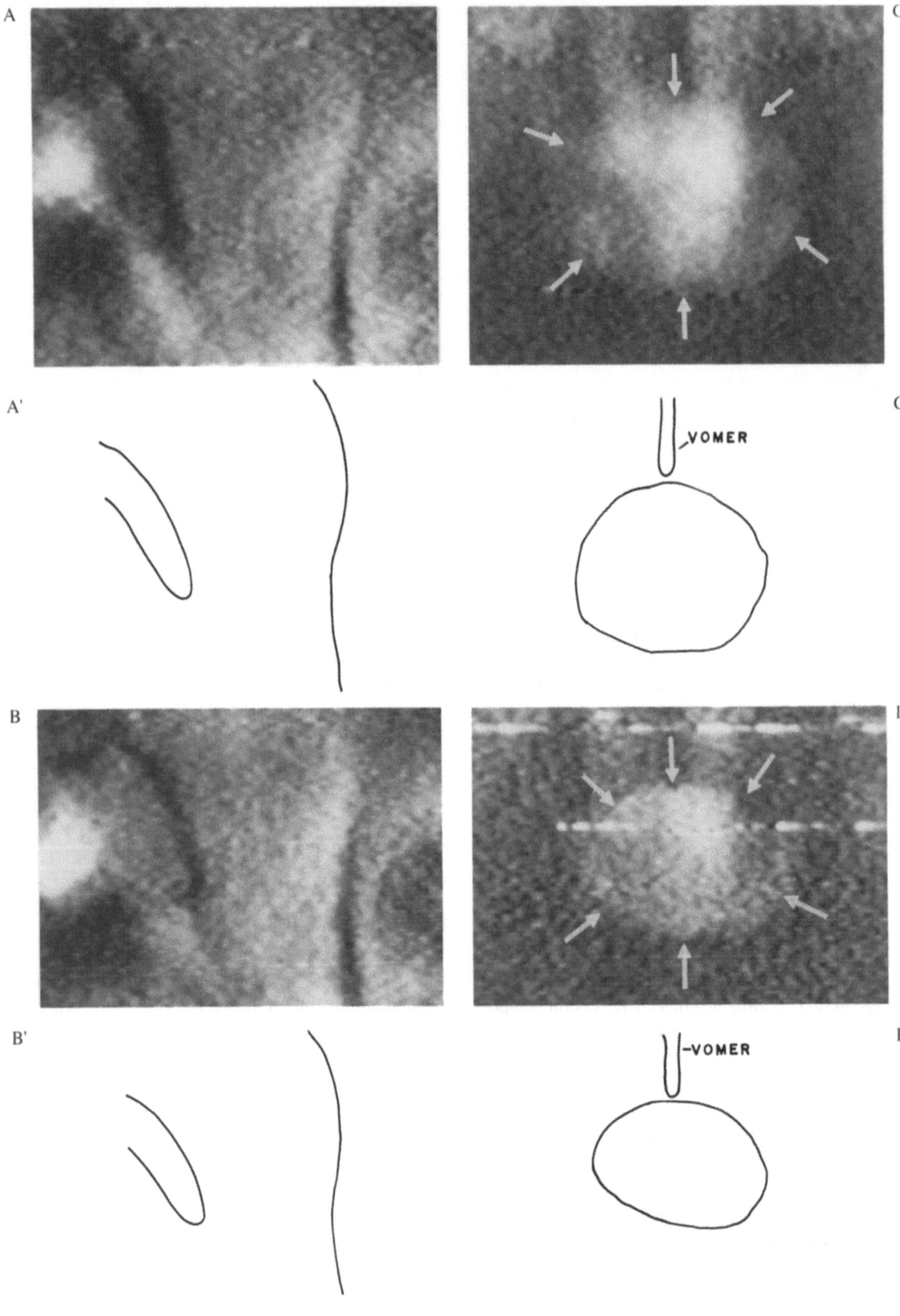
A
C
A'
C'
VOMER
B
D
B'
D'
VOMER

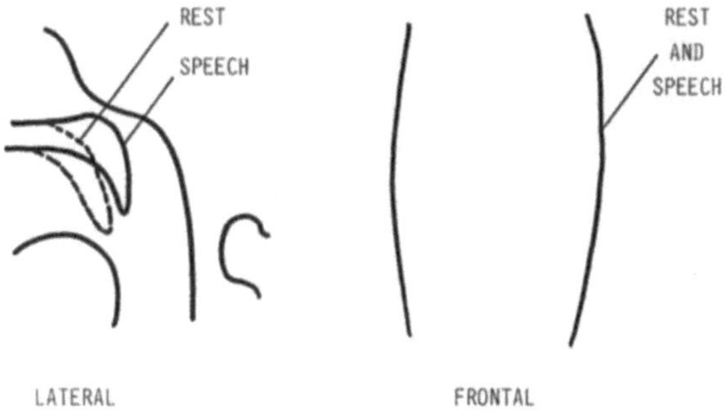

FIGURE 8.13. Tracing of videofluoroscopic image of a patient during speech showing velar motion in lateral view but no lateral pharyngeal wall motion in frontal view. Reproduced with permission from Shprintzen RJ, Rakoff SJ, Skolnick ML, Lavorato AS: Incongruous movements of the velum and lateral pharyngeal walls. Cleft Palate J 14:148–157, 1977.

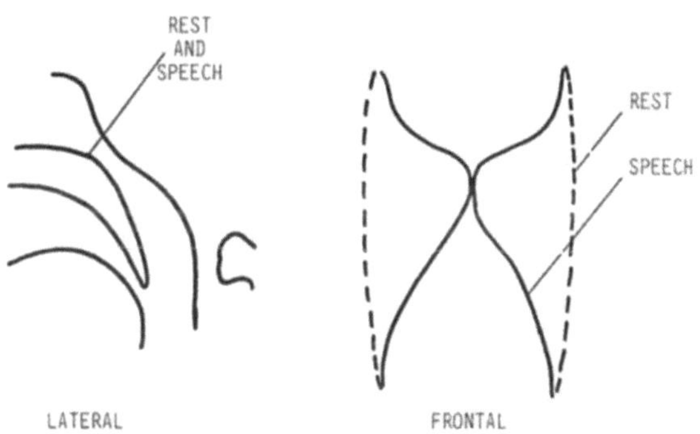

FIGURE 8.14. Tracing of videofluoroscopic image of a patient during speech showing no velar motion on lateral view but good lateral pharyngeal wall motion in frontal view. Reproduced with permission from Shprintzen RJ, Rakoff SJ, Skolnick ML, Lavorato AS: Incongruous movements of the velum and lateral pharyngeal walls. Cleft Palate J 14:148–157, 1977.

◁ FIGURE 8.12. Severely hypernasal patient with repaired cleft palate that has no significant palatal or pharyngeal wall movements during phonation. Compare lateral and base views during breathing (A-A', C-C') and phonation (B-B', D-D'). Arrows define barium-coated margins of velopharyngeal portal on base view. Reproduced with permission from Skolnick ML, McCall GN, Barnes M: The sphincteric mechanism of velopharyngeal closure. Cleft Palate J 10:286–305, 1973.

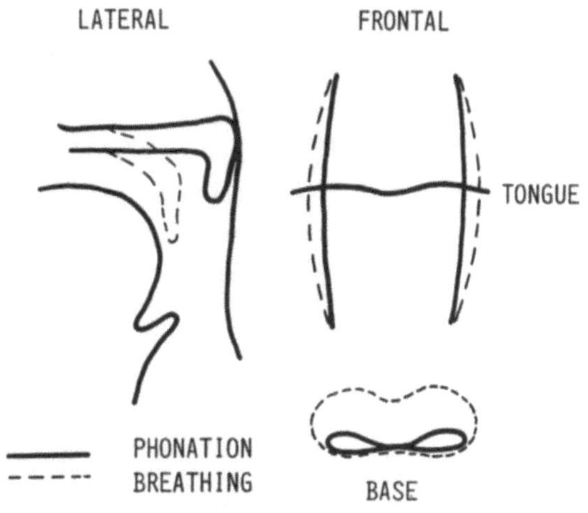

FIGURE 8.15. Central portal closure with lateral portal inadequacy. Although lateral view shows midsagittal velopharyngeal closure, lateral view alone is inadequate for completely evaluating velopharyngeal movements. Frontal view shows reduced lateral wall motion. Base view confirms central closure and lateral portal openings.

mechanism on frontal and lateral views, but cannot be satisfactorily studied via the base view, then oblique views should be used to examine the posteriolateral aspects of the velopharyngeal portal that may not be seen on the frontal and lateral views (see Chapter 4 for details). In these patients one should also carefully look for the presence of barium blowing through the velopharyngeal portal as a sign of portal inadequacy. This barium-blowing sign is usually seen when the area of inadequacy is small. When a non-nasal sound is uttered by the subject, the air rushing through the inadequate region mixes with the barium within the opening and produces a bubble of barium in a manner similar to the bubbles produced when a child dips a small metal ring in a solution of soapsuds and blows through the ring. When the pharyngeal gap becomes much larger, then the membrane of barium-coated mucus may not be suspended across the opening and, thus, bubbles of barium blowing through the portal may not be appreciated. (See Chapter 6 for further details.)

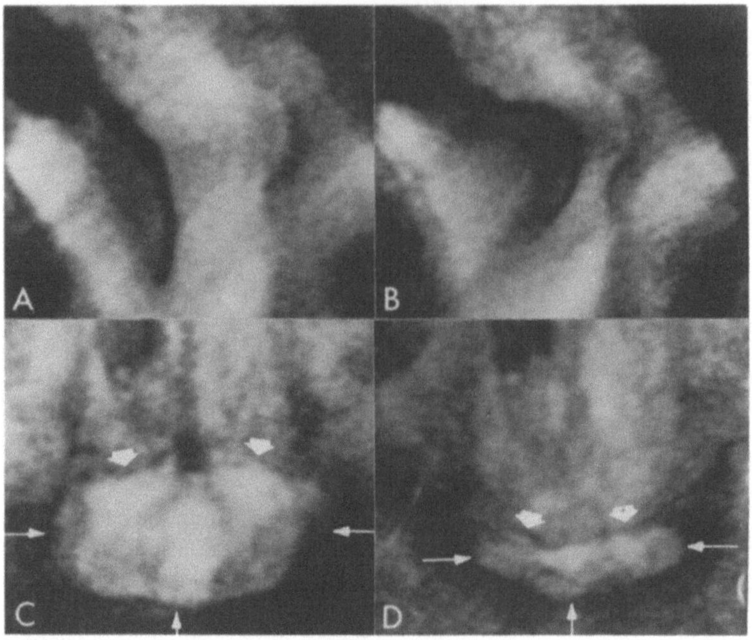

FIGURE 8.16. Lateral and base views during breathing (A, C) and phonation (B,D). Lateral view (B) shows good velar motion. Base view (D) shows almost no medial movement of lateral pharyngeal walls. Velum—arrowheads; pharyngeal walls—arrows, on base view.

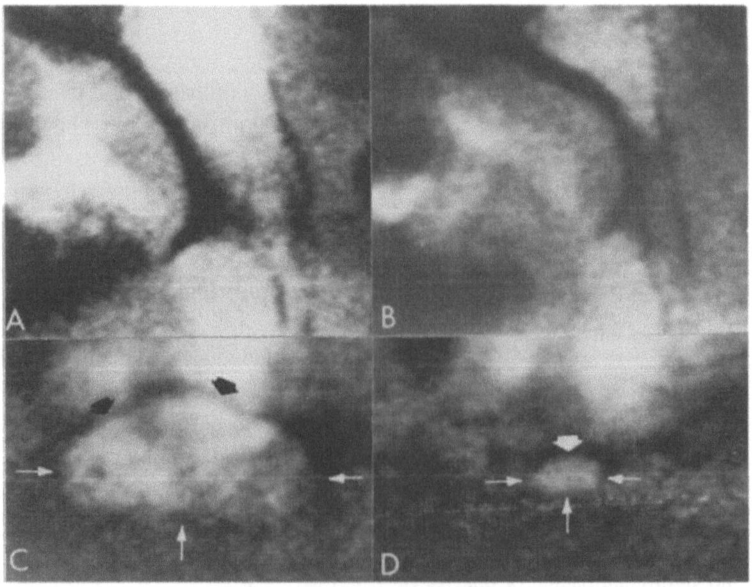

FIGURE 8.17. Lateral and base views during breathing (A, C) and phonation (B, D). Lateral view shows good velar motion (similar to Fig. 8.16B). However, base view (D) shows very good lateral wall motion. Velum—arrowheads; pharyngeal walls—arrows, on base view.

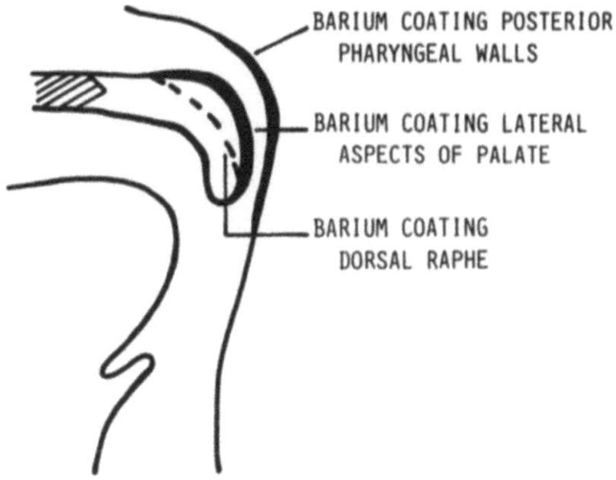

FIGURE 8.18. Patient with prominent raphe in dorsum of palate either representing site of primary repair or submucous cleft palate. Lateral view following nasal instillation of barium shows a double barium coating on the velum: one line representing barium in the raphe and the other representing barium coating the more lateral aspects of the velum.

References

1. Skolnick ML, McCall GN, Barnes M: The sphincteric mechanism of velopharyngeal closure. Cleft Palate J 10:286, 1973
2. Glaser (Cohn) ER: A multiview videofluoroscopic study of Passavant's ridge in patients with and without cleft palate. Unpublished doctoral dissertation, University of Pittsburgh, 1980
3. Skolnick ML: Velopharyngeal function in cleft palate. Clin Plast Surg 2:280, 1975
4. Shprintzen RJ, Rakoff SJ, Skolnick ML, Lavorato AS: Incongruous movements of the velum and lateral pharyngeal walls. Cleft Palate J 14:148, 1977

9 Passavant's Ridge

Passavant's ridge is a much misunderstood and controversial pharyngeal structure. In 1863 and 1869, Gustav Passavant (1, 2) reported his observations of a transverse bulge on the posterior pharyngeal wall in a patient with an unrepaired cleft palate. Passavant believed this bulge to be essential for normal speech:

> This "cross-roll" is of the greatest importance for speech; it has nothing to do with cleft palate and is a physiological process. For every word we speak it is an essential condition for normal pronunciation as it produces closure of the palate; otherwise a nasal pronunciation occurs (2).

Common misconceptions are as follows:

1. The ridge is essential to normal speech in both cleft and noncleft individuals.
2. The presence of a ridge signals the presence of velopharyngeal insufficiency.
3. The ridge is a static, ever-present structure in the pharynx.
4. The ridge is always too low in the pharynx to assist in velopharyngeal closure.
5. The ridge is never contributory to velopharyngeal closure.

Some of the major controversies surrounding ridge function are as follows:

1. Is Passavant's ridge a compensatory structure that develops in response to velopharyngeal insufficiency?
2. What is the tissue/muscular composition of Passavant's ridge?
3. Does the ridge appearance change over time, and when actually does the ridge form?

In light of the controversy and misunderstanding surrounding ridge function, we have elected to devote an entire chapter to the topic. We will consider the appearance, significance, and possible origins of Passavant's ridge and demonstrate how multiview videofluoroscopy can be useful in documenting ridge activity.

What Is Passavant's Ridge?

We define Passavant's ridge as a "localized soft-tissue projection that extends in an arc or crescent along the lateral and posterior nasopharyngeal walls." Passavant's ridge is not a permament projection, but rather a "functional" structure, which appears during speech and disappears during quiet breathing (Fig. 9.1). Ridge function during deglutition is controversial and will be addressed later in the chapter. The ridge has also been reported to emerge during other activities involving the velopharyngeal closure, such as whistling and blowing.

Passavant's ridge is not be confused with the broad or generalized anterior movement of the posterior pharyngeal wall during speech that lacks any localized region of movement. The difference between the generalized and discrete forward pharyngeal movement is shown in Figure 9.2.

Prevalence

Passavant's ridge has been reported in individuals both with and without hypernasality and/or radiographic evidence of velopharyngeal insufficiency.

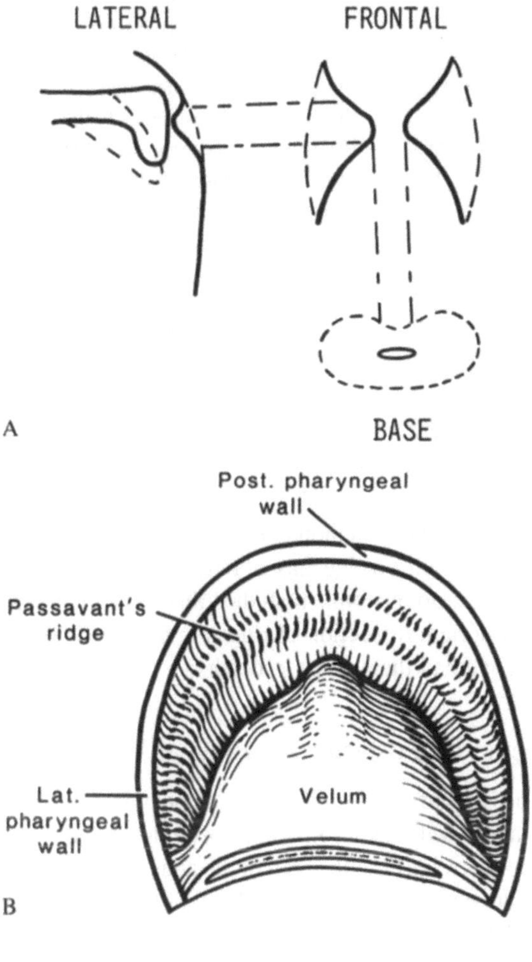

FIGURE 9.1. (A) Passavant's ridge: appearance during phonation (solid lines), disappearance during breathing (dotted lines). (B) Three-dimensional view of Passavant's ridge: looking down on velopharyngeal portal from above with nasopharyngeal cut-away.

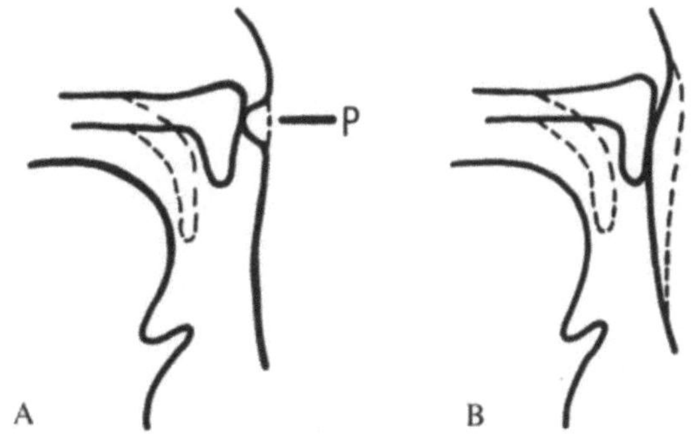

FIGURE 9.2. Lateral views distinguishing (A) Passavant's ridge (P), localized anterior movement of nasopharyngeal wall from (B) broad generalized anterior movement of posterior nasopharyngeal wall. (Solid lines, phonation; dotted lines, breathing.)

Passavant (1), using an intraoral mirror, was only able to visualize the ridge in 1 out of 50 normal subjects. However, he hypothesized that a ridge was present in all normal speakers.

It is now widely agreed that Passavant's ridge is not a prerequisite part of velopharyngeal closure for normal speech. Calnan (3) observed a ridge in only 20% of 20 normal adult speakers, and Fletcher (4) identified a ridge during speech in 30% of ten noncleft children.

It is also known that a Passavant's ridge is not present in all or even in a majority of patients with a repaired cleft palate. Reports of ridge occurrence during speech in these patients range from 9.5% (5) to 50% (6). We conducted our own survey of 88 patients with speech symptoms of VPI (7): 72% had clefts and 28% did not. Of these 88 patients, 25% had Passavant's ridges. Patients with and without ridges did not differ significantly with regard to speech characteristics or prior history of speech therapy.

Thus, one cannot assume that the presence or absence of a ridge is necessarily associated with the presence or absence of velopharyngeal insufficiency. The occurrence of Passavant's ridge in noncleft patients with normal speech is difficult to fit with the view that Passavant's ridge is a compensatory mechanism. However, although it could be said that some noncleft individuals would be hypernasal if a ridge were not present, this hypothesis is virtually impossible to investigate.

What Causes Passavant's Ridge to Appear

The anatomical and physiological bases of Passavant's ridge have been subjects of debate since Passavant first described the ridge. There are several major points of disagreement.

First, there is controversy as to whether the ridge is caused by a passive folding of mucosa initiated by the lifting action of muscles external to the ridge, by muscular contraction of fibers underlying the ridge mucosa, or by a combination of the two.

A B

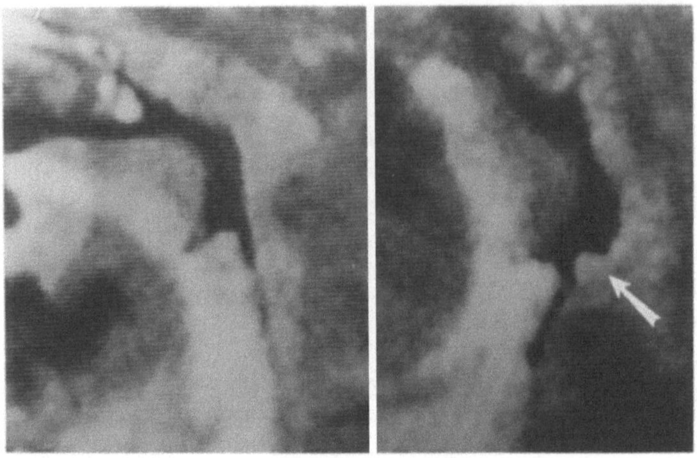

FIGURE 9.3. Passavant's ridge (arrow) in lateral view: (A) Neutral versus (B) hyperextended head position during phonation of sustained /s/. Note the increased prominence of the ridge in hyperextension. Reproduced with permission from Ellen R. Glaser(-Cohn) et al.: The dynamics of Passavant's ridge in subjects with and without velopharyngeal insufficiency—A multiview videofluoroscopic study. Cleft Palate J 16:24–33, 1979.

However, the bulk of the literature agrees that Passavant's ridge is generated by muscle fibers which course around the posterior pharyngeal wall in a horizontal direction to enter the palate. There remains disagreement as to whether these fibers are part of the palatopharyneus muscle (6, 8-10), the superior constrictor muscle (1, 11-17), or both.

Our observations (7, 18) support the involvement of fibers of the posterior pharyngeal wall, which enter the velum for the generation of Passavant's ridge. For instance, we have seen ridges appear after surgery for maxillary advancement, which suggests that the stretching of fibers of the posterior pharyngeal wall entering the velum may produce a ridge. Furthermore, on the lateral view videofluoroscopic projection, Passavant's ridge is often more prominent when the head is in a hyperextended rather than in a neutral position (Fig. 9.3). We hypothesize that the observed increase in the prominence of the ridge results from the deepening of the pharynx that occurs when the head is shifted into hyperextended posture (19,20). We further believe that the increase in pharyngeal depth results in a stretching of those fibers of the posterior pharyngeal wall that enter the velum, thus increasing the prominence of the ridge. Evidence that the ridge decreases with head flexion also supports this view, as does the observation of RW Pigott, MD (personal communication, 1984) that some ridges diminish or disappear after palatal pushback surgery. Based upon these observations, we hypothesize that there are muscle fibers that are continuous between the ridge and the palate, and activities stressing these fibers accentuate the ridge.

A conclusive demonstration of the composition of Passavant's ridge has not been made because there are no anatomical studies that have included postmortem dissection of ridges in patients known to have had them prior to death.

Level on Posterior Pharyngeal Wall

The level of Passavant's ridge on the posterior pharyngeal wall has traditionally been described as it appears relative to the arch of the atlas. While many wrote that the ridge is at the level of the atlas (3, 6, 8, 12, 16, 22), in a cinefluoroscopic study of six subjects, Carpenter (21) found the ridge to be moderately above the level of the tubercle of the atlas. Harrington (23) noted that the ridge may vary its position relative to the arch of the atlas and may appear from 3 mm above the arch to 7 mm below it.

Other investigators have related the ridge's location to one or more velar landmarks. Honjo et al. (24) studied ten subjects with ridges, using static and dynamic radiography, and reported that ridges were located 4.4 mm to 14.2 mm *below* the hard palate, with a mean of 9.5 mm. Passavant (2) described the ridge as appearing at the level of the base of the uvula, a description that Calnan (6) later criticized as inadequate because ridge level relative to the uvula may vary with maturation.

We studied Passavant's ridges in 43 patients using multiview videofluoroscopy (18) and considered the level of Passavant's ridge relative to its relationship to the velum during speech (Fig. 9.4). We found that 80% of the ridges were positioned

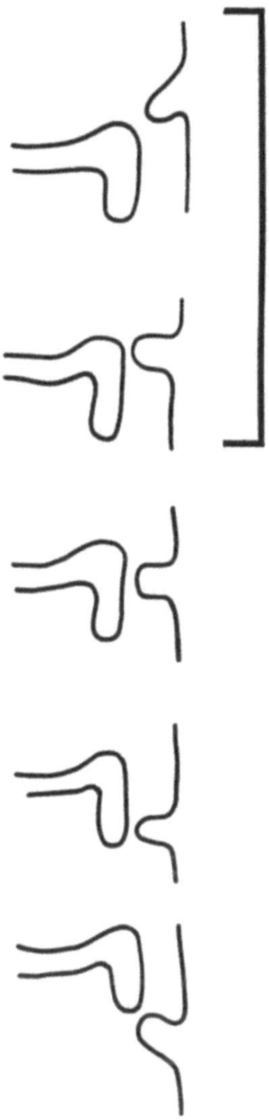

FIGURE 9.4. Lateral views showing level of Passavant's ridge relative to velum during speech in 43 subjects. (A) In 5% of the patients the ridge was above or at the level of the eminence. (B) In 58% of patients the ridge was opposite the vertical part of the velum. (C) In 26% the ridge was at the uvula level. (D) In 12% it was below the uvula level.

with sufficient height on the posterior pharyngeal wall to assist in constriction or closure. Results were as follows:

5%—opposite velar eminence
58%—opposite vertical portion of velum
26%—opposite uvula
12%—below uvula

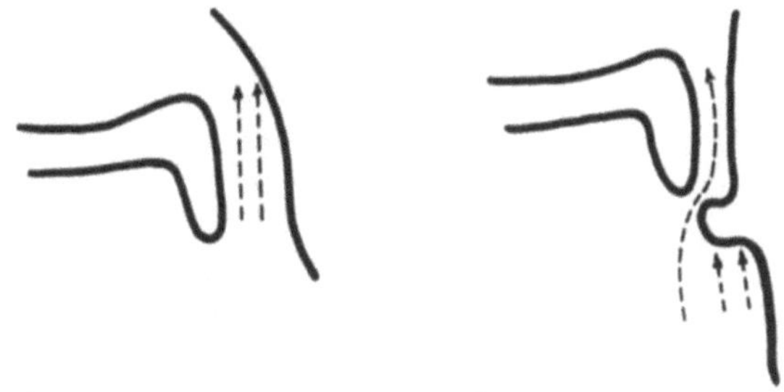

FIGURE 9.5. Hypothetical role of Passavant's ridge when located below the uvula. (A) No Passavant's ridge; air flow has no barrier. (B) Passavant's ridge, which could reduce hypernasality by impeding flow of air. Reproduced with permission from Ellen R. Glaser(-Cohn) et al.: The dynamics of Passavant's ridge in subjects with and without velopharyngeal insufficiency—a multiview videofluoroscopic study. Cleft Palate J 16:24–33, 1979.

These results disagree with the assumption of other authors (3, 6, 14, 26, 27) that Passavant's is usually too low in the pharynx to assist in velopharyngeal closure. Futhermore, we suggest that a ridge which appears below the uvula may still assist in reducing hypernasality by serving as a partial barrier to airflow during speech, and/or by damping (Fig. 9.5) sound, particularly if the ridge is large relative to the velopharyngeal gap.

Although the location of Passavant's ridge was variable across subjects, it was relatively constant within each patient.

Orientation of Ridge

We also studied the orientation of Passavant's ridge in the 25 patients (Fig. 9.6). In 18 subjects (82%), the ridge was perpendicular to the velum. In two subjects (9%), the ridge was pointed in an upward direction oriented toward the velar eminence, whereas in two other subjects (9%), the ridge was above the velum and oriented in a downward direction (25).

Association with Lateral Pharyngeal Wall Motions

Intraoral observations of Passavant's ridge have shown it to be crescent-shaped and extending from the posterior to the lateral pharyngeal walls (1, 2, 8, 12, 32). We made the same observations, using

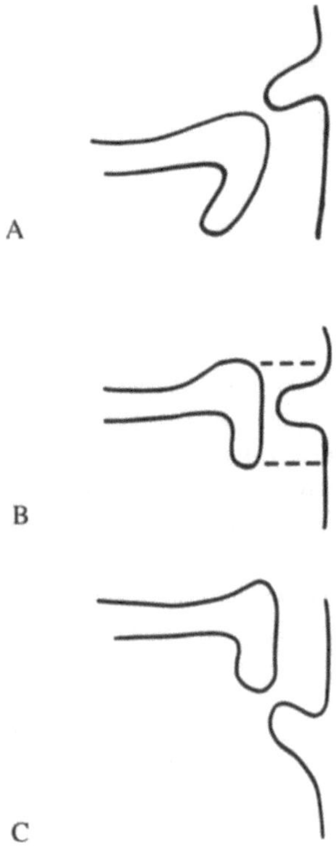

◁ FIGURE 9.6. Lateral views showing orientation of Passavant's ridge to velum in 25 subjects. In 82% the ridge (B) was oriented perpendicular to the velum; located within the region of the dotted lines, in 9% (C) it was below the uvula and pointed upward, while in 9% (A) it was above the eminence and pointing downwards toward it.

FIGURE 9.7. Passavant's ridge during multiview videofluoroscopy. Lateral, frontal and base views during breathing (A, C, E) and phonation (B, D, F). Velopharyngeal insufficiency is present during phonation manifested by a small central velopharyngeal opening. Horizontal black arrowhead indicates ridge on lateral view (B). Double black arrowheads indicate medial movements of ridge involving lateral pharyngeal walls (D). Opposing arrowheads and arrows reveal crescent shape of ridge on base view (F). White oval above arrowheads is central velopharyngeal opening. Opposing rows of white arrows show lateral pharyngeal walls at rest on frontal view (C). Reproduced with permission from Skolnick ML: Velopharyngeal function in cleft palate. Clin Plast Surg 2:280-297, 1975.

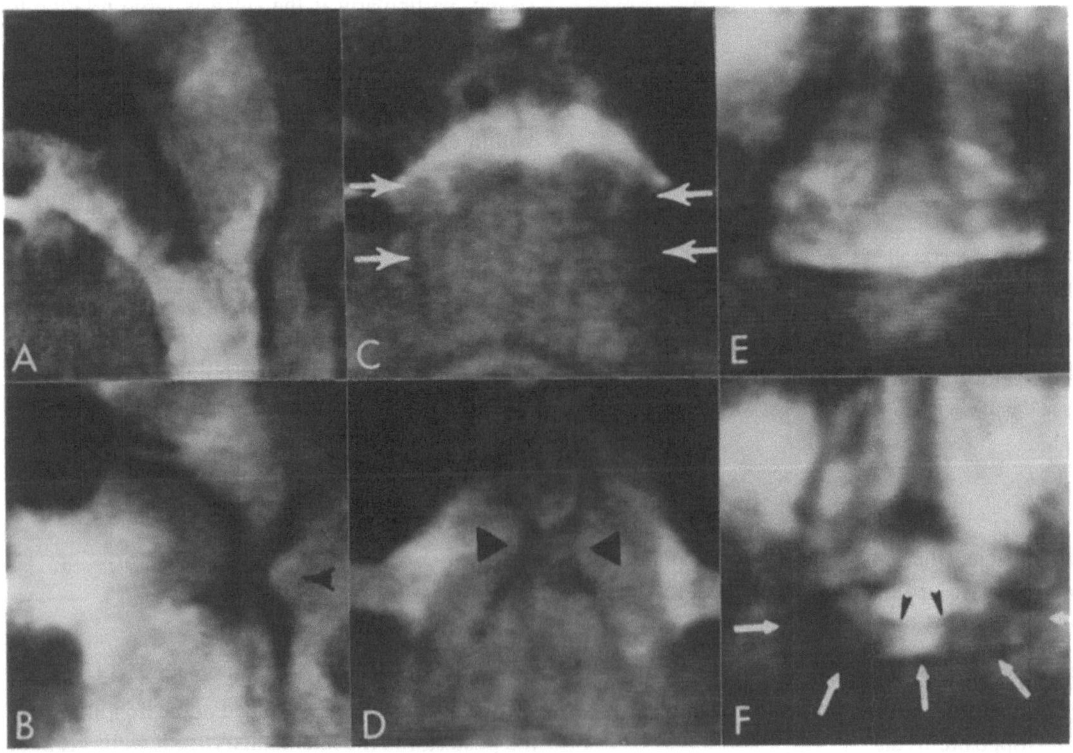

multiview videofluoroscopy (18, 25, 29). In our studies, a Passavant's ridge seen on the lateral videofluoroscopic projection was usually associated with active, localized shelflike lateral pharyngeal wall motion on the frontal view and a circular closure pattern on base view (Fig. 9.7).

The high incidence of circular closure in patients with ridges (32 of 43 cases, or 74%) (18) is in sharp contrast to a much lower incidence of circular closure in subjects *without* Passavant's ridge (10% of 80 non-nasal anatomically normal cases and 20% of 500 hypernasal cases) (30).

As expected from the observations, ridge activity was seen to be grossly synchronous with lateral pharyngeal wall activity. That is, the lateral pharyngeal walls moved in concert with Passavant's ridge activity as seen on the base view.

Passavant's Ridge and Velar Activity

The issue of the exact relationship of ridge movement to velar movement has been a central one in the attempt to understand both the functional signficance and anatomical bases of Passavant's ridge activity. That the ridge moves in synchrony with the velum was observed by Glaser-(Cohn) et al. (18) and Skolnick et al. (29) as well as by Passavant (1), Bosma (15), Barton (16) and Carpenter (21). This view was disputed by Calnan (6), who wrote that ridges moved slowly and fatigued.

Carpenter (21) reported that, like the velum, Passavant's ridge maintained a stable phonatory position during continuous speech unless interrupted by a nasal consonant, and that the ridge reached a phonatory position simultaneously with the velum although the ridge lagged behind the palate in the initiation of movement. Carpenter thus concluded that the ridge functions much as the velum does. These observations were supported by Glaser-Cohn et al. (18), who also reported that ridge activity reflected articulatory competency. For example, one of our patients had active ridge and velar activity for all speech sounds except the /s/, which was misarticulated. We have not observed the isolated fatigue of the ridge reported by Calnan, unless of course the entire velopharyngeal mechanism was seen to fatigue.

A

B

C

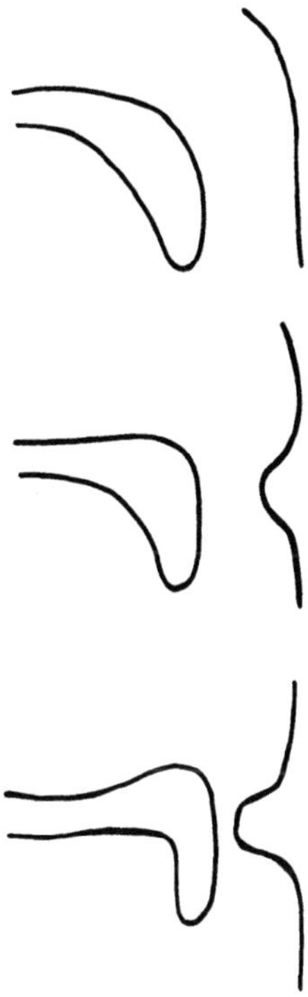

FIGURE 9.8. Lateral views showing enlarging Passavant's ridge with increasing velar elevation. (A) Quiet breathing. (B) Phonation of low vowel. (C) Phonation of plosive consonant.

Thus, observers of ridge activity can expect the ridge to move grossly in concert with the velar component of the sphincter. As does velar activity, ridge size appears to vary according to the speech sound, and the overall degree of velar activity (Fig. 9.8).

Clinical Significance of Passavant's Ridge

In our study of 43 patients (18), we found that Passavant's ridge did, in some individuals, contribute to closure. The ridge was in a primary position to achieve constriction in 37% of this group, and actually achieved closure in 7% (Fig. 9.9A). The ridge occurred at a lower, secondary site of narrowing in 28% of this patient group (Fig. 9.9C), increased the length of narrowing or closure in 23% (Fig. 9.9B) and was located below the uvula in 12% (Fig. 9.9D). As noted previously, whereas others contend that the subuvular ridge does not influence

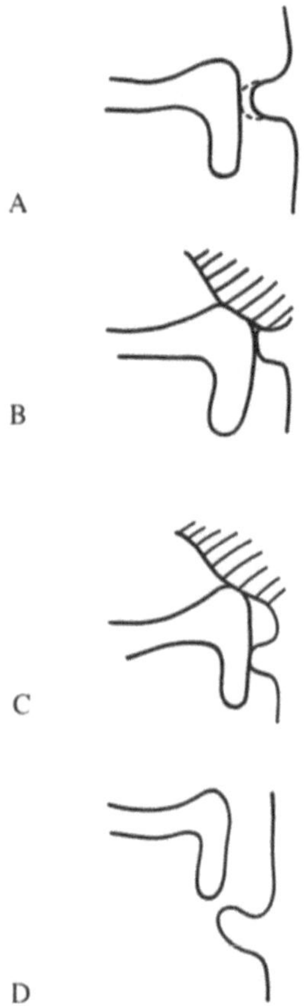

FIGURE 9.9. Contribution of Passavant's ridge to velopharyngeal closure in 43 subjects. Lateral views. Ridge was in primary position to achieve closure (opposite vertical portion of velum) in 37% and actually achieved closure in 7% (A). Ridge increased zone of narrowing or closure in 23% (B). Ridge was a secondary site of closure in 28% (C); primary closure was between velum and adenoids. Ridge was too low to achieve closure in 12% (D), but may have reduced degree of hypernasality.

speech, we hypothesize that a large ridge may slightly alter the airflow or absorb sound (Fig. 9.5).

Use of Barium

We have found that a barium sulfate coating of the nasopharynx is valuable, if not essential, to videofluoroscopic visualization of the ridge. In a previous study (31) of 28 ridges visible with a barium sulfate coating, 25 were not detected when barium was not used (Fig. 9.10).

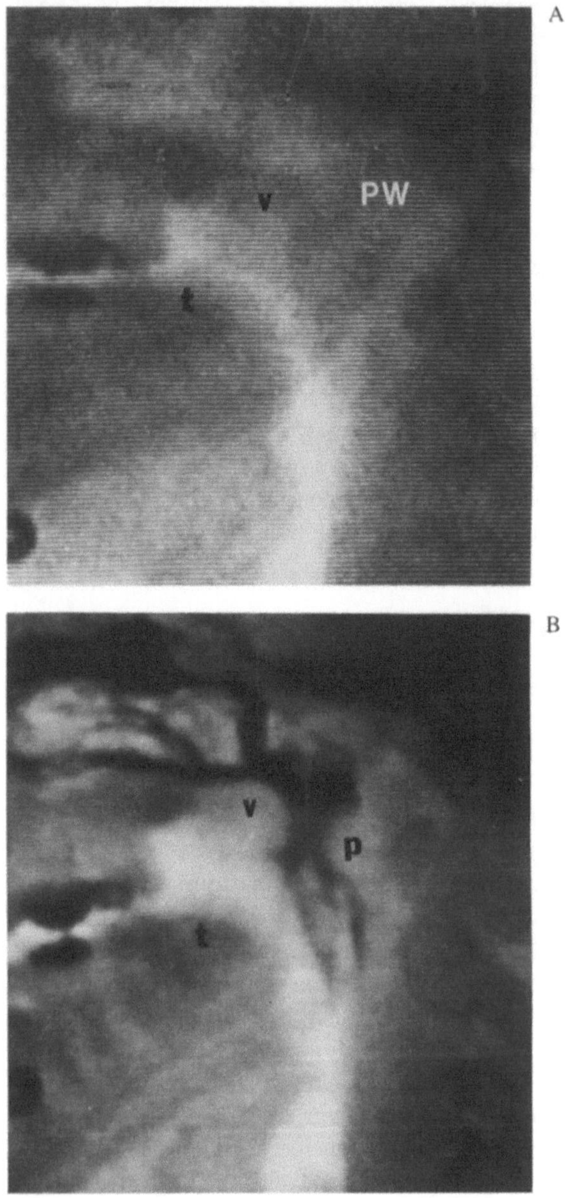

FIGURE 9.10. Appearance of Passavant's ridge on lateral view without (A) and with (B) nasal barium coating during phonation of "s." Without use of barium, velum (*V*) blends into posterior pharyngeal wall (*PW*). With use of barium, a distinct Passavant's ridge (*P*) is seen as well as a midsagittal VP gap (filled with barium) between the velum and ridge. Reproduced with permission from Glaser(-Cohn) ER et al.: The dynamics of Passavant's ridge in subjects with and without velopharyngeal insufficiency—a multiview videofluoroscopic study. Cleft Palate J 16:24–33, 1979.

Swallowing Versus Speech

We compared the position and forward movement of the ridge that is seen during speech with posterior pharyngeal wall bulging during swallowing (25). As can be seen in Figure 9.11, the superior boundary of the ridge was higher during speech than during swallowing. This observation is consistent with the hypothesis of Shprintzen et al. that during speech the uppermost fibers of the superior constrictor are activated, which does not occur during reflexive swallowing (28). Shprintzen et al. observed that maximum lateral pharyngeal wall activity takes place at a lower site for swallowing than for speech. We also observed that anterior movement of the posterior pharyngeal wall occurred at a lower level for swallowing than for speech. The forward movement of the posterior pharyngeal wall also tended to be greater during swallowing than during the speech tasks. Finally, the ridge seen during swallowing was less discrete than the ridge seen during speech (Fig. 9.11).

We therefore recommend that the use of the term "Passavant's ridge" be limited to speech functions and not include swallowing.

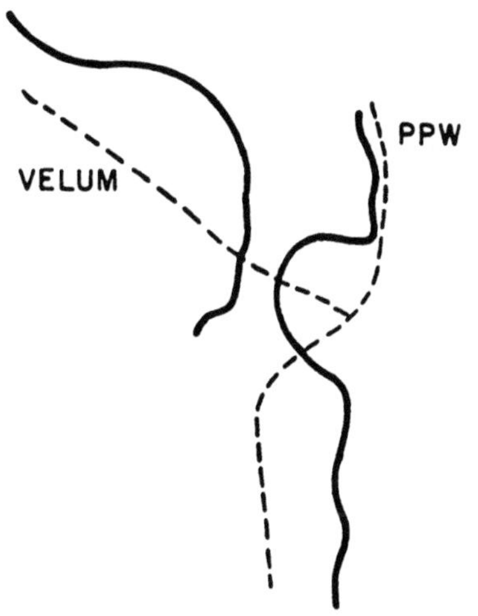

FIGURE 9.11. Swallowing versus speech activity with Passavant's ridge in the lateral videofluoroscopic view. Solid lines show speech "s" sound; dotted lines show swallowing. During speech velum elevates and localized pharyngeal movement occurs while during swallowing velum elongates and broad pharyngeal movement occurs at a lower level than with speech. Reproduced with permission from Glaser(-Cohn) ER. A multiview videofluoroscopic study of Passavant's ridge in patients with and without velopharyngeal insufficiency. Unpublished doctoral dissertation, University of Pittsburgh, 1980.

Clinical Implications

As a result of our observations, we believe that there is a clinical need to determine the presence, location, and possible significance of Passavant's ridge within the velopharyngeal valve of *each* patient. It is no longer valid to assume that all Passavant's ridges are noncontributory to velopharyngeal closure. Furthermore, patients with ridges should not be presumed to have different speech characteristics than those without ridges. Thus, we believe that the mere presence of a ridge, in the absence of both speech stigmata of velopharyngeal insufficiency and a demonstrable velopharyngeal gap, is not symptomatic of a clinically significant problem necessitating treatment. Each Passavant's ridge should be evaluated within the total context of the velopharyngeal closure mechanism for that individual patient.

In patients with ridges requiring a pharyngeal flap or prosthetic treatment, the flap or prosthesis should be positioned to take full advantage of the contribution of the ridge, provided that the maximum velar and lateral pharyngeal wall motion occurs at the level of the ridge.

References

1. Passavant G: Ueber die Verschliessung des Schlundes beim Sprechen. Arch Heilkd 3:305, 1863 (translated by Jose Paltan, 1977)
2. Passavant G: Ueber die Verschliessung des Schlundes beim Sprechen. Virchows Arch [A] 46:1, 1869, cited in Townsend (8)
3. Calnan, J: The error of Gustav Passavant. Plast Reconstr Surg 13:275, 1954
4. Fletcher SG: A cinefluorographic study of the movements of the posterior wall of the pharynx during speech and deglutition. MS Thesis, University of Utah, 1957
5. Massengill R, Walker T, Pickrell K: Characteristics of patients with a Passavant's pad. Plast Reconstr Surg 44:268, 1969
6. Calnan J: Modern views on Passavant's ridge. Br J Plast Surg 13:275, 1954

7. Cohn E, Skolnick ML: Insights into the function of Passavant's ridge. Paper presented at American Cleft Palate Association Annual Meeting, 1985
8. Townsend RH: The formation of Passavant's bar. J Laryngol 55:154, 1940
9. Negus VE: The mechanism of swallowing. Proc R Soc Med 36:85, 1942
10. Ardan GM, Kemp FH: Radiologic investigation of pharyngeal and laryngeal palsy. Acta Radiol 46:446, 1956
11. Sappey MPC. Traité d'anatomie descriptive 4, 1874, cited in Calnan (6).
12. Wardill WEM: Cleft palate—results of operation for cleft palate. Br J Surg 16:127, 1928
13. Whillis J: A note on the muscles of the palate and the superior constrictor. J Anat 65:92, 1930
14. Veau V, Borel-Maissony S: Mem Acad Chir 69, 1943, cited in Calnan (6).
15. Bosma JF: A correlated study of the anatomy and motor activity of the upper pharynx by cadaver dissection and cinematic study of patients after maxillofacial surgery. Ann Otol Rhinol Laryngol 62:51, 1953
16. Barton RT: The clinical significance of the ridge of Passavant. Surg Gynecol Obstet 101:317, 1955
17. Lavorato A: Normal lateral pharyngeal wall motion during velopharyngeal functioning: A cinefluorographic study. Unpublished doctoral dissertation, University of Pittsburgh, 1976
18. Glaser(-Cohn) ER, Skolnick ML, McWilliams BJ, Shprintzen RJ: The dynamics of Passavant's ridge in subjects with and without velopharyngeal insufficiency—A multiview videofluoroscopic study. Cleft Palate J 16:24, 1979
19. McWilliams BJ, Musgrave RH: The influence of head position upon velopharyngeal closure. Cleft Palate J 5:117, 1968
20. Lloyd, RS, Pruzansky S, Subtelny JD: Prosthetic rehabilitation of a cleft palate patient subsequent to multiple surgical and prosthetic failures. J Prosthet Dent 7:216, 1975
21. Carpenter M: A preliminary study of Passavant's pad within the velopharyngeal mechanism. Unpublished MA thesis, University of Iowa, 1966
22. Shelton RL, Brooks AR, Youngstrom KA: Articulation and patterns of palatopharyngeal closure. J Speech Hear Disord 29:390, 1964
23. Harrington R: A study of the mechanism of velopharyngeal closure. J Speech Hear Disord 9:325, 1944
24. Honjo I, Kojima M, Kumazawa T: Role of Passavant's ridge in cleft palate speech. Arch Otorhinolaryngol 211:203, 1975
25. Glaser(-Cohn) ER: A multiview videofluoroscopic study of Passavant's ridge in patients with and without velopharyngeal insufficiency. Unpublished doctoral dissertation, University of Pittsburgh, 1980
26. Pigott RW: The nasendoscopic appearance of the normal palatopharyngeal valve. Plast Reconstr Surg 43:19, 1969
27. Pigott RW, Bensen JF, White FD: Nasendoscopy in the diagnosis of velopharyngeal incompetence. Plast Reconstr Surg 43:141, 1969
28. Shprintzen RJ, McCall GN, Skolnick ML, Lencione R: Selective movements of the lateral aspects of the pharyngeal walls during velopharyngeal closure for speech, blowing, and whistling in normals. Cleft Palate J 12:51, 1975
29. Skolnick ML, McCall GN, Barnes B: The sphincteric mechanism of velopharyngeal closure. Cleft Palate J 10:186, 1973
30. Croft CB, Shprintzen RJ, Rakoff SJ: Patterns of velopharyngeal valving in normal and cleft palate subjects: A multiview videofluoroscopic and nasendoscopic study. Laryngoscope 91:265, 1981
31. Cohn ER, Rood SR, McWilliams BJ, Skolnick ML, Abdelmalek LR: Barium sulphate coating of the nasopharynx in lateral view videofluoroscopy. Cleft Palate J 21:72, 1984
32. Casey DM, Emrich LJ: Passavant's ridge in patients with soft palatectomy. Cleft Palate J 25:72, 1988

10

Patterns of Velopharyngeal Closure in Patients with Pharyngeal Flaps

In this chapter we will discuss the movements of the velopharyngeal portal during closure or attempted closure in patients with pharyngeal flaps, and describe the appearance of the pharyngeal flap and its effects upon behavior of the velopharyngeal portal.

Causes of Velopharyngeal Insufficiency

Velopharyngeal insufficiency is caused by a failure of the velopharyngeal portal to achieve complete closure during phonation. The resultant space or gap between the palate and pharyngeal wall allows air to escape into the nasal cavity. In cleft palate patients this space may be present because of (1) deficient tissue along the midline of the palate from a submucous cleft palate, or primary repair of a cleft palate, (2) reduced movement of the palate and/or pharyngeal walls, and/or (3) velopharyngeal mislearning. Velopharyngeal insufficiency can also occur in anatomically normal patients due to neuromotor abnormalities. In the neurologically compromised patient, the palate cannot properly elevate or elongate enough to effect closure, or there is reduced movement of the pharyngeal walls. A failure to achieve VP closure can also occur in patients with anatomic discrepancies between the size of the palate and the depth of the nasopharynx. If palatal length is too short relative to the pharyngeal depth, closure cannot be effected even though the palate and pharyngeal walls can move adequately .

The Pharyngeal Flap

A common method of treating patients with velopharyngeal insufficiency is to create a pharyngeal flap. This surgical procedure involves the construction of a soft-tissue bridge between the pharyngeal wall, usually the posterior wall, and the velum. The flap is designed to fill the opening in the velopharyngeal portal that remains even after the patient has made a maximum attempt to achieve palatal-pharyngeal closure. The surgeon usually forms a pharyngeal flap by raising a tongue of tissue from the posterior pharyngeal wall and suturing this tissue to the velum. The flap can either be superiorly (Fig. 10.1) or inferiorly based (Fig. 10.2) (1) although the superiorly based flap, which has been performed at our institution, is much more common. In a superiorly based flap, the base of the flap is cephalad and the flap is lifted upward from the pharyngeal wall, whereas in the inferiorly based flap the reverse is the case. To be effective in eliminating velopharyngeal insufficiency during speech, the flap should be wide enough to fill the gap between the lateral pharyngeal walls at their maximum medial movement (Fig. 10.3). If, however, the flap does not fulfill these conditions, air can escape around the sides of the flap, usually resulting in velopharyngeal insufficiency despite the presence of the flap (Fig. 10.4). The degree of insufficiency, however, may be less than it was prior to the pharyngeal flap operation.

It is important that the pharyngeal flap not so alter the pharyngeal anatomy that the patient cannot breath through his nose. Therefore, the optimal flap

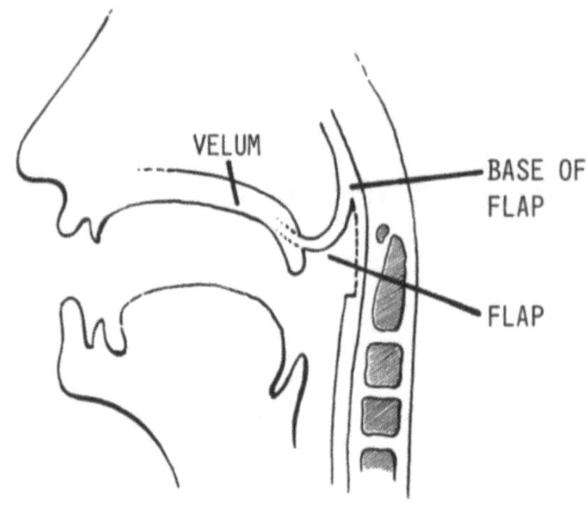

FIGURE 10.1. Superiorly based pharyngeal flap.

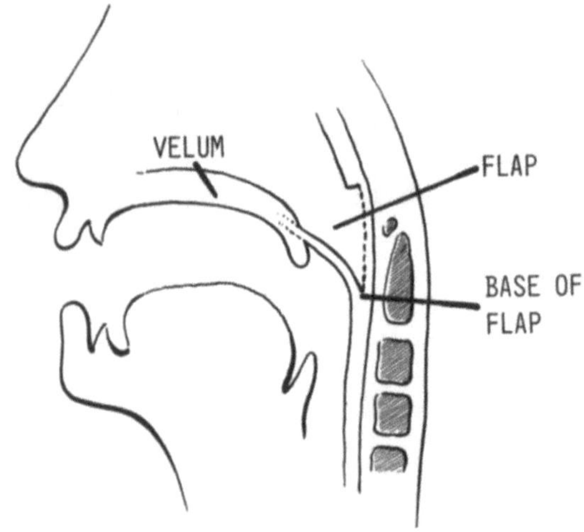

FIGURE 10.2. Inferiorly based pharyngeal flap.

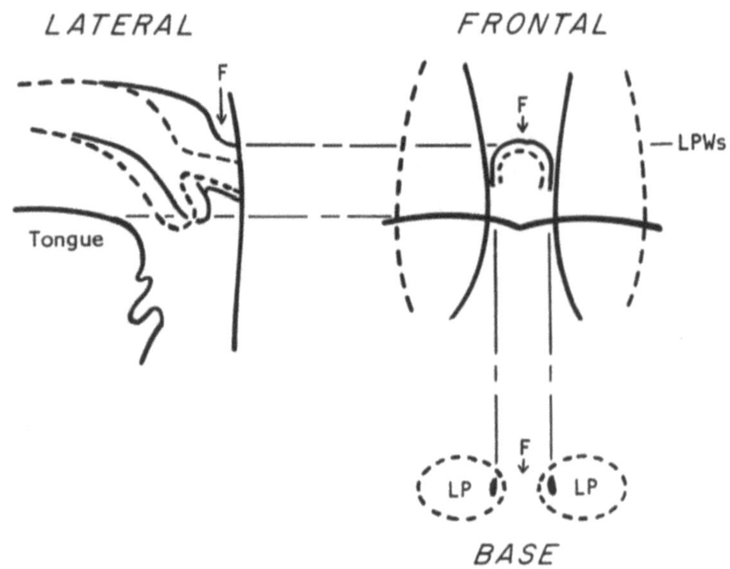

FIGURE 10.3. Pharyngeal flap effectively closes velopharyngeal portal during phonation. Lateral pharyngeal walls touch sides of flap on frontal view. Lateral portals completely close on base view. (*F*, pharyngeal flap; *LPWs*, lateral pharyngeal walls; *LP*, lateral portals; solid line, phonation; dotted line, breathing.) Reproduced with permission from Skolnick ML, McCall GN: A radiographic technique for demonstrating the causes of persistent nasality in patients with pharyngeal flaps. Br J Plast Surg 26:12–15, 1973.

should not be so wide that it completely occludes the nasopharynx and prevents nasal breathing. When the lateral pharyngeal walls are at rest (patient not speaking or uttering nasal phonemes), there should be spaces (lateral portals) between the pharyngeal walls and lateral edges of the flap (Fig. 10.3). If the flap is made so wide as to prevent the formation of lateral portals, then hyponasal speech and mouth breathing will occur (Fig. 10.5). However, hypernasal patients with no medial movement of the lateral pharyngeal walls may require a very wide flap to eliminate velopharyngeal insufficiency. These patients should be forwarned that hyponasal speech and mouth breathing may be unavoidable sequelae to the operation. Thus, the amount of medial movement of the lateral pharyngeal wall should always be considered prior to fabricating the flap (2).

Choosing the Surgeon

It is important to remember that constructing a pharyngeal flap is an approximate art. Although it appears easy to surmise from a videofluoroscopic study what the width and position of the flap should be, the surgical result may not always match the clinical desire.

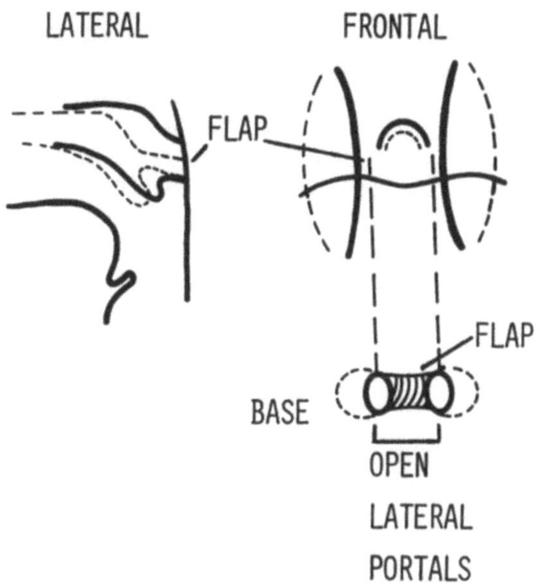

FIGURE 10.4. Velopharyngeal insufficiency persists in spite of pharyngeal flap. Both lateral portals remain open because either the flap was not wide enough or the pharyngeal walls did not have enough medial movement for the edges of the flap and pharyngeal walls to touch each other. (Solid line, phonation; dotted line, breathing.)

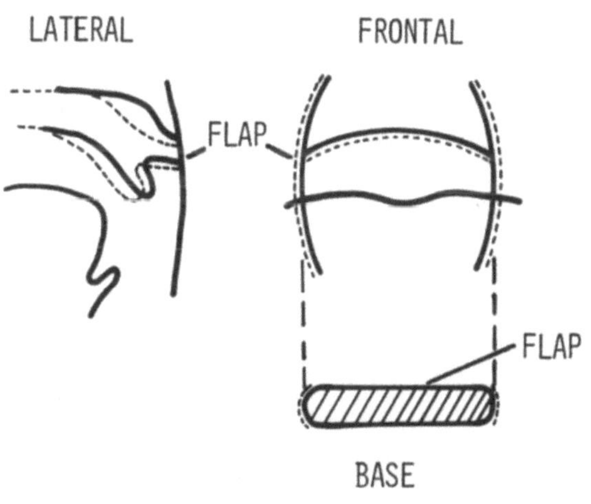

FIGURE 10.5. Excessively wide pharyngeal flap completely occludes velopharyngeal portal even during breathing. Speech is hyponasal and patient cannot breathe through the nose. (Solid line, phonation; dotted line, breathing.)

Not only do different surgeons perform different types of pharyngeal flap procedures, but even the same type of operation in the hands of different surgeons may result in flaps of different widths and positions. Therefore, if a particular patient needs an unusually narrow or unusually wide pharyngeal flap, we believe that the patient should be referred to a surgeon whose pharyngeal flap best matches the clinical need of the patient. This is not to say that surgeons cannot vary the width of their pharyngeal flaps or cannot perform different types of flap operations for different clinical needs. We are simply pointing out that constructing a pharyngeal flap is a fine art, and that better results may be achieved by matching a given surgeon's art with a given patient's needs.

Postoperative Changes in the Flaps

There are a variety of methods of performing the pharyngeal flap operation and the postoperative results may show significant differences depending upon the type of surgery performed.

Some flaps maintain a relatively constant width during healing whereas others may progressively narrow. A flap that maintains its width is usually lined on its raw surface with tissue from the palate; a flap that narrows is unlined. This lining is designed to prevent the flap from shrinking or tubing upon itself with healing.

Whether or not the flap changes the level at which it was originally connected to the posterior pharyngeal wall depends to some extent upon: (1) whether the flap was superiorly or inferiorly based; and (2) whether the under surface of the flap or the donor site on the posterior pharyngeal wall is covered with tissue or left open. In a superiorly based pharyngeal flap, the site of attachment to the posterior pharyngeal wall has a tendency to migrate inferiorly with healing, whereas the reverse is true with an inferiorly based pharyngeal flap. When the raw surface of the flap is lined with palatal tissue, this migration with healing may be reduced. Another way that the surgeon can reduce migration of the flap attachment to the pharyngeal wall during healing is by closing the exposed donor site on the pharyngeal wall from which the flap was raised. Closure of the donor site also has another benefit. It helps to narrow the size of the velopharyngeal portal and thereby helps to bring the lateral pharyngeal

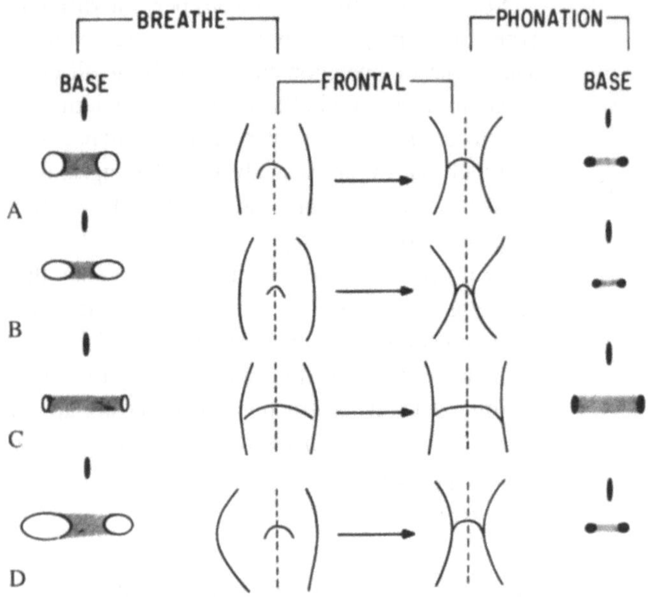

FIGURE 10.6. Patterns of velopharyngeal closure depending upon the width and position of the flap, and movement of lateral pharyngeal walls. (A) Average width flap and average amount of pharyngeal wall movement. (B) Narrow flap and increased wall movement. (C) Wide flap and reduced wall movement. (D) Asymmetrically positioned pharyngeal walls at rest with excessive medial movement of more laterally placed wall during phonation. Reproduced with permission from Skolnick ML: Velopharyngeal function in cleft palate. Clin Plast Surg 2:280–297, 1975.

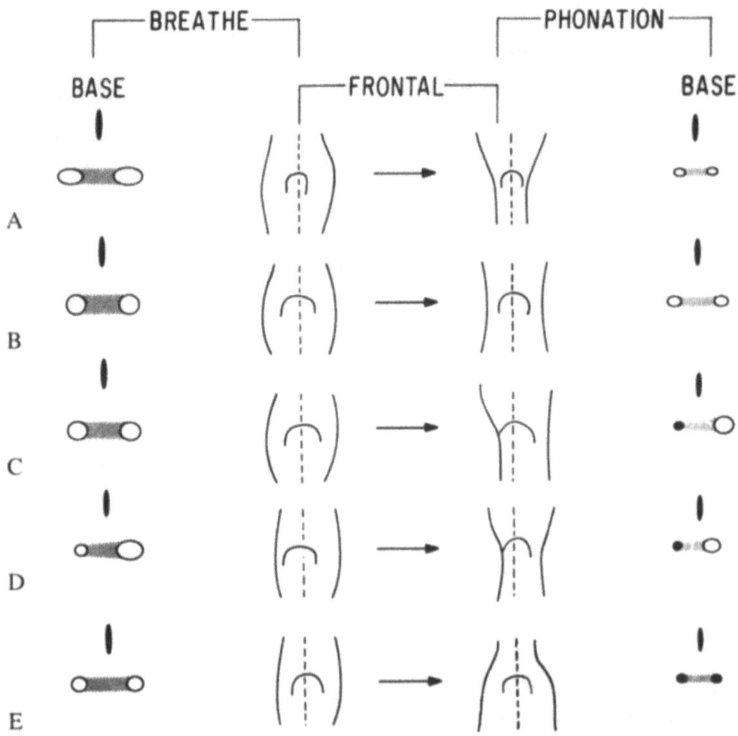

FIGURE 10.7. Patterns of velopharyngeal insufficiency that can occur following pharyngeal flap surgery. (A) Narrow-width flap with average medial movement of the lateral pharyngeal walls. (B) Average-width flap with reduced wall movement. (C) Average-width flap with reduced movement of one lateral wall. (D) Flap asymmetrically placed with average movement of both walls. (E) Flap below level of maximum medial movement of pharyngeal walls. Reproduced with permission from Skolnick ML: Velopharyngeal function in cleft palate. Clin Plast Surg 2:280–297, 1975.

walls closer together so that during medial movement with phonation they can more readily press against the lateral margins of the pharyngeal flap to effect closure.

How the Flap Closes the Portal During Speech

For velopharyngeal closure to be effective following pharyngeal flap surgery, the pharyngeal flap must fill the gap between the lateral pharyngeal walls at their maximum medial excursions during phonation. This concept is the key to understanding velopharyngeal portal closure with a pharyngeal flap. There are a variety of ways to achieve this effect. There may be a narrow pharyngeal flap (Fig. 10.6B) in a patient with excellent medial movement of the lateral pharyngeal walls, so that the edges of the pharyneal walls touch against the margins of the flap. At the other extreme, closure can be achieved when the pharyngeal flap is so wide as to touch the lateral pharyngeal walls even when they are at rest (Fig. 10.5). However, this situation results in hyponasal speech and mouth breathing. In between these two extremes is a range of combinations of flap width and pharyngeal wall medial movement that results in effective closure (Fig. 10.6A,C). Furthermore, the flap may be asymmetrically placed in the portal, either intentionally or accidentally as a result of healing, but closure may nonetheless be effectively achieved because the lateral pharyngeal walls demonstrate enough medial movement to reach thc margins of the flap, with one wall obviously exercising greater medial movement than the other (Fig. 10.6D) (3).

Flap Failures

The pharyngeal flap operation will be inadequate if the marriage between flap width and effective medial movement of the pharyngeal walls is not achieved. Once again, we will refer to sketches of the frontal and base views to illustrate this concept (Fig. 10.7). Several possibilities can occur. First, the pharyngeal flap may be of average width, but the pharyngeal walls may show reduced medial movement, thereby resulting in bilateral portal inadequacy (Fig. 10.7B). Second, the pharyngeal flap may be abnormally narrow (perhaps having tubed in on itself during healing) in a subject with good pharyngeal wall movement (Fig. 10.7A). Third, the pharyngeal flap may have been made intentionally wide in a patient with poor pharyngeal wall movement, but the pharyngeal walls still do not have adequate movement to reach the edges of the flap. Fourth, the flap may be asymmetrically positioned, probably from a shift during healing, resulting in unilateral portal incompetence (Fig. 10.7D). Fifth, it is possible that the surgeon did not appreciate the fact that one pharyngeal wall had reduced movement as compared with the other and he placed the flap symmetrically in the center of the portal, assuming equally good movement of both lateral portals. The result is again unilateral inadequacy (Fig. 10.7C). Sixth, the pharyngeal flap attains an inappropriate level in the nasopharynx with reference to the level of maximum medial movement of the pharyngeal walls. In this scenario, the flap is usually at a level below the maximum medial movement of the pharyngeal wall and this abnormally low level may result from inferior migration of a superiorly based pharyngeal flap with healing (Fig. 10.7E).

It is important to note that one cannot appreciate an abnormally low flap attachment from the base view. In fact, the base view can erroneously create the impression that an inadequate flap effectively closes the portal because on base view one cannot distinguish between structures located at different vertical levels. The shadows of their margins project as if they were all located at the same level. Therefore, it is important to remember that another view, the frontal view, is required to document the level at which the pharyngeal flap is located with reference to the maximum medial movement of the pharyngeal walls. (See *Frontal View* for details.)

Analysis of Videofluoroscopic Images

Lateral View Without Barium

We perform the lateral view first without barium and then repeat it following the nasal instillation of barium just as we do in patients without pharyngeal flaps. The nonbarium lateral view is necessary because excessive barium may mix with mucus above the flap and produce a variety of confusing shadows, in some cases even obscuring the flap.

The lateral view examination should always begin with the patient breathing quietly through his nose and with his mouth closed. During this maneuver, the velum descends anteriorly to open the nasopharynx. In the subject without a pharyngeal flap, the velum descends to touch the posterior surface of the tongue (Fig. 10.8A), whereas in a patient with a pharyngeal flap the flap tethers the velum and prevents its contact with the posterior surface of the tongue (Fig. 10.8B). Therefore, one sees an air space between the velum and the tongue, and a soft-tissue line representing the pharyngeal flap that runs either horizontally or diagonally from the posterior pharyngeal wall to the posterior surface of the velum. Unless the palate is unusually scarred from prior surgery, the palate will always touch the posterior surface of the tongue during quiet nasal breathing in someone without a pharyngeal flap, whereas in a patient with a flap the palate does not touch the tongue. If the palate does not descend to touch the tongue, one should suspect the presence of a flap, even if the pharyngeal flap is not seen. It is important to emphasize this point because at times the patient does not give a history of having a pharyngeal flap, and the presence of a flap is first demonstrated during initial nonbarium examination in the lateral view.

Although a brief recording in quiet nasal breathing is all that is needed in this view, one may also wish to get a brief recording of speech as well, especially if one suspects that the patient will not tolerate the nasal instillation of barium.

Lateral View with Barium

The lateral view with barium demonstrates the level of palatal attachment to the pharyngeal wall, and shows the amount of palatal movement and midsagittal narrowing of the velopharyngeal portal. On occasion one may even see velopharyngeal closure above the flap level. This view may also reveal the presence of any palatal or flap fistula or incomplete closure of the lateral portals by the presence of barium blowing above the level of the flap. However, with the exception of barium blowing as a sign of portal incompetence or fistula, the lateral view gives no information about effectiveness of lateral portal closure.

The pharyngeal flap arises from the posterior pharyngeal wall in the nasopharynx, usually slightly below the level of the eminence. On rare occasions the flap origin from the posterior pharyngeal wall may be in the oropharynx, at or below the level of the superior surface of the tongue. Observations of palatal movement during phonation are usually of little importance on the lateral view if the patient has satisfactory velopharyngeal closure during speech. There is usually no contact between the palate and posterior pharyngeal wall above the level of the flap because the reason for performing the flap operation is to close the central portion of the velopharyngeal portal to compensate for the failure of the palate to effect this closure on its own.

There are two exceptions, however, in which it is important to observe palatal-pharyngeal relationships above the flap. First, when one sees velopharyngeal closure above the level of the flap in patients with non-nasal speech, the question should be raised as to whether the pharyngeal flap is really necessary or effective in helping to close the portal. In these patients the level of the pharyngeal flap should also be carefully observed on frontal view in order to determine whether the flap is at or below the maximum movements of the pharyngeal walls and, therefore, whether or not the flap really contributes to closure of the portal (Fig. 10.9). Second, in patients with hypernasal speech following pharyngeal-flap surgery, the palate may show good movement with a distinct eminence formation, but the palate may appear to be unable to reach the posterior pharyngeal wall because it is restricted by the tethering affect of the pharyngeal flap. In such situations, the pharyngeal flap usually has a low attachment, perhaps even as low as the oropharynx level, and appears to prevent adequate posterior and superior movement of the palate (Fig. 10.10). If, in such a subject, the frontal view confirms the observation that the pharyngeal flap is below the level of greatest medial movement of the lateral pharyngeal walls, thereby indicating that the flap is of little or no help in closing the portal, then consideration should be given to cutting the flap to improve palatal motion. If the flap is cut, the palate may become mobile enough to satisfactorily close against the posterior pharyngeal wall. Such closure may occur even if, prior to flap surgery, the patient had a central region of velopharyngeal insufficiency. Just the procedure of attaching the flap to the palate may have increased the soft-tissue bulk along the posterior surface of the palate enough to result in closure of the portal.

FIGURE 10.8. Lateral views comparing positions of velum during breathing in patients (A) without and (B) with pharyngeal flaps. Without the flap, the velum touches the tongue. With the flap present the velum is tethered and cannot descend to touch the tongue.

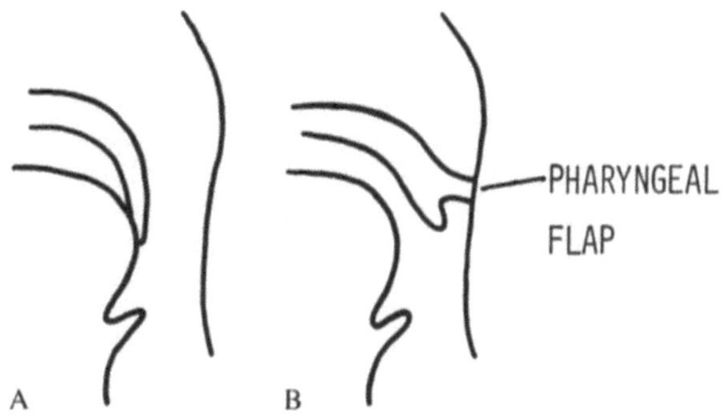

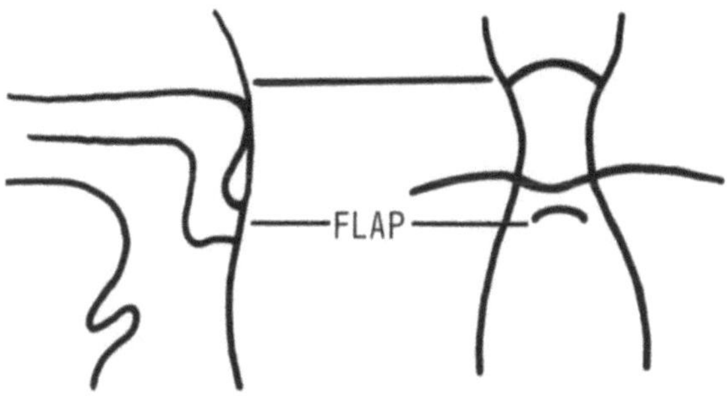

FIGURE 10.9. Velopharyngeal closure on lateral and frontal views above level of pharyngeal flap.

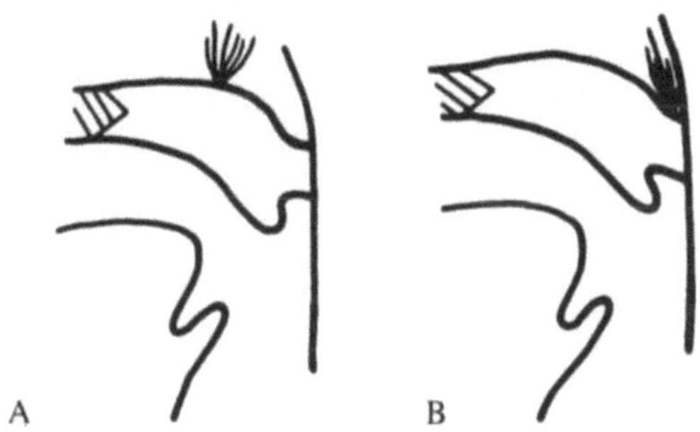

FIGURE 10.11. (A) Barium blowing above palate indicates palatal fistula. (B) Barium blowing above level of flap indicates either a fistula through the flap or inadequacy of lateral portal(s).

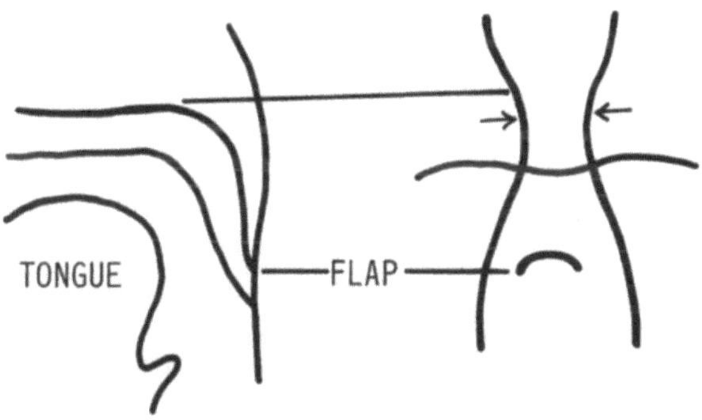

FIGURE 10.10. Low attachment of flap to pharyngeal wall at oropharyngeal level. Maximum medial movement of pharyngeal walls occurs in nasopharynx, considerably above flap level. Flap appears to tether the elevating movements of the velum.

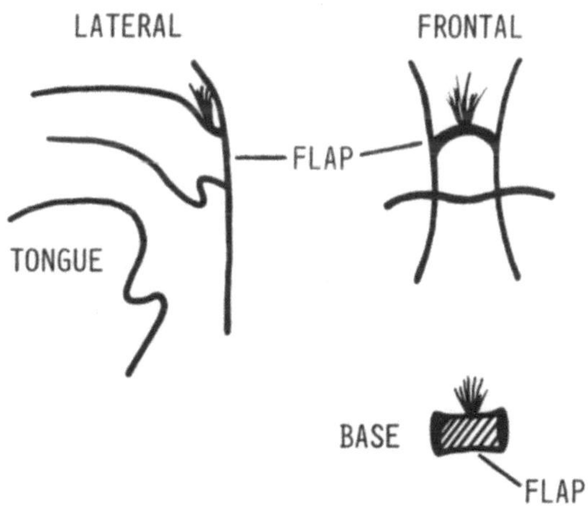

FIGURE 10.12. Barium blowing through fistula in pharyngeal flap.

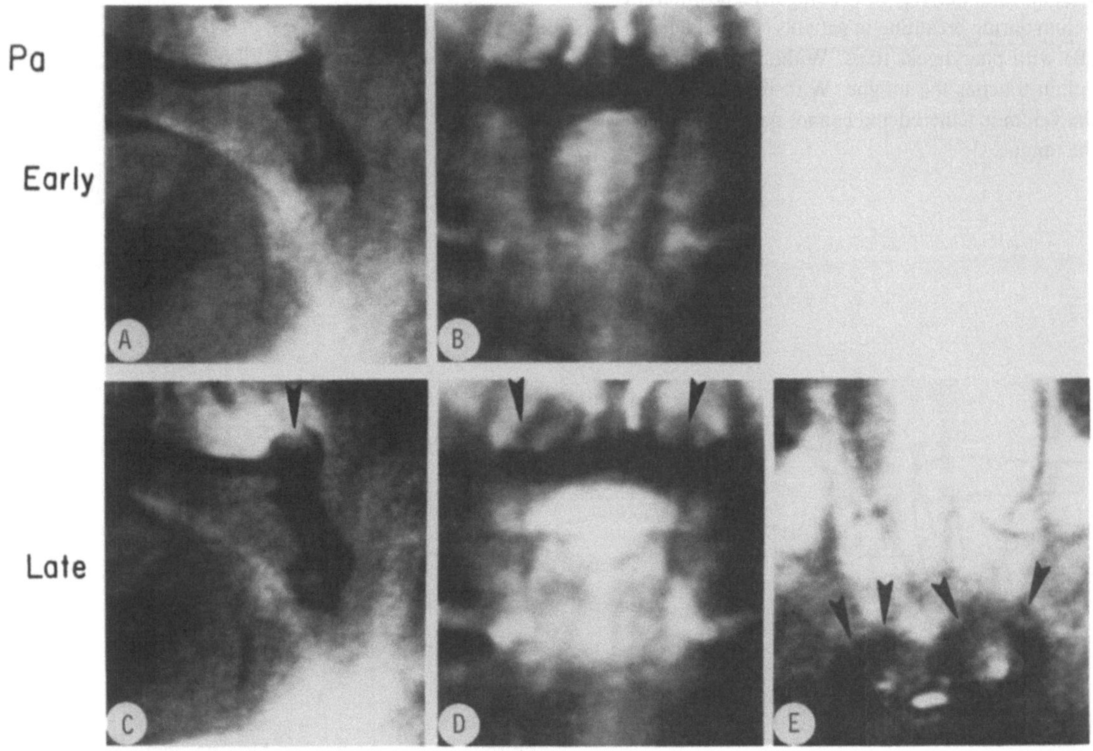

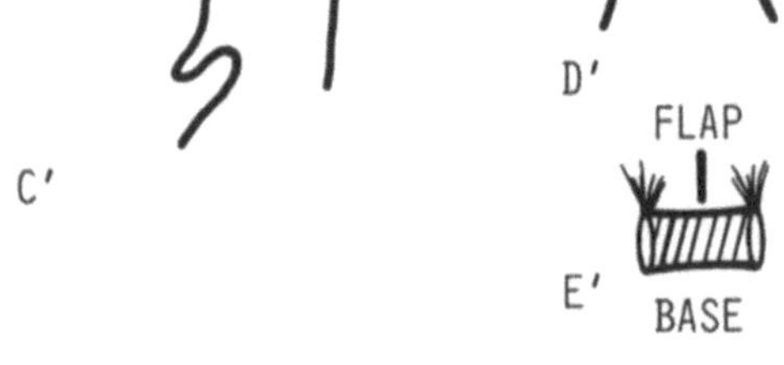

FIGURE 10.13. Barium blowing through inadequate lateral portals. Patient phonating "pa." Lateral (A) and frontal (B) views during early phase of "pa." Lateral (C), frontal (D), and base (E) during late phase of "pa." (Barium blowing, vertical arrows. C', D', E' are sketches corresponding to C, D, E.)

Fistula and Lateral Portal Inadequacy

The lateral view may also document a fistula in the palate or flap or reveal lateral portal inadequacy by demonstrating the blowing of barium through the fistula or the inadequate portals. Such blowing is most often seen during the phonation of consonants that require the build-up of high amounts of pressure within the oropharynx, for example, the "S," "P," or "K." If the barium is seen blowing through the region of the soft palate, then the defect is a fistula in the palate (Fig. 10.11A). However, if the site of barium blowing is in the region of the pharyngeal flap (Fig. 10.11B), it can either represent a fistula through the flap (Fig. 10.12) or barium blowing through an inadequate lateral portal (Fig. 10.13). Usually this sign, whether involving a palatal fistula or an inadequate lateral portal, is best seen when the opening is relatively small. If the opening is relatively large, then a layer of barium-coated

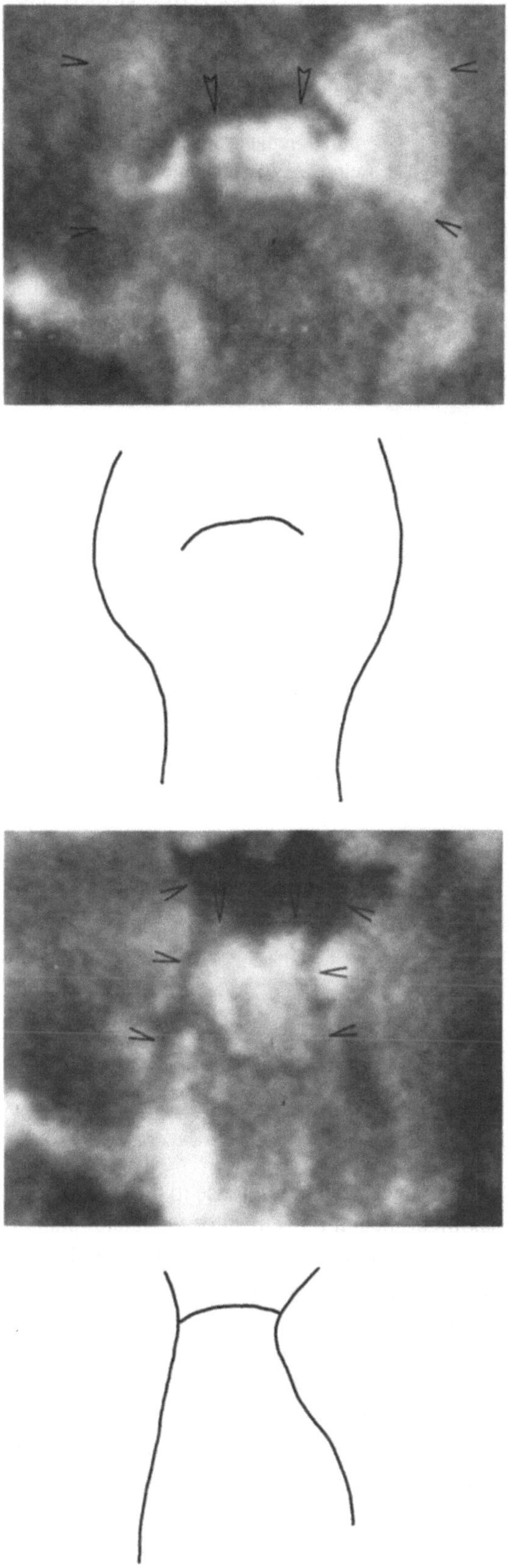

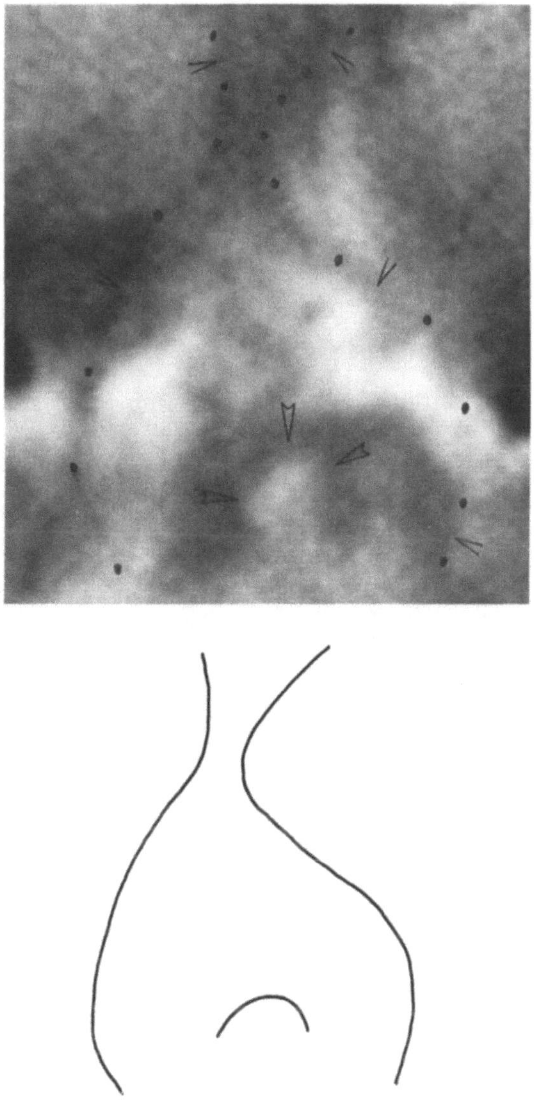

FIGURE 10.15. Frontal view (A, A') during phonation. Flap is below level of maximum medial movement of lateral pharyngeal walls. (Vertical arrowheads, flap; dots and horizontal arrowheads, lateral pharyngeal walls.) Reproduced with permission from Skolnick ML, McCall GN: Velopharyngeal competence and incompetence following pharyngeal flap surgery: Videofluoroscopic study in multiple projections. Cleft Palate J 9:1–12, 1972.

◁ FIGURE 10.14. Frontal view: breathing (A, A') and phonation (B, B'). Flap is at level of maximum medial movement of lateral pharyngeal walls. Vertical arrowheads, flap; horizontal arrowheads, lateral pharyngeal walls. Reproduced with permission from Skolnick ML, McCall GN: Velopharyngeal competence and incompetence following pharyngeal flap surgery: Videofluoroscopic study in multiple projections. Cleft Palate J 9:1–12, 1972.

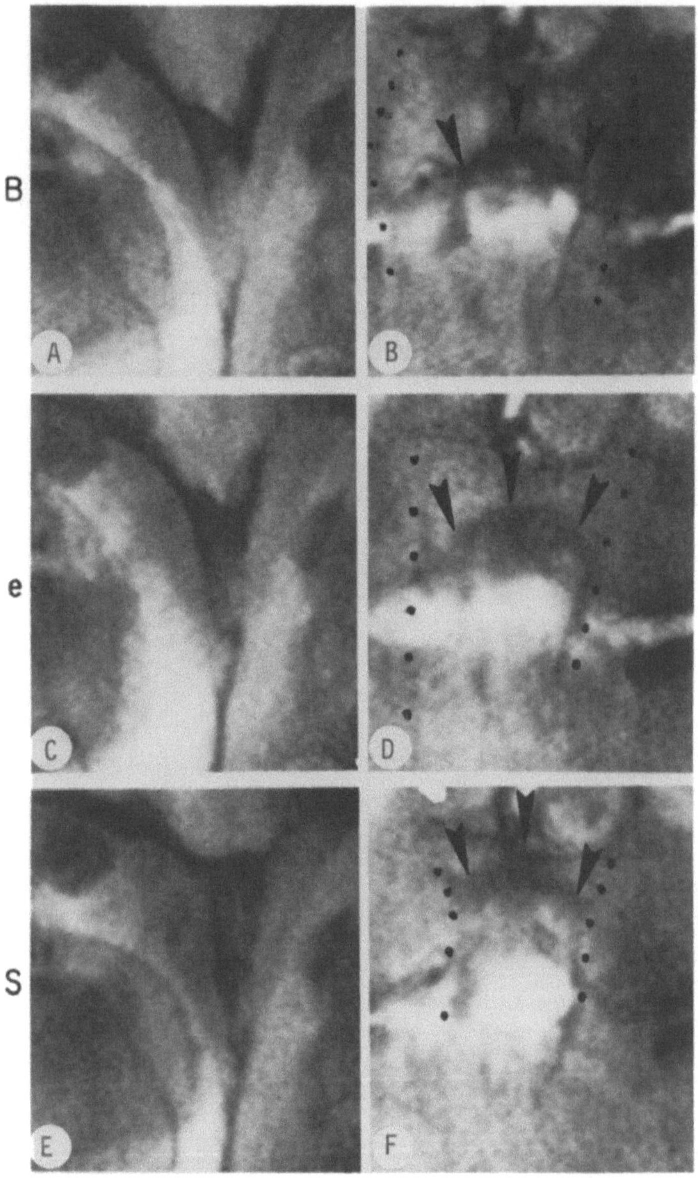

FIGURE 10.16. Asymmetrically positioned lateral pharyngeal walls on either side of flap during breathing (A = lateral, B = frontal views). During phonation of "e" (C = lateral, D = frontal) right portal remains open while left portal closes. With phonation of "s" (E = lateral, F = frontal) more forceful closure occurs and both portals close. Note: Flap level is more cephalad with phonation (F) than with breathing (B). Vertical arrowheads, flap; vertical dots, lateral pharyngeal walls.

mucus may not bridge the gap so that barium will not form a bubble as the air blows through the opening. The frontal or base view is required to distinguish whether the site of barium blowing represents a fistula through the flap or inadequacy of one or both lateral portals (Figs. 10.12, 10.13).

Frontal View

The frontal view is used to evaluate the amount of pharyngeal wall movement, to demonstrate the width of the pharyngeal flap, and to determine the level of the pharyngeal flap in reference to the region of the maximum medial movement of the lateral pharyngeal walls. Therefore, one can tell whether the pharyngeal flap is properly positioned in the vertical dimension so that the lateral pharyngeal walls touch against the lateral edges of the flap during phonation (Fig. 10.3, 10.14). The frontal view will also show if the flap has been placed at an inappropriate level either above or below the region of maximum medial movement of the pharyngeal walls (Fig. 10.15).

The palate may not always be well visualized on the frontal view because its upper surface may not

A B C D

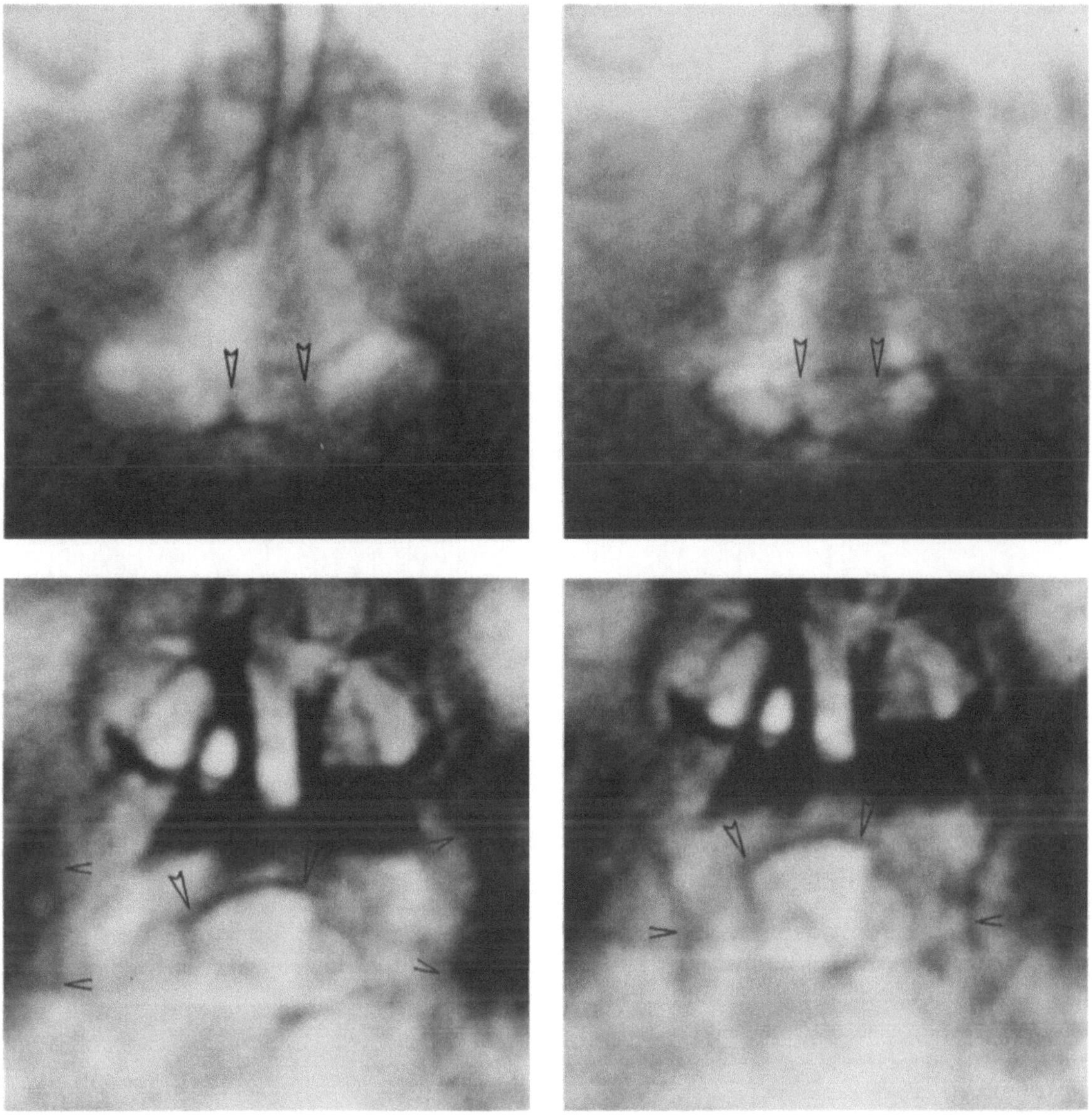

FIGURE 10.17. Bilateral portal inadequacy in patient with narrow flap and good lateral pharyngeal wall motion. Breathing: base (A), frontal (C). Phonation: base (B), frontal (D). In all views, nasal barium is used. (Vertical arrowheads, flap; horizontal arrowheads, lateral pharyngeal walls.) Reproduced with permission from Skolnick ML, McCall GN: Velopharyngeal competence and incompetence following pharyngeal flap surgery: Videofluoroscopic study in multiple projections. Cleft Palate J 9:1–12, 1972.

be tangential to the x-ray beam. The pharyngeal flap, however, will usually be seen if barium adequately coats its superior surface (Fig. 10.14A,B). This surface is seen as a flat or convex superior structure that may move up during phonation of non-nasal sounds and down during quiet breathing (Fig. 10.16). One can also determine whether the pharyngeal walls are symmetrically placed on either side of the flap or not, and how the lateral portals respond during phonation (Fig. 10.16). When the portals narrow but do not close because of reduced lateral pharyngeal wall motion, there is persistent velopharyngeal insufficiency. Two examples of bilateral inadequacy are shown in Figure 10.17 (narrow flap with good pharyngeal wall motion), and in Figure 10.18 (wide-flap and reduced pharyngeal wall motion). In both cases the flap failed to fill the gap between the lateral pharyngeal walls at their

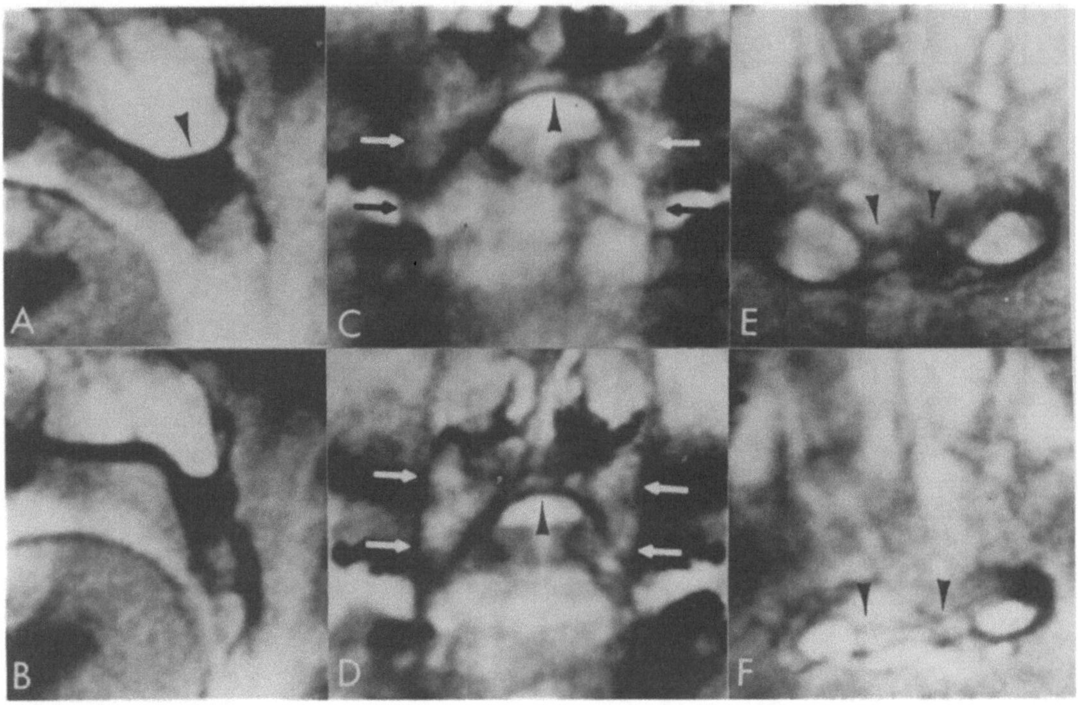

FIGURE 10.18. Bilateral portal inadequacy in patient with wide flap and reduced lateral pharyngeal wall motion. Breathing: lateral (A), frontal (C), and base (E) views. Phonation: lateral (B), frontal (D), base (F) views. (Vertical arrowheads, flap; horizontal arrows, lateral pharyngeal walls.) Reproduced with permission from Skolnick ML, McCall GN: A radiographic technique for demonstrating the causes of persistent nasality in patients with pharyngeal flaps. Br J Plast Surg 26:12–15, 1973.

maximum medial excursions. However, when the lateral pharyngeal walls completely close the portals, one can see contact between the surfaces of the lateral walls and the lateral margins of the flap (Fig. 10.14, 10.16).

If in spite of what appears to be adequate contact of the lateral pharyngeal walls against edges of the flap, the patient has hypernasal speech, there are two possibilities. Most commonly, the entire portal is incompletely closed but one cannot appreciate the separation between the pharyngeal wall and flap because the separation is narrow and there is enough barium to fill and obscure the gap. However, air can readily escape through the gap and carry barium in front of it to produce bubbling above the inadequate portal(s) (Fig. 10.13). Less commonly, a portion of the portal closes so that the pharyngeal walls touch the margins of the flap along part but not all of its extent. On frontal view a barium line is formed by the region of partial lateral wall to flap contact. Unless there is barium blowing through the residual opening, this incomplete closure is not appreciated on frontal view. However, on the base view the residual opening is visualized (Fig. 10.23).

Base View

During quiet breathing, the pharyngeal flap appears as a central band or structure connecting the posterior pharyngeal wall to the palate. On the fluoroscopic image one sees what appears to be a double-barrel shotgun viewed on end. Each open lateral portal represents one barrel of the shotgun and the pharyngeal flap represents the metal separating the two barrels (Fig. 10.19, 10.20). As we have already described earlier in this chapter, the pharyngeal flap can range in width from a thin band (Fig. 10.21A) to a broad sheet (Fig. 10.21B) that may completely occlude the lateral portals during quiet breathing. At this extreme end of the spectrum (Fig. 10.21B), the excessively wide flap is actually obturating the nasopharynx and the patient is unable to breathe through the nose. If one sees what appears to be

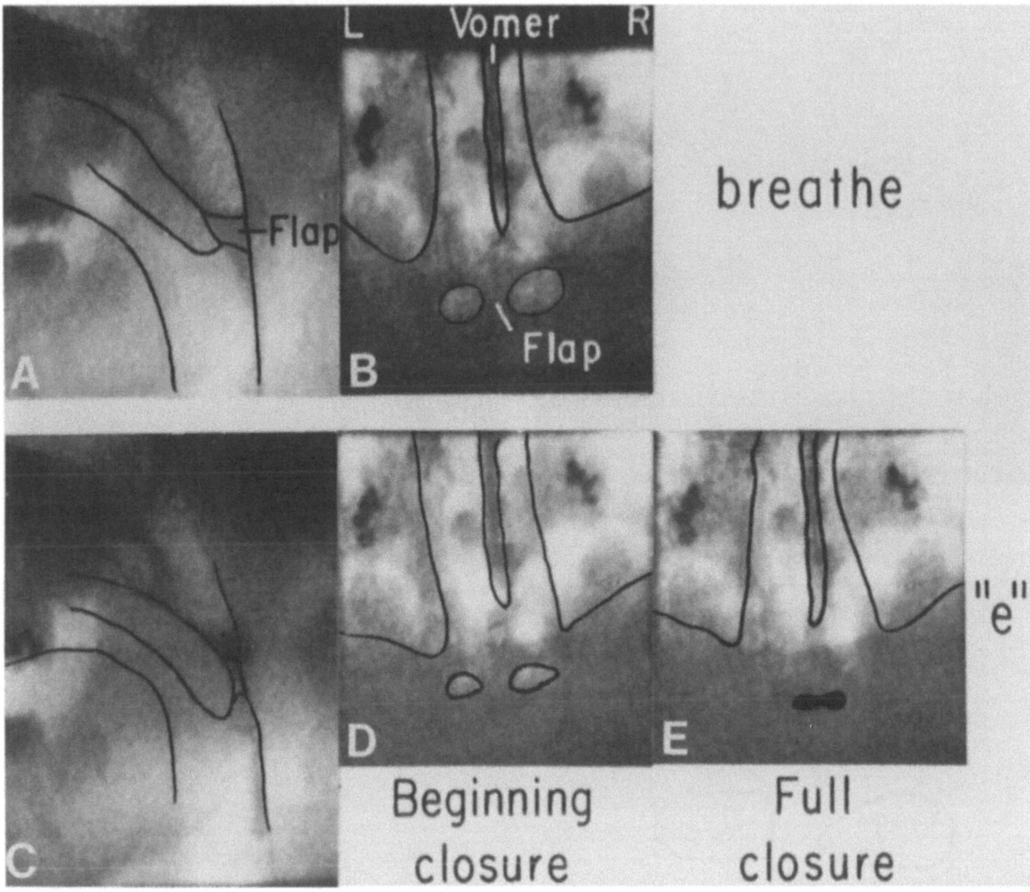

FIGURE 10.19. Bilateral portal closure with pharyngeal flap. Breathing: (A) lateral and (B) base views. Phonation: (C) lateral and base in (D) early and (E) full closure. Note that in full closure flap and closed portals blend together.

such a broad flap during quiet breathing, and is not sure whether one or both portals can open at all, the patient should be instructed to sniff through his nose. This forceful inhalation may demonstrate a narrow opening through a portal or portals not be seen during quiet breathing. It is important also to point out that quiet breathing must be quiet breathing through the nose. If the patient makes no attempt to breathe through the nose, then even if the lateral portals can open slightly, one would not see such an opening because no air is directed through them.

Having determined the width and position of the flap during quiet breathing, one now observes the appearance of the lateral portals during various phonation maneuvers. Unlike the frontal view, on the base view one can see all margins of the portal at once and observe how these margins move medially and centrally to close the individual lateral portals. At maximal closure, the portals appear either as two parallel and sagittally oriented oval slits produced by the barium-coated lateral pharyngeal walls squeezing against the sides of the flap (Fig. 10.21B) or as two circles when each portal closes in a circular manner (Fig. 10.21A).

Portal Inadequacy

When portal inadequacy exists, one may see varying degrees of constriction of one or both lateral portals compared with the rest state. The extent of constriction depends upon the degree of reduced pharyngeal wall movement, flap width, and flap position (Fig. 10.7). Examples of still-frame photographs from videotapes show bilateral inadequacy with a narrow flap (Fig. 10.17), a wide flap (Fig.

A B

A' B'

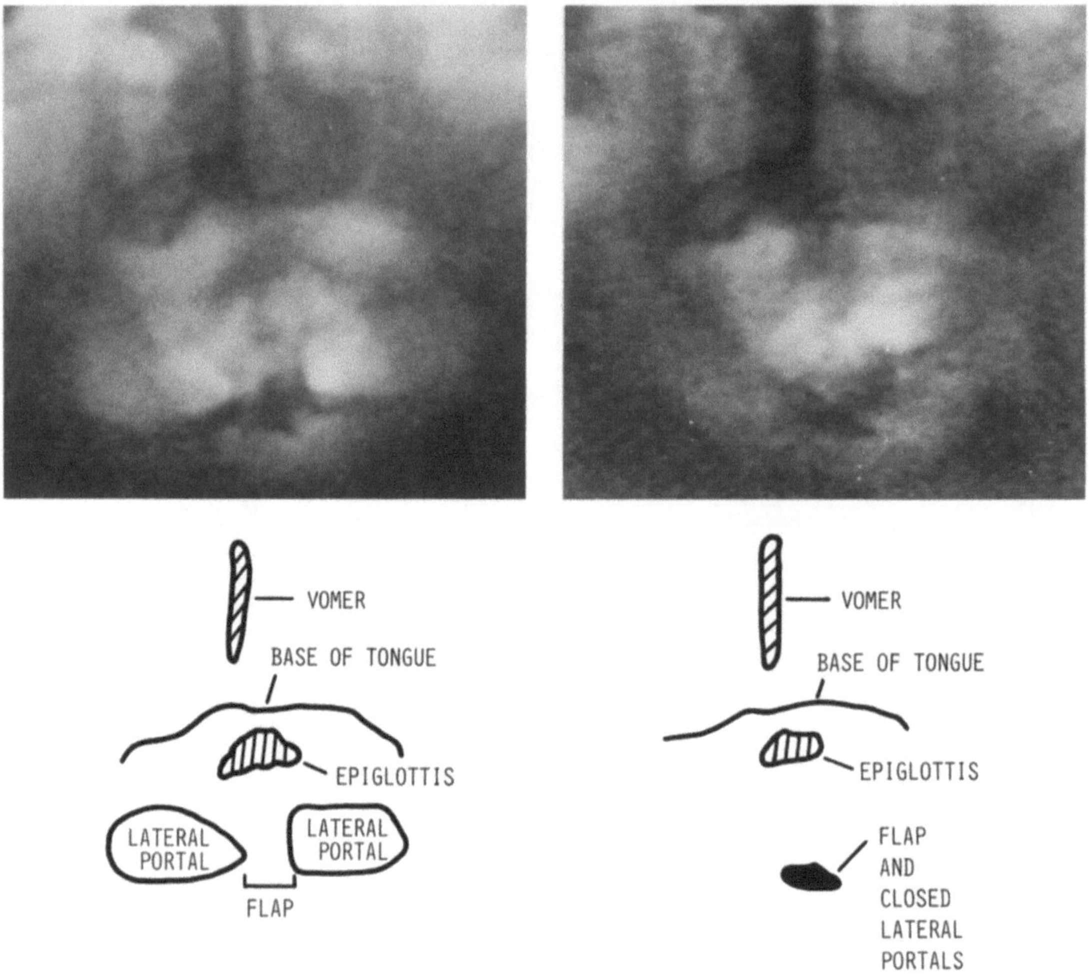

FIGURE 10.20. Base views: Pharyngeal flap with bilateral portal closure during phonation. (A, A') during breathing and (B, B') during phonation. Note that barium-coated base of tongue and epiglottis project on the base view anterior to the velopharyngeal portal and should not be mistaken for the portal.

A

B

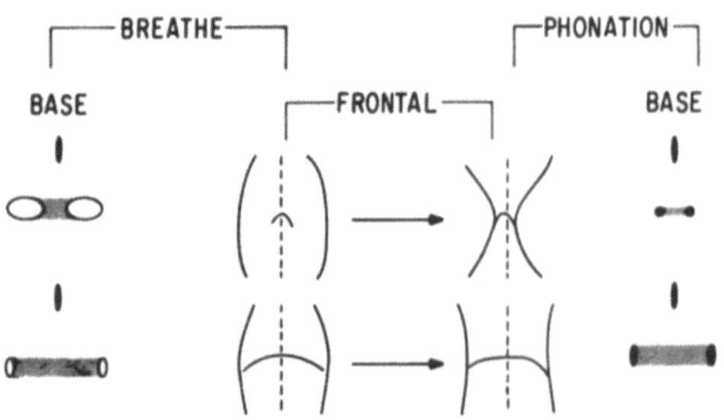

FIGURE 10.21. Comparison between (A) narrow and (B) wide flaps during breathing and phonation. Modified from Skolnick ML: Velopharyngeal function in cleft palate. Clin Plast Surg 2:280–297, 1975, Fig. 12.

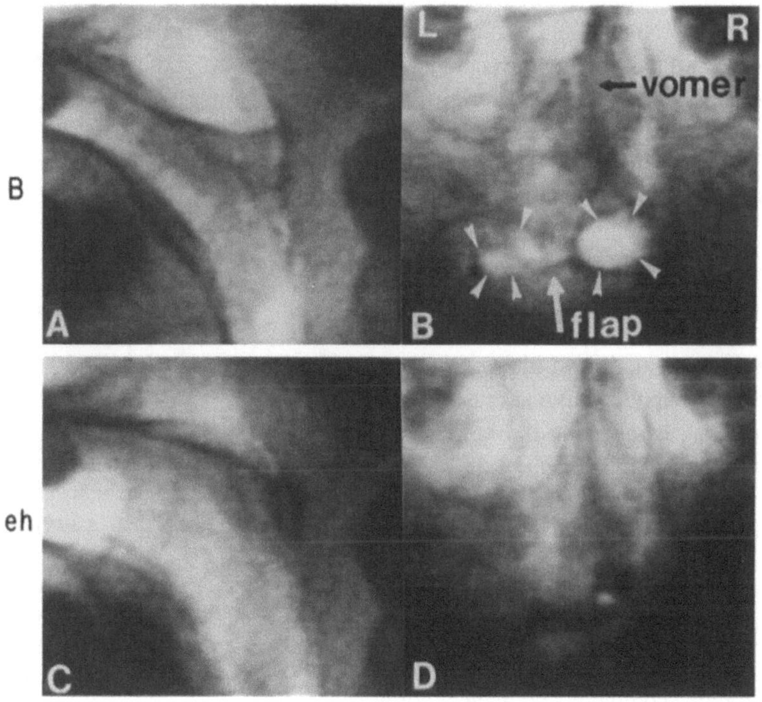

FIGURE 10.22. Unilateral portal inadequacy. Pharyngeal flap positioned off center, closer to portal that closes during phonation. Lateral and base views during breathing (A, B) and phonation (C, D). Note that flap is to right of midline (vomer is midline reference). (Margins of portals, arrowheads; right, *R;* left, *L.*)

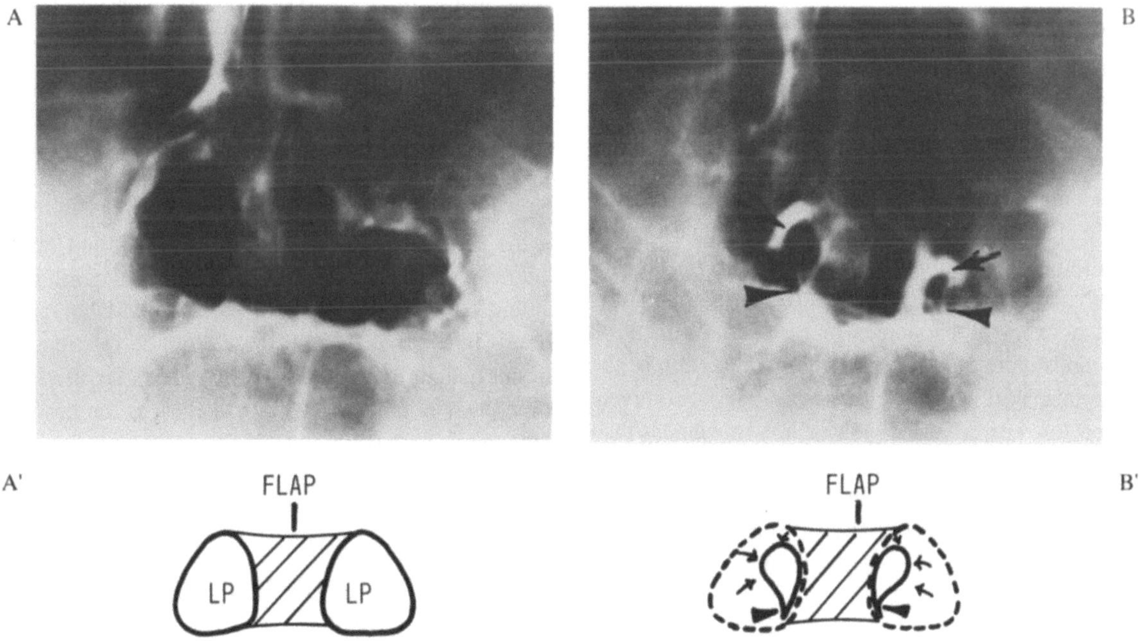

FIGURE 10.23. Partial closure of each lateral portal seen on base view. A, A') breathing. B, B') phonation. Posterior aspects of lateral portals have closed (arrowheads) while more anterior portions of portals remain open (arrows). (*LP*, lateral portals.)

10.18), and unilateral inadequacy (Fig. 10.22). If a small gap remains in the portal, then one may observe barium blowing through the portal in a manner similar to that seen on the frontal and lateral views (Fig. 10.13). Occasionally, as described under *Frontal View,* one may see localized closure of a part of each of the portals while the remainder narrows but remains open (Fig. 10.23).

False Appearance of Closure

Before attempting to interpret images in the base view, we repeat that one must be sure the pharyngeal flap is indeed at the same level as the maximum medial movement of the pharyngeal walls. If the flap is at a lower level, there may appear to be portal narrowing or closure on the base view, whereas in reality there is inadequacy (Fig. 10.24). The only possible tip-off to the presence of such an abnormal flap position on the base view would be the observation that barium appears to be blowing through a completely competent portal.

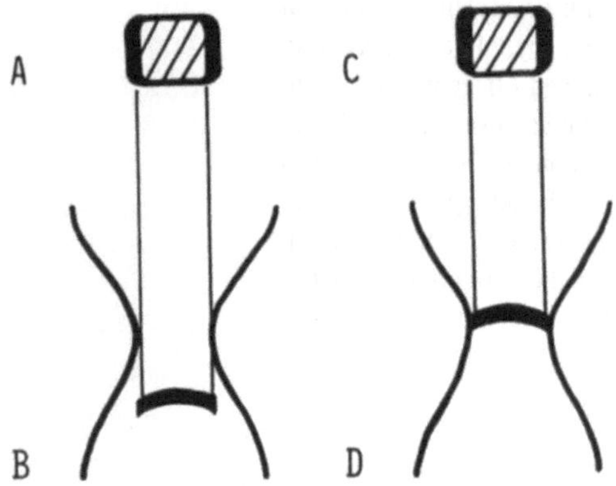

FIGURE 10.24. Example of misleading information presented on base view. During phonation, there appears to be closure of the lateral portals against the margins of the flap on both base views (*A* and *C*). However, frontal view (*B*) shows that flap is actually below level of maximum medial movement of the lateral pharyngeal walls and that there is a gap between the flap and lateral walls for air to pass through. Superimposition of edges of flap upon barium-coated lateral walls on base view (*A*) give impression of portal closure. Frontal view (*D*), which coincides with base view (*C*), shows that the lateral walls touch the lateral margins of the flap to close the lateral portals. Thus, frontal views are necessary for interpreting the base views.

Analysis of Rapid Movements

At times the portal moves so rapidly from a fully open to a completely closed position during phonation of non-nasal phonemes that one has difficulty in appreciating the way in which its margins move to effect closure. Of course, one could play the videotape back in slow motion and try to analyze the changing movements on the individual frames. Another way to analyze the movements during closure is to observe how they move during phonation of a series of phonemes in isolation. With nasal sounds there may be only slight contraction of the portals. With vowels, closure may not be complete and low vowels usually show less movement than higher vowels. Thus, by observing different phonemes one can see the margins of the portal frozen in different degrees of partial closure so as to better appreciate how these margins move.

Congruence of Dimensions Among Multiple Views

In Chapter 7 we discussed the need for cross-correlating dimensions of the velopharyngeal portal seen on one view with the dimensions seen on another view in order to insure the accuracy of positioning and interpretation. The same rules apply with studies of flaps. The width of the flap as measured on the base view should equal that seen on the frontal view. Likewise, the width of the lateral portals should be the same between base and frontal views, and the anterio-posterior dimensions of the lateral portals should be the same on both the base and lateral views (Figs. 10.3, 10.4). If these cross-correlations do not match, then one of the two views has been improperly obtained and most likely it is the base view that is malpositioned.

Effect of Pharyngeal Flap Surgery on Lateral Pharyngeal Wall Movement

In our discussion of patients who have persistent inadequacy of the lateral portals following pharyngeal flap surgery, we have indicated that the problem has persisted because the flap is of inadequate width to fill the gap between the lateral pharyngeal walls at their maximum medial excursions. We have not addressed the question of whether the flap surgery itself could affect lateral pharyngeal wall motion, especially in a detrimental manner.

In order to attempt to answer this question, a study was performed by Shprintzen, McCall and Skolnick (4) in which 100 patients were studied prior to and at least 6 months following pharyngeal flap surgery with videofluoroscopy in multiple views. The degree of lateral pharyngeal wall motion was rated on a 4-point scale. Poor motion was defined as no or only slight movement. Fair motion was a medial excursion of the lateral walls less than one quarter of the distance toward the midline. Good motion was defined as a medial excursion of between one quarter and and one half of the distance toward the midline. Excellent motion was the movement of the lateral walls greater than one half of the distance toward the midline. The pattern of lateral wall motion, specifically regarding the vertical extent of the medial motion versus the position of the pharyngeal flap, was also observed.

Results of the study reveal that: (1) In 84 of 100 patients no change in the degree of lateral pharyngeal motion occurred between pre- and postoperative studies. (2) In seven patients who had fair, good, or excellent lateral wall motion preoperatively, a marked reduction in wall motion occurred postoperatively. All of these patients had an extremely wide pharyngeal flap that restricted medial movement of the lateral pharyngeal walls (Fig. 10.25). (3) In seven patients who had poor or fair lateral wall motion preoperatively, pharyngeal wall motion improved postoperatively (Fig. 10.26). (4) In two patients the vertical extent of the region of medial motion of the lateral walls changed. Preoperatively, these patients showed discreetly localized regions of medial movement. Postoperatively, the flaps were below the regions of medial movement seen on the preoperative studies and the medial movement was much broader, enabling the walls to touch the edges of the flap (Fig. 10.27).

Although 7/100 patients (7%) showed an increase in lateral wall motion following flap surgery, it was not possible to predict from the preoperative videofluoroscopy which patients would show increased wall motion postoperatively. These patients also had speech therapy following surgery. It is not known what effect the speech therapy had in increasing the lateral wall movement.

The main conclusion of the study was that flap surgery did not alter the degree of lateral pharyngeal wall motion in the great majority (84%) of the patients. Therefore, the surgeon should attempt to design the flap to fill the gap between the lateral pharyngeal walls at their maximum medial excursions rather than expect the flap operation to increase the medial movement of the lateral pharyngeal walls.

When to Perform a Videofluoroscopic Study After Pharyngeal Flap Surgery

Because a videofluoroscopic study involves radiation exposure and is therefore not innocuous, we believe that the study should only be performed when there is a valid clinical indication. In patients who have received a pharyngeal flap, we believe that the prime indication for a postoperative videofluoroscopic study is that the patient continues to exhibit speech that is symptomatic of VPI and that the *patient* (or family in the case of a child) *is dissatisfied* with the current state of the speech. This latter point is crucial because if treatment is not desired, there is *no* purpose in repeating the videofluoroscopic study. However before performing the videofluoroscopic study, one should wait between 6 months to 1 year following the flap surgery to allow for complete healing.

Another, though less common indication for a post-flap fluoroscopic study is the situation in which the flap is so wide that it obturates the nasopharynx causing mouth breathing and hyponasal speech. The information obtained through videofluoroscopy could help the surgeon decide how to modify the flap so as to provide some opening of the lateral portals while minimizing the risk of recurrent hypernasality.

It is important when interpreting a post-flap videofluoroscopic study to review the preoperative study at the same time. In this way one can determine what effect the flap operation had on movement of the lateral pharyngeal walls. In patients who have a wide and obturating flap, review of the preoperative study is especially vital prior to undertaking flap revision. In order to determine how much the flap could be narrowed and still have the lateral pharyngeal walls touch the edges of the flap to close the lateral portals, one should know how mobile the walls were preoperatively.

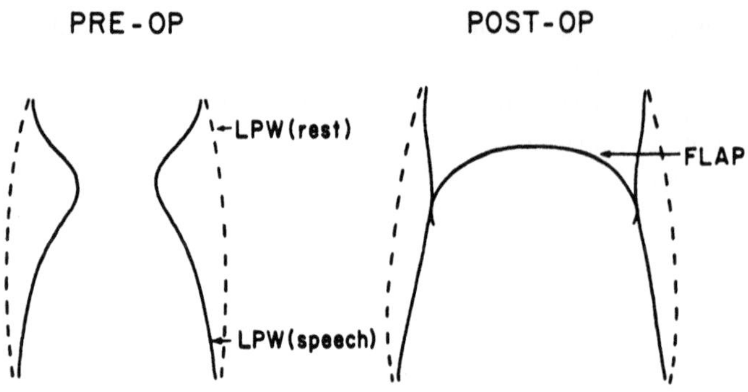

FIGURE 10.25. Reduced pharyngeal wall motion postoperatively because side flap limits medial movement of lateral pharyngeal walls (*LPW*). Frontal view. Reproduced with permission from Shprintzen RJ et al.: The effect of pharyngeal flap surgery on the movement of lateral pharyngeal walls. Plast Reconstr Surg 66:570–573, 1980.

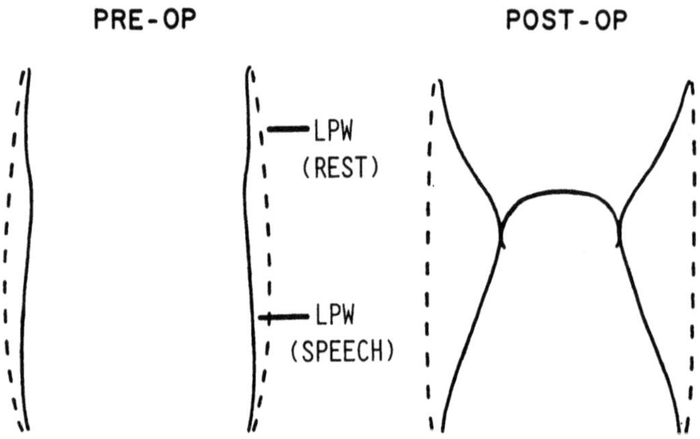

FIGURE 10.26. Increased medial movement of lateral pharyngeal walls (*LPW*) following pharyngeal flap surgery. Frontal view. Reproduced with permission from Shprintzen RJ et al.: The effect of pharyngeal flap surgery on the movement of the lateral pharyngeal walls. Plast Reconstr Surg 66:570–573, 1980.

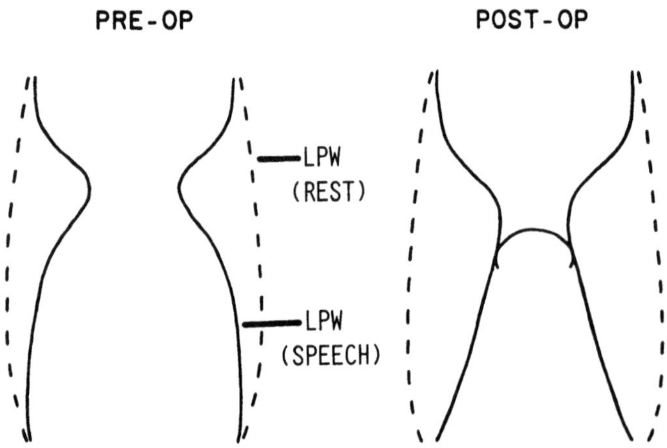

FIGURE 10.27. Change of pattern of lateral pharyngeal motion (*LPW*) from localized region of medial motion preoperatively to greater vertical extent postoperatively. Frontal view. Reproduced with permission from Shprintzen RJ et al.: The effect of pharyngeal flap surgery on the movement of the lateral pharyngeal walls. Plast Reconstr Surg 66:570–573, 1980.

References

1. Yules RB, Chase RW: Pharyngeal flap surgery: A review of the literature. Cleft Palate J 6:303, 1969
2. Shprintzen RJ, Lewin ML, Croft CB, Daniller AI, Argamaso RV, Ship AG, Strauch B: A comprehensive study of pharyngeal flap surgery: Tailor-made flaps. Cleft Palate J 16:46, 1979
3. Skolnick ML, McCall GN: Velopharyngeal competence and incompetence following pharyngeal flap surgery: Videofluoroscopic study in multiple projections. Cleft Palate J 9:1, 1972
4. Shprintzen RJ, McCall GN, Skolnick ML: The effect of pharyngeal flap surgery on the movements of the lateral pharyngeal walls. Plast Reconstr Surg 66:570, 1980

11 Fundamental Principles of Normal Speech Production

The purpose of this chapter is to describe fundamental principles of normal speech production. This will be done to provide a framework for the discussion of speech symptoms of velopharyngeal insufficiency in Chapter 12, and the description of the speech protocol used in the videofluoroscopic evaluation in Chapter 14.

We hope that the nonspeech pathology readers of this text will study these chapters with an eye toward utilizing the information in the execution and interpretation of the videofluoroscopic examination. Just as it is important for the speech-language pathologist to appreciate the procedures employed by the radiologist, so is it helpful for the radiologist to have a basic understanding of normal and pathologic speech production. Our experience indicates that an informed radiologist can significantly shorten the time and the radiation exposure necessary to obtain a diagnostic videofluoroscopic study. Furthermore, a solid knowledge of speech production is essential to the accurate interpretation of the completed study. We offer the following information with the belief that it will aid in the planning and evaluation of the videofluoroscopic study.

The Roles of Thought and Language

The primary purpose of the videofluoroscopic examination is to ascertain the ability of the palate to elevate and separate the oral and nasal cavities during speech. However, we must recognize that underlying systems of *thought* and *language* must function in order to produce *speech*.

Let us consider what occurs before a normally hearing four-year-old can answer his mother's question, "What would you like for lunch?" The child must have the intellectual capacity and *receptive language* capability to *understand* the question. The child must not only attach the appropriate meaning to each of the words (the *semantic* component), but he must also understand the rules for their grammatical usage (the *syntactic* component). Furthermore, he must recognize the question as a request for information to which he responds (the *pragmatic* component).

Before the child answers the question, it will be necessary for him to formulate or think of an answer. *Thought* is essentially an internal representation of one's experience, and may take the form of visual images, muscle imagery, or language. Perhaps our hypothetical child will form the visual image of a peanut butter and jelly sandwich. In order to communicate this culinary craving to his caretaker, he must be able to summon forth the appropriate vocabulary (the *semantic* component), use grammar (the *syntactic* component), and enlist language to express his needs (the *pragmatic* component). Thus, the child demonstrates competency in utilizing *expressive language* as a carrier for thought.

When the child has a message he wishes to communicate, audible speech is most commonly used as the method of transmission. Alternative sources of message transmission may incorporate sign language, gestures, writing, or communication devices capable of producing audible or graphic or pictorial messages.

Knowledge of the rules of language production and understanding of language is often referred to as the speaker's *language competence*. This differs from the actual expression of thought and language: the speaker's *performance*. In general, *performance* is thought to be inferior to the user's *competence*. This explains why many of us find it easier to read a book than to write one, or to think of an experience than to describe it to a friend. Each of us has most likely been unable at one time or another to verbalize the name of a person or an object we are thinking of. Another common and similarly uncomfortable experience is finding ourselves at a "loss for words" to express a complex abstract idea. These are normal happenings when they occur sporadically and to a mild degree.

However, the independence of language and thought is dramatically demonstrated in individuals who have suffered aphasia: language impairment due to brain injury. Such problems are most commonly seen in an adult population, and are not often encountered in a cleft palate clinic. Patients with aphasia may have difficulties in understanding language (receptive aphasia) or in expressing their thoughts (expressive aphasia) (1). These conditions may be complicated by difficulty in the motor sequencing of speech (*apraxia*) or in muscle weakness or flaccidity (*dysarthria*).

Why then is it important for the radiologist to appreciate the difference between the disorders of thought, language, and speech? Occasionally a patient will be referred with hypernasality who also has difficulty in the formulation and/or understanding of language. The instructions given to this patient, as well as the speech protocol, must be individualized to obtain optimal speech performance.

It is the responsibility of the speech-language pathologist to obtain information regarding the intellectual, language, and speech capabilities of a patient prior to an x-ray evaluation, and to determine if an x-ray evaluation is indeed appropriate for the patient. For example, there may be limited benefit in doing a videofluoroscopic study of a severely intellectually-impaired individual with minimal language if the poor palatal movement is a reflection of dysfunction in the more basic and underlying systems of speech production. In such a case, it is our experience that surgical or prosthetic management of the VPI often produces less than optimal speech results.

Another commonly encountered clinical entity necessitating the need to appreciate these concepts is the language-impaired child. Such a child may be unable to produce vocabulary and syntax that are commensurate to his age and nonverbal intellectual abilities. It is important that we not confuse this deficiency with the inability to physically produce speech and thus to utilize the velopharyngeal valve. The speech-language pathologist will be able to administer a diagnostic test battery to quantify the extent of the language deficit in this patient.

A subtle example of the need to distinguish between speech and language production was posed in the following case. A seven-year-old had normal intelligence, yet failed to utilize the phoneme /s/ at the ends of words to denote pluralization. However, the /s/ was used appropriately at the beginning and in the middle of words. When he omitted the /s/, his palate did not elevate as expected. This absence of palatal elevation was a symptom of an *expressive language* deficit, the omission of /s/ to mark plurals, rather than a demonstration of velopharyngeal insufficiency. Speech therapy focusing on the teaching of the pluralization rule solved the problem.

The speech-language pathologist has the needed clinical skills to assist in the assessment of the individual with mental retardation, or the intellectually normal individual with language dysfunction. These patients often require a tailormade, individualized speech protocol for the videofluoroscopic examination, in order to elicit the best speech they are capable of producing. The formulation of such a protocol will be described further in Chapter 14.

Development of Speech and Language

Most normal children develop the ability to understand and produce speech and language. This development follows a general sequence that has approximate ages assigned to each of the stages, with a good bit of overlap between the stages.

It is also important to reiterate that thinking provides a foundation for language, and that we can only talk about what we know. Thus, the child's intellectual capabilities, as well as his opportunities to learn, will affect his language development.

The process involved in the miraculous acquisition of language is complex and mysterious. It appears that children do not initially acquire the adult's rules and adopt them in the form they have been presented. Instead, children seem to seek and perceive patterns, form hypotheses about linguistic rules, and apply them to match these hypotheses. Thus, children may initially use the word "dada" to refer to *all* adult males, before the word becomes specific to their own male parent. Throughout the course of speech development they may omit consonant sounds they have not yet mastered, and simplify words so that "banana," "bottle," and "baby" all sound like /ba/. Very young children's syntactic systems also reflect their attempts to simplify and approximate the adult system, as in the sentence "No go bed." These stages in the development of speech and language are somewhat predictable, although much individual variation exists.

Many investigators believe that children with cleft palate may lag behind the noncleft child in speech sound development (2-5). This is not generally attributable to any difference in intellectual capabilities, but may be a result of physical limitations of the oral mechanism prior to surgical repair, the impact of the surgical experience, and/or the hypothesized negative consequence of persistent middle-ear disease with its resultant mild-to-moderate conductive hearing loss on speech acquisition and production. The speech-language pathologist will have the necessary expertise to assess and document the young child's level of speech and language development. This information will be critical in determining whether speech problems reflect velopharyngeal incompetence and in formulating an individualized speech protocol for the videofluoroscopic evaluation.

Hearing

Role in Acquisition of Speech/Language

It stands to reason that we must hear others' speech in order to learn language and that a profound sensorineural hearing loss will negatively affect speech acquisition. The sense of hearing not only provides the child with the opportunity to learn speech and language, but also enables him to appreciate the extent to which speech can be used to influence the actions of others. Hearing also allows the child to compare his sound production to that of the adult model, or to his internalized representation of adult speech.

Feedback Function

Another function of hearing is to provide feedback to the speaker, so that he can consciously, or unconsciously, monitor and make improvements in his speech. If a severe hearing loss is acquired after language is learned, the patient's speech may deteriorate as a result of insufficient auditory feedback. (Auditory feedback is not the only type of feedback from which the speaker derives information. Proprioceptive feedback, for example, provides one with a sense of movement and position of the articulators. Many of us are familiar with the odd sensation associated with the deprivation of proprioceptive feedback via an injection of local anesthetic during a visit to the dentist.)

Audition seems to play a critical role in providing feedback about palatal function. Because we are not able to "feel" the position of the velum during ongoing speech production, we must rely upon auditory feedback to achieve an oronasal resonance balance for nonhypernasal speech production. Patients with severe congenital hearing loss in the absence of clefting or observable muscle insufficiency may have hypernasal speech because they lack audtory feedback.

The relationship between the degree of hearing loss and the magnitude of the resulting speech problem is an imperfect one. However, in general, the more normal a person's hearing, the better speech is likely to be (6).

Speech Production

The videofluoroscopic x-ray study is primarily conducted to determine the competency of oral and velopharyngeal movement during *speech* production. We will describe three mechanisms that have an impact upon the production of speech: *phonation, resonance,* and *articulation.* These three aspects can be isolated as separate areas of study. However, it is important to remember that they are

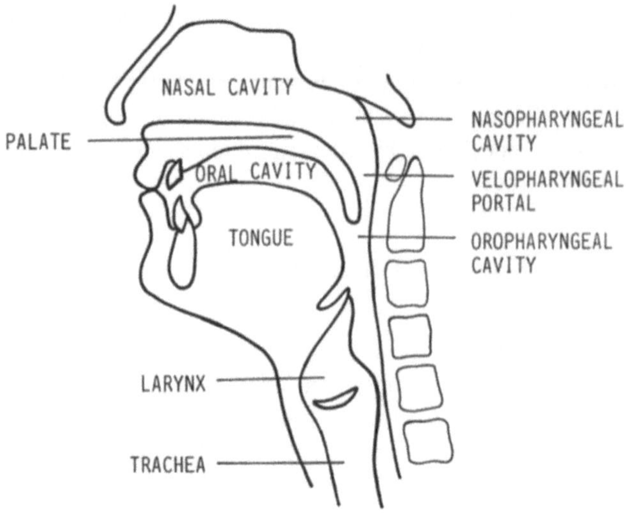

FIGURE 11.1. Resonating cavities of the vocal tract.

intimately related and coordinated in the normal speaker. We will assume, for the forthcoming discussion, that language, thought, and audition are not disordered and that speech has been acquired normally. Pathologic speech production will be considered in a subsequent chapter.

Phonation

The sounds of speech may be dichotomized into two groups. *Voiced,* or phonated sounds, such as vowels and voiced consonants (e.g., "b," "d," and "g"), are produced as a result of the vibration of the vocal folds of the larynx. This vibration is initiated as a result of a build-up and the pulselike release of subglottal air. *Unvoiced* sounds (e.g., "p," "t," and "k") are not, by their nature, characterized by vocal fold vibration. Instead, air from the lungs passes through the open, nonvibrating larynx and is modified by articulatory movements of the tongue, lips, jaw, and pharynx (7).

Resonance

Both voiced and unvoiced sounds are modified by the resonance characteristics of the vocal tract. *Resonance* is a term that refers to the vibratory response of the vocal tract to sound.

The vocal tract is comprised of the supralaryngeal air passages, which span the area from above the vocal folds to the lips (Fig. 11.1). Resonance occurs in the air-filled spaces of the buccal cavities (the cheeks), the larynx, trachea, and most important, in the oral and the pharyngeal cavities. The nasal cavity also serves as a resonator when the palate is in the rest position for quiet breathing and during the production of *nasal* consonant sounds and nasalized vowels (i.e., "m" in "mom," "n" in "no," and ŋ in "ring"). At other times, the soft palate elevates and separates the oral and nasal cavities. Movements of the articulators further alter the acoustic resonance characteristics of the vocal tract (1).

Too much nasal resonance is termed *hypernasality,* and too little nasal resonance, *hyponasality.* These speech problems are discussed in greater detail in Chapter 12.

Articulation

The term *articulation* is used by speech-language pathologists to refer to movements of the vocal tract that produce the sounds of speech (7). Speech articulators include the lips, mandible, palate, tongue, and the pharynx.

In the following sections, we will describe aspects of articulatory production, including vowel, consonant, and diphthong classification. (This information is derived from references 7-11.) We will also consider the dynamic nature of speech articulation and how it relates to velopharyngeal function. This information is included as background reference material for upcoming chapters describing the speech problems associated with velopharyngeal dysfunction and the speech protocol for the x-ray evaluation.

TABLE 11.1. The sounds of American English based upon the International Phonetic Alphabet.

Consonant sounds	Key words
p	pay
b	bay
m	may
w	we
ʍ	why
f	fan
v	van
θ	this
ð	thin
t	to
d	do
n	no
l	low
r	red
s	sip
z	zip
ʃ	shop
ʒ	rouge
j	yes
k	key
g	go
ŋ	ring
h	he
Consonant Combinations (Affricates)	**Key Words**
tʃ	chew
dʒ	gem

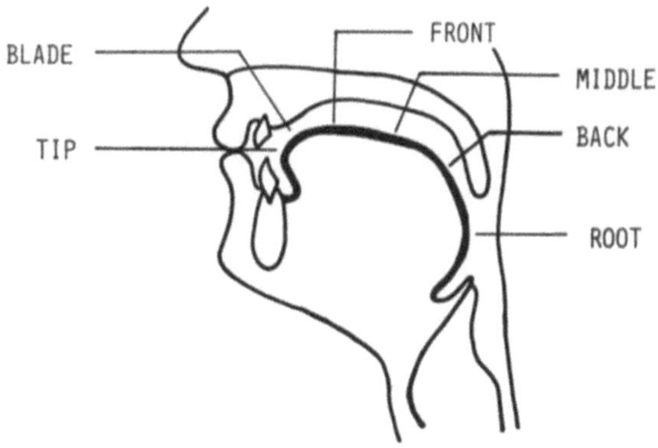

FIGURE 11.2. Functional divisions of the tongue.

The International Phonetic Alphabet

Throughout this chapter, we will utilize the International Phonetic Alphabet (IPA) to designate consonants, vowels, and diphthongs, Designed in the late 1800s, the IPA enables us to accurately refer to a sound or phoneme. The IPA eliminates the potential for confusion by assigning *one symbol* to each distinctive sound, as shown in Table 11.1. For example, the letter "c" used in English represents an /s/ sound, (e.g., circle) or a /k/ sound (e.g., circle). Examples of phonemes in word contexts are provided for the reader when the sounds they represent are not immediately obvious (7).

Vowel Articulation

Vowel sounds are produced when the sound energy produced by the vibrating vocal folds escapes through a relatively unoccluded oral-pharyngeal tract. Acoustically recognizable differences in vowels primarily occur as a result of the positions of the lips, tongue, and mandible. Vowels are also classified according to the duration of their production and the degree of muscular tension. Differences between vowels can also be described with regard to velar, lateral pharyngeal wall, and posterior pharyngeal wall position. The primary determinants of vowel differentiation are as follows:

Lip Position

Vowels are described as being round (lips positioned to form circular opening), as in the vowel /u/ (e.g., in the word "who"), or unrounded (lips forming oval shape), as in the vowel /i/ (e.g., in the word "key"). Not only is the lip rounder in the vowel /u/, but the lips protrude and elongate the oral tract.

Tongue Position

The tongue is a dynamic articulatory structure which is influential in the production of all vowels and consonants except for: /m/, /p/, /b/, /v/, and /f/. Functional divisions of the tongue are shown in Figure 11.2. Tongue position for vowels is traditionally considered relative to two dimensions in the oral cavity. First, the tongue may be described as being "high," or "low" in the oral cavity (Fig. 11.3A).

high vowels

/i/	/ɪ/	/u/	/ʊ/
(beat)	(bit)	(boot)	(book)

low vowels

/æ/	/ɑ/
(bat)	(bottle)

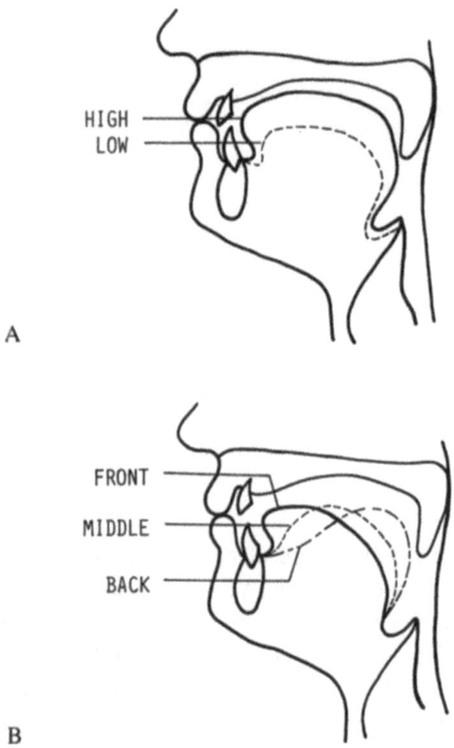

FIGURE 11.3. Tongue position for vowel production results from the combination of vertical and horizontal tongue placement. Vertical placement (A) may be high or low. Horizontal placement (B) may be front, middle or back.

Vowels with lingual height somewhere between "high" and "low" are described as "mid" vowels:

/e/	/ɛ/	/ɝ/
(bait)	(better)	(Bert)
/ʌ/	/ɚ/	/ə/
(but)	(butter)	(again)
/o/	/ɔ/	
(boat)	(bought)	

Secondly, vowels may be described with respect to the anterior-posterior lingual position (Fig. 11.3B). *Front* vowels are made with the tongue forward in the mouth), *back* vowels with the tongue posterior to the resting position, and *central* vowels with the tongue somewhere inbetween.

front vowels

/i/	/ɪ/	/ə/	/ɛ/	/æ/
(beat)	(bit)	(bait)	(better)	(bat)

central vowels

/ɝ/	/ʌ/	/ɚ/	/ə/
(Bert)	(but)	(butter)	(again)

back vowels

/u/	/ʊ/	/o/	/ɔ/	/ɑ/
(boot)	(book)	(boat)	(bought)	(bottle)

Tongue Tension and Duration of Production

Vowels are also described according to the degree of tongue muscular tension and the duration of sound production. Tense vowels (those produced with increased muscular tension within the tongue), such as /i/ as in "ski," are generally produced with more effort and a longer duration relative to the lax vowels. Lax vowels (those produced with reduced muscular tension) involve less muscular tension and a shorter relative duration.

tense

/i/	/e/	/ɝ/	
(beat)	(bait)	(Bert)	
/u/	/o/	/ɔ/	/ɑ/
(boot)	(boat)	(bought)	(bottle)

lax

/ɪ/	/ɛ/	/ɚ/
(bit)	(better)	(butter)
/ə/	/ʌ/	/ʊ/
(again)	(but)	(book)

tense/lax

/æ/
(bat)

To summarize, we have described ways in which to describe a vowel according to:

1. the relative position of the lips or tongue
   a. rounded or unrounded (lips)
   b. high, mid, or low (tongue)
   c. frontal, central, or back (tongue)
2. the degree of muscular tension and duration of the production

Diphthong Articulation

Diphthongs are sequences of two vowel sounds. Three of the five diphthongs found in the English language are "true diphthongs" used to differentiate words:

/aɪ/ ("buy")
/aʊ/ ("bough")
/ɔɪ/ ("boy")

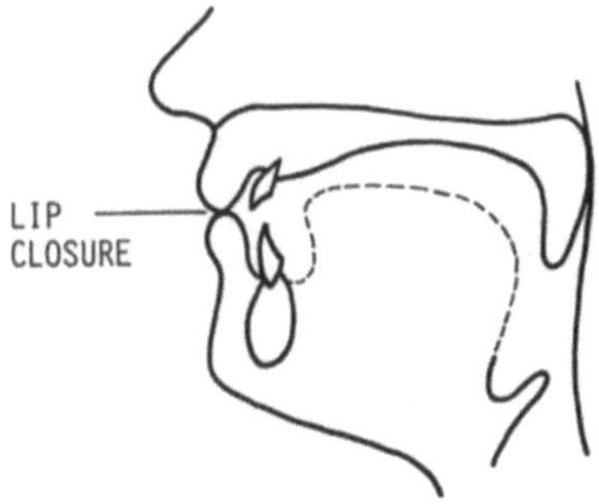

FIGURE 11.4. Example of tongue placement for bilabial sound /p/ or /b/. Dotted line indicates that the tongue position is not fixed but that it varies according to the influence of the associated vowel.

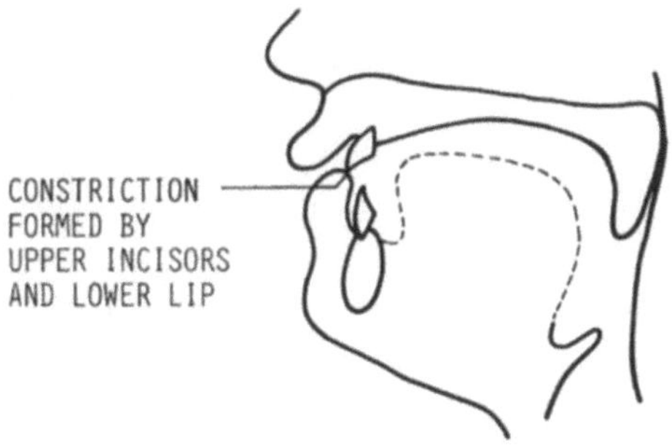

FIGURE 11.5. Example of tongue placement for labiodental sound /v/ or /f/. Dotted line indicates that the tongue position is not fixed but that it varies according to the influence of the associated vowel.

TABLE 11.2. Classification of the consonant sounds of English

	Place of production						
	Bilabial	Labio-dental	Lingua-dental	Lingua-alveolar	Lingua-palatal	Lingua-velar	Glottal
Stop	b p			d t		g k	ʔ
Fricative		v f	ð θ	s z	ʒ ʃ		h
Affricate					dʒ tʃ		
Nasal	m			n		ɔ	
Glide	w*				j	ʍw*	
Lateral				l			
Rhotic					r		

*Bilabial/velar

Adapted from the works of: Bernthal and Bankson (7), Calvert (10), and Tiffany and Carrell (11).

The other two diphthongs /eɪ/ ("bay") and /oʊ/ ("bow") are stressed variants of the vowels /e/ and /o/, respectively.

Like vowels, diphthongs serve as nuclei for syllables and are produced with an open vocal tract. Unlike vowels, they are characterized by a progressive change in the shape of the vocal tract as it assumes the form necessary to produce the first vowel in the pair, and then the second.

Consonant Articulation

Consonants are made with a complete or partially constricted vocal tract, whereas vowels are produced with an open vocal tract. Consonants and vowels also differ with regard to their function within a syllable. Generally, syllables contain one vowel (V) nucleus to which one or more consonants (C) may be added to form different syllable shapes. Common syllable shapes are as follows:

CV	"so"
VC	"at"
CVC	"tag"
CCVC	"stop"
CCVCC	"drags"

Consonants are described by the location or *place* of the complete or partial vocal tract closure and the *manner* of closure. Consonants are also characterized according to the presence or absence of vocal-fold vibration during their production. Table 11.2 shows the place of articulation, manner of production, and voicing for each consonant sound.

Place

The *place of articulation* refers to the site where two opposing structures achieve a localized constriction of the vocal tract.

Bilabial sounds are formed by closure at the lips (i.e., /b/; /p/) (Fig. 11.4).

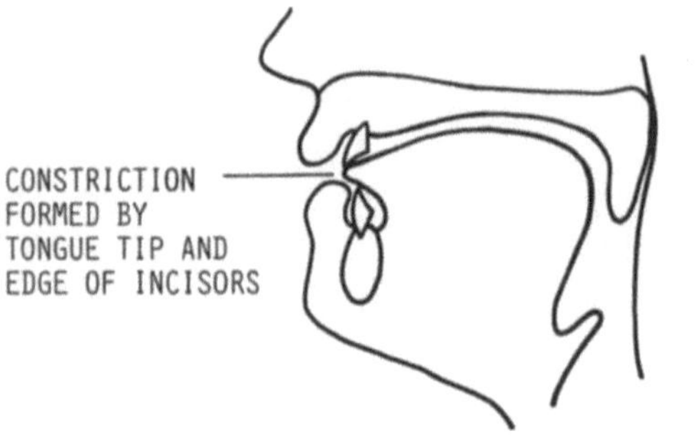

FIGURE 11.6. Example of tongue placement for linguadental sound / θ / (thin) or / ð / (this).

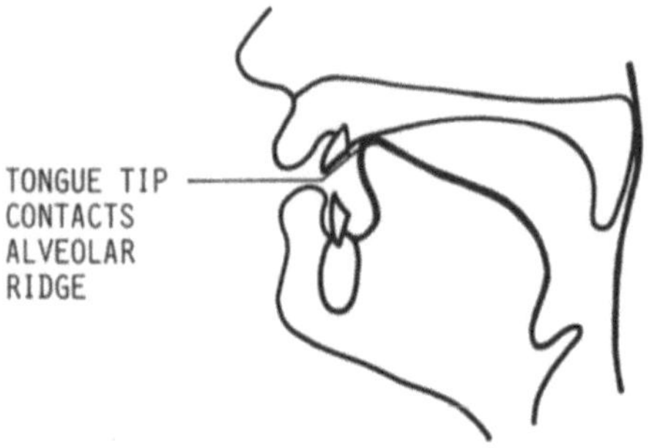

FIGURE 11.7. Example of tongue placement for lingua-alveolar sounds /t/ or /d/ that involves tongue-alveolar contact.

Labiodental sounds involve contact between the upper central incisors and the lower lip (i.e., /v/; /f/) (Fig. 11.5).

Linguadental, also referred to as *dental* or *interdental* sounds, are produced with contact between the tip of the tongue and the upper central incisors (i.e., / θ / (thin); / ð / (this) (Fig. 11.6).

Lingua-alveolar or *alveolar* sounds (i.e., /t/, /d/, /z/, /s/, /n/) are characterized by contact (Fig. 11.7) or approximation (Fig. 11.8) of the tip of the tongue and the alveolar ridge. Although the majority of speakers produce these sounds by raising the tip of the tongue to the maxillary alveolar ridge, some normal speakers have lingual contact with the mandibular alveolar ridge.

Linguapalatal, also referred to as *palatal* sounds, are produced via the constriction between the blade of the tongue and the hard palatal area behind the alveolar ridge (i.e., /dʒ/ (judge); / tʃ/ (church); /r/, /j/ (yes) (Fig. 11.9).

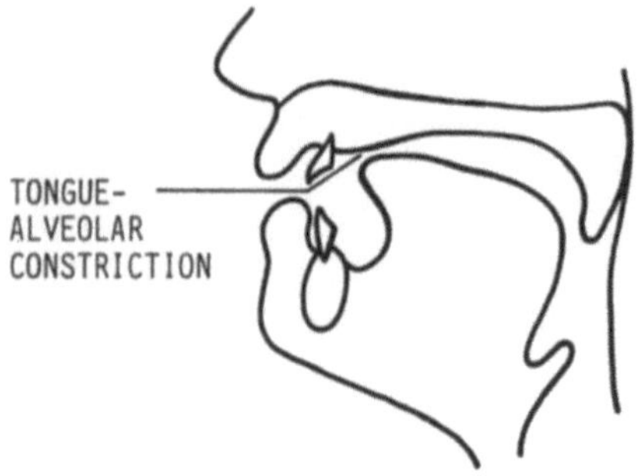

FIGURE 11.8. Example of tongue placement for lingua-alveolar sound /s/ that involves tongue-alveolar constriction.

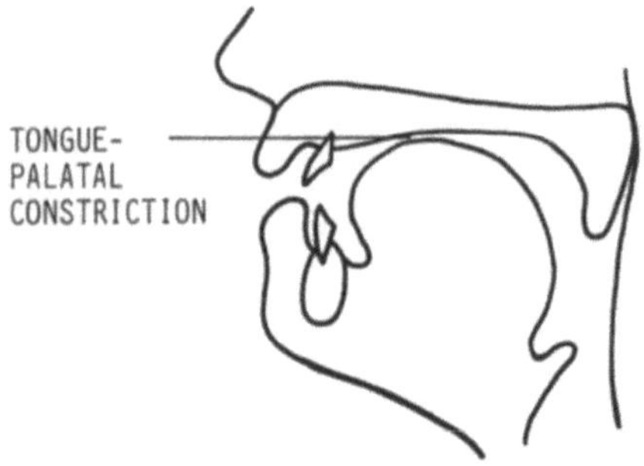

FIGURE 11.9. Example of tongue placement for linguapalatal sound / ʃ / (ship) or / ʒ / (rouge). Site of tongue-velar constriction varies with the adjacent vowel. Constriction site is more anterior for front vowels and more posterior for back vowels.

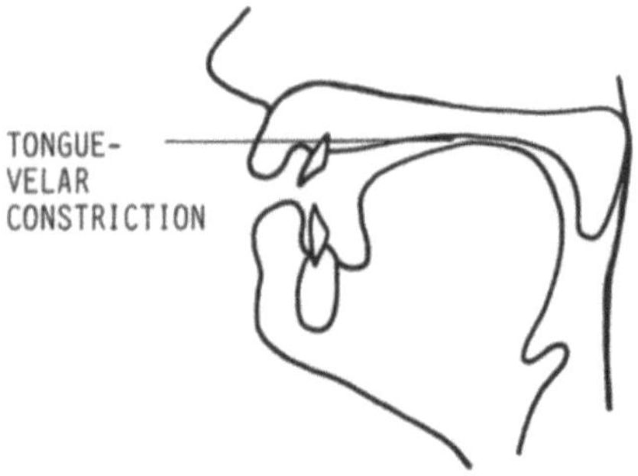

FIGURE 11.10. Example of tongue placement for linguavelar sound /g/ or /k/. Site of linguavelar constriction varies with the adjacent vowel. Constriction site is more anterior for front vowels and more posterior for back vowels.

Linguavelar, or *velar* phonemes (i.e., /g/, /k/, /ŋ/ (ring) referred to in other chapters as "ng") are articulated even more posteriorly in the oral cavity, between the dorsum of the tongue and the velum of the palate (Fig. 11.10).

Glottal sounds (i.e., the glottal stop /ʔ/ (bottle); /h/ (who) are produced at the level of the vocal folds.

Manner

Manner of articulation refers to the way in which sounds are produced in the vocal tract.

Plosives or *stops* are formed when two structures completely block the flow of air through the vocal tract (i.e., /b/, /p/, /d/, /t/, /k/, /g/, and /ŋ/). Air pressure builds up behind the closure and is then released, sometimes resulting in an audible burst of air, as in the words "pit" and "two."

Fricatives, less commonly referred to as *spirants,* are produced when the two articulators are in close approximation so that friction is created in the air stream moving through the narrow constriction. There are nine fricative sounds in English (v, f, ð (thin), θ (that), z, s, ʒ (rouge), ʃ (should), h*). Four of the *fricatives* are characterized by a high-frequency hissing sound, and are called *sibilants* (z, s, ʒ (rouge), ʃ should).

Affricates are sounds that have characteristics of first the stop and then the fricative. English has two affricates: /tʃ/ (church) and /dʒ/ (judge). Affricates are considered by some to be sibilants.

Nasals are consonants that are produced with velopharyngeal opening so that air passes into the nasal cavity (i.e., m, n, ŋ (ring). These sounds are also characterized by the presence of constriction somewhere in the oral cavity. They are sometimes referred to as *nasal* stops.

Sounds that are produced with approximation in the vocal tract, but, unlike fricatives, are not characterized by friction, are called *approximants* or *frictionless continuants.* There are four such sounds in English: /l/, /r/, /w/, and /j/ (yes). The /l/ sound is often referred to as a *"lateral"* sound, as it is the only sound in English for which air is emitted laterally and is directed over the front of the tongue.

*Sometimes classified as a glide.

Voicing

Sounds are said to be *voiced* if the vocal folds are vibrated during production (e.g., /b/ and /d/). If the vocal folds do not vibrate during sound production (e.g., /p/ and /t/), the resultant sounds are considered to be *voiceless* or *unvoiced.*

All of the vowels and diphthongs are voiced. True *consonants* or *obstruents* (stops, fricatives, and affricates) occur in pairs of a voiced sound and an unvoiced cognate (e.g., /b/ and /p/; /t/ and /d/). The cognates share the same site of articulation and manner of production.

Sound Influence

Our descriptions of the characteristics of English speech sounds might tempt us to think of the sounds of our language in isolation. However, it is important to recognize that normal conversational speech is comprised of a stream of speech sounds, and that these sounds are influenced and modified by their neighboring sounds.

The type of sound influence most commonly alluded to is *coarticulation.* A strict definition of coarticulation is that "two articulators are moving at the same time for different phones" (8). Coarticulation occurs when a speaker, saying "dew" (du), rounds his lips for the /u/ while the tongue is still moving for /d/. It also occurs in the word "simple" when the speaker lowers the palate in anticipation of the /m/ while moving the tongue to produce the /s/.

Another type of sound influence may occur when characteristics of the production of a specific speech sound alters (i.e., undergoes adaptation or partial assimilation) or completely changes a neighboring sound. Partial assimilation or adaptation is exemplified by the /p/ sound in the word "pit" (p^hit). The /p/ is aspirated (p^h) as a result of the plosive nature of the /t/ consonant that follows it. This is termed *anticipatory forward* or *right-to-left assimilation.* When assimilation occurs in the opposite direction, it is known as *retentive, left-to-right, backward,* or *carryover assimilation.* In the word "dogs," for example, the /s/ is spoken as a /z/. This sound change occurs as a result of the influence of the preceding voiced /g/ sound. Unlike the prior example (p^hit), in which characteristics of the neighboring sound were assumed, *complete* assimilation, or sound change, has occurred. Edwards and Shriberg (9) state that regressive assimilation is more common than progressive assimilation.

Assimilation processes may occur in sounds that are adjacent or contiguous, or, when sounds are distant or noncontiguous.

The concepts of coarticulation, adaptation, and assimilation are important ones to consider when selecting the speech sample for the videofluoroscopic examination, as discussed in Chapter 14.

Both the phonetic composition of the speech protocol and the rate with which it is spoken will result in changes in velopharyngeal and oral function. An understanding of the normal variation attributable to sound influence is essential to the proper diagnosis of pathologic function.

Suprasegmentals

Suprasegmentals, or prosodic elements, are speech characteristics that are superimposed on speech sounds, syllables, words, or phrases. Examples of suprasegmentals are stress, intonation, loudness, pitch level, juncture, and speaking rate (7). It should be noted that changes in suprasegmentals may affect both the timing characteristics and the competency of velopharyngeal functioning (1).

References

1. Borden GJ, Harris K: Speech Science Primer: Physiology, Acoustics and Perception of Speech. Baltimore: Williams & Wilkins, 1983
2. Bzoch KR: Articulatory proficiency and error patterns of preschool cleft children and normal children. Cleft Palate J 2:340, 1965
3. Philips BJ, Harrison RJ: Articulation patterns of preschool cleft palate children. Cleft Palate J 6:245, 1969
4. McWilliams BJ, Musgrave RH: Diagnosis of speech problems in patients with cleft palate. Br J Disord Commun (Spring): 26, 1977
5. Van Demark DR, Morris HL, Vandehaar C: Patterns of articulation abilities in speakers with cleft palate. Cleft Palate J 16:23, 1979
6. Ling D: Speech and the Hearing-Impaired Child: Theory and Practice. Washington, DC: AG Bell, 1976
7. Bernthal JE, Bankson NW: Articulation Disorders. Englewood Cliffs, NJ: Prentice Hall, 1981
8. Newman PW, Creaghead NA, Secord W: Assessment and Remediation of Articulatory and Phonological Disorders. Columbus, OH: Charles E Merrill Publishing Company, 1985
9. Edwards NL, Shriberg LD: Phonology: Applications in Communicative Disorders. San Diego: College Hill Press, 1983
10. Calvert DR: Descriptive Phonetics. New York: Brian C Decker, 1980
11. Tiffany WR, Carrell J: Phonetics: Theory and Application. New York: McGraw-Hill Book Company, 1977

12
Speech Patterns of Velopharyngeal Dysfunction

The purpose of this chapter is to acquaint the non-speech-language pathologist with speech characteristics that are symptomatic of velopharyngeal insufficiency (VPI). Our intent is not to furnish the reader with an all-encompassing description of "cleft palate speech" but to broadly define issues and terms.

There are a multitude of factors that may impact upon speech. These include, but are not limited to: motivation and learning, hearing, cognitive abilities, phonetic context and complexity, rate of speech, respiratory effort, patient motivation, fatigue, nasopharyngeal tissue characteristics, dental and skeletal structure, and neurologic status. Complicating the issue further is the observation that variability in speech may be the rule rather than the exception for many patients.

An in depth discussion of each of these factors and their interrelationships is beyond the scope of this text. For a more complete examination of cleft palate speech, the reader is referred to the text by McWilliams et al. (1). In addition, perceptual training in the diagnosis of speech symptomatic of VPI is available on an audio series, produced by McWilliams and Philips (2).

Some "Not-So-Uncommon" Misconceptions

The following section presents some misconceptions we have heard about the speech of individuals with clefting.

Misconception 1: All Individuals with Repaired Cleft Palate Have Abnormal Speech

Nowadays, with improved diagnosis, surgical and orthodontic techniques, and early and aggressive management of otitis media, many more patients will have socially acceptable, if not completely normal speech.

Misconception 2: All Patients with a Submucous Cleft Palate Will Have Hypernasal Speech

Research has shown that the majority of the patients with a bifid uvula, palpable notching of the posterior nasal spine, and/or a blue raphe will have *normal*, not hypernasal speech (3, 4)

Misconception 3: Patients with a Cleft Lip Will Have Hypernasal Speech

In general, the presence of a cleft lip in a person with an anatomically and neurologically intact palate will not result in resonance problems. However, articulation errors may be present if clefting of the alveolus results in missing teeth or malocclusion, or if a short, tight upper lip precludes lip closure for bilabial sound production (1).

Misconception 4: There Is a One-to-One Relationship Between Structure and the Speech Product

Bloomer (5) stressed the imperfect relationship between oral structure and speech. He discussed several possible relationships between the two and the therapeutic implications:

1. Normal structure–normal function. Intervention is not needed.
2. Normal structure–abnormal function. Speech therapy should be considered rather than structural change.
3. Abnormal structure–compensatory function. In this instance, the patient may adopt speech behaviors that compensate for the normal structure. There are several treatment choices. If the compensation does not create other problems such as vocal cord nodules, does not call attention to the speaker, and is effective, the decision may be made not to intervene. However, when the compensation is ineffective, structural improvement may be required. Even after the necessary structural changes are accomplished, speech therapy may be required to eliminate the maladaptive habits.

While it is generally accepted that the severity of the speech problem is related to the degree of VPI, the correlation is an imperfect one. This is not surprising in light of the dual importance of both structure and function. Morris (6), for example, in a review of research, found a correlation coefficient of 0.50 between the velopharyngeal gap and articulation proficiency. Multivariate studies by Van Demark (7) and Fletcher (8) provide evidence that articulation errors in children are influenced by maturation and learning, as well as by velopharyngeal incompetency.

Thus, it can be seen that the relationship between velopharyngeal structure and function is complex and is influenced by many factors.

Misconception 5: If a Patient's Speech Is Hypernasal, Speech Therapy Will Correct the Problem

In general, speech therapy cannot overcome the effects of VPI. An exception to this may be the patient with inconsistent velopharyngeal closure. Some such patients may improve resonance by using the technique devised by Shprintzen et al. (9).

Speech Symptoms

Previously, we described the development, production, and definition of "normal" speech. However, to adequately select the speech protocol for the videofluoroscopic evaluation, it is also necessary to become familiar with the pathologic speech that is characteristic of VPI.

Hypernasality

Hypernasality is a deviant resonance characteristic that occurs on vowels, vocalic consonants, glides, or liquids (10). (Speech classifications are discussed in Chapter 11.) This is primarily caused by failure of the palate to separate the oral and nasal cavities during non-nasal consonant production (all consonants except /m/, /n/, /ng/). Instead, sound energy enters the nasal airways, and speech sounds as if it is "coming through the nose."

It is important to appreciate that the perception of hypernasality is based upon features of the entire vocal tract rather than just the presence and size of the VP gap. One such feature may be tongue posture, although there is a lack of consensus on the nature of the influence. Although McDonald and Koepp-Baker (11) contend that a high-riding, posteriorly positioned tongue may block the oral airstream, accentuating the oral-nasal resonance imbalance, Trost (10) speculates that "speakers with velopharyngeal insufficiency often use the back of the tongue to aid in velopharyngeal valving so that it becomes obligatory for lingual articulatory maneuvers to be shifted posteriorly."

Other features which might influence the perception of hypernasality include the degree of respiratory effort, the extent of mouth opening, the structure of the nasal cavities, and the effects of vocal-tract-tissue composition on resonance. Discussion of these topics is beyond the scope of this book. The reader is referred to works by Falk and Kopp (12) and McWilliams et al. (1).

Hyponasality

Hyponasality is a reduction of the normal nasal resonance that occurs on /m/, /n/, and /ng/. When hyponasality is severe, these nasal phonemes may sound like, but not identical to, /b/, /d/, and /g/, respectively.

Even though the presence of hyponasality suggests an absence of VPI, such is not always the case. Obstruction of the posterior nasal airway or adenoid enlargement in the nasopharynx may prevent air escape from an incompetent velopharyngeal valve (2).

Mixed Hypo-Hypernasality

The presence of both hyper- and hyponasality in the same patient may be partly explained by a partial obstruction of the nasal airways. The hyponasality is a consequence of the obstruction's impact on the /m/, /n/, and /ng/ sounds. Because the obstruction is only partial, air is able to escape into the nasal passages, causing hypernasality (2).

Cul-de-sac Resonance

This variation in resonance occurs when air enters the nasal airway, but cannot escape because of an anterior nasal blockage. Cul-de-sac resonance has a muffled sound quality.

Nasal Emission

Nasal escape of air normally accompanies production of /m/, /n/, and /ng/. However, nasal air flow that occurs on stop, affricate, and fricative phonemes may be associated with poor VP function.

There are many varieties of nasal emission described in the literature:

1. Visible, inaudible nasal emission: This may be detected on a cold mirror held under the nose, but is not perceptible to the listener.
2. Visible, audible nasal emission: The air flow traveling through the nasal cavities creates a noise which signals the presence of VPI.
3. Nasal turbulance: This term, introduced by McWilliams (13), refers to a severe form of audible nasal escape, which is produced when the rush of air meets increased resistance in the nasal airways.

It should be noted that pathologic, audible nasal emission does not always occur because of structural or neurologic limitations of the velopharyngeal valve. Audible nasal emission may instead occur in the presence of VP closure because of an underlying articulatory problem, which may be solved by intensive speech therapy (10, 14, 15). The videofluoroscopic x-ray can be used to detect and document this problem (16).

It is often helpful for the speech-language pathologist to instruct the patient in the proper production of the misarticulated sound during the course of the x-ray examination. (Instruction should be delivered when the x-ray is turned off.) A competent VP response to instruction confirms that the VPI is related to articulatory, rather than structural causes.

Articulation

Lingual Posture and Movement

The lingual movement and posture of many speakers with cleft palate can differ from those of noncleft speakers, a finding that is even more frequent for patients with a past or present history of VPI.

There are two major patterns of this lingual dysfunction: tongue-pharyngeal contact which obturates the space between the palate and posterior wall, or tongue-palatal contact. In the latter case, the tongue actually appears to lift the palate toward the site of VP constriction (17).

The second pattern of lingual dysfunction, that of tongue positioning that is excessively low and forward, is also observable on videofluoroscopy, but is not as dramatic in appearance. This is probably why fewer investigators (18, 19) have reported this behavior.

Error Types and Patterns

Much has been written about types and patterns of errors in speakers with cleft palate. While it is beyond the scope of this text to detail all of these findings, the following summary statements have been included to indicate general trends:

1. Speakers may evidence sound distortions (e.g., phoneme nasalization), substitutions (i.e., substitutions of one phoneme for another), or omissions (i.e., failure to articulate a phoneme). There is disagreement in the literature as to which of these error types are most common. McWilliams (20, 21) reported that of 1,814 misarticulated sounds, most were distortions (1,436), followed by omissions (335) and substitutions (43). According to Spriestersbach et al. (22), however, omissions were most prevalent, followed by substitutions and distortions.
2. Speakers with cleft palate are more likely to articulate plosives correctly, followed by nonsibilant fricatives and sibilants (23). Fricatives and affricates have the highest rate of misarticulation (24). The /s/ sound is the phoneme that cleft-palate speakers most frequently and consistently misarticulate (20, 21).
3. In cleft-palate speakers, sounds that require lingual contacts (e.g., /s/) are more often defective than sounds requiring lip movement (e.g., /w/) (26). Sounds made with posterior lingual activity (e.g., /k/) are more often disordered than those made with tongue-tip contact (e.g., /t/) (25).
4. Voiceless sounds are more likely to be misarticulated than their cognates. (Cognates are sounds that only differ in the presence of voicing, such as the unvoiced /p/ and voiced /b/) (20-22).
5. Unless nasalization is a problem, speakers with cleft palate generally articulate vowels correctly (26). However, when compared to noncleft speakers, they are more likely to have problems with producing intelligible vowels (27) and have more errors in vowel production (24).
6. There is a relationship between the vowel sound and perceived nasality. High vowels (e.g., /i/ as in beat) are perceived as more nasal than low vowels (e.g., /æ/ as in battle), and front vowels (e.g., /i/ as in beat) are more nasal than back vowels (e.g., /u/ as in boot) (28).
7. Cleft-palate speakers are more likely to misarticulate sounds in blends (two sounds together, as in "bl") than a single sound (one sound, as in "b") (22).
8. Speakers with cleft palate tend to misarticulate sounds inconsistently. McWilliams et al. (1) hypothesized that this inconsistency may be related to both dental malocclusion and VPI.
9. Children with cleft palate have slower articulatory development than noncleft speakers (24, 29) and are at greater risk for disordered articulation (1, 30). However, the articulation of cleft speakers does tend to improve with age (1, 31).
10. Speakers with cleft palate may have weak consonant production because of reduced intraoral air pressure (32).

Compensatory Articulations

Compensatory articulations are believed to be the speaker's attempts to achieve oral-nasal valving (Fig. 12.1). The speaker employs the correct "manner" of articulation (e.g., plosive, affricate, glide, etc.) but shifts the "place" of articulation from the expected site (33).

In general, cleft-palate speakers tend to shift the "place" of articulation in a backward direction. This posterior shifting, referred to by some as "backing," was documented by Powers (34), Brooks et al. (35, 36), and Lawrence and Philips (19).

Investigators believe that "compensatory" or "gross" articulations, like the glottal stop and pharyngeal fricative, reflect the speaker's attempt to create as successful a valve as possible, at a site other than the velopharyngeal portal (37, 38).

The following is a description of some of the most common types of compensatory articulations.

Nasal Snort

The nasal, or nasopharyngeal snort described by Morley (32), is often associated with the /s/ phoneme and other fricatives. The auditory perception is that of excessive force of production, and represents the introduction of air through an initially closed velopharyngeal valve. Intraoral pressure mounts until the velopharyngeal valve is forced open, culminating in a brief, but noisy escape of air into the nasal airways (1).

Posterior Nasal Fricative

The posterior nasal fricative is characterized by audible nasal emission accompanied by audible frication, which (10) is created by "velar flutter" activity. Some patients also evidence "lingual assistance" in the form of simultaneous linguavelar articulation, and/or posterior lingual positioning. The posterior nasal fricative is notably present in noncleft palate patients with VPI, specifically in pho-

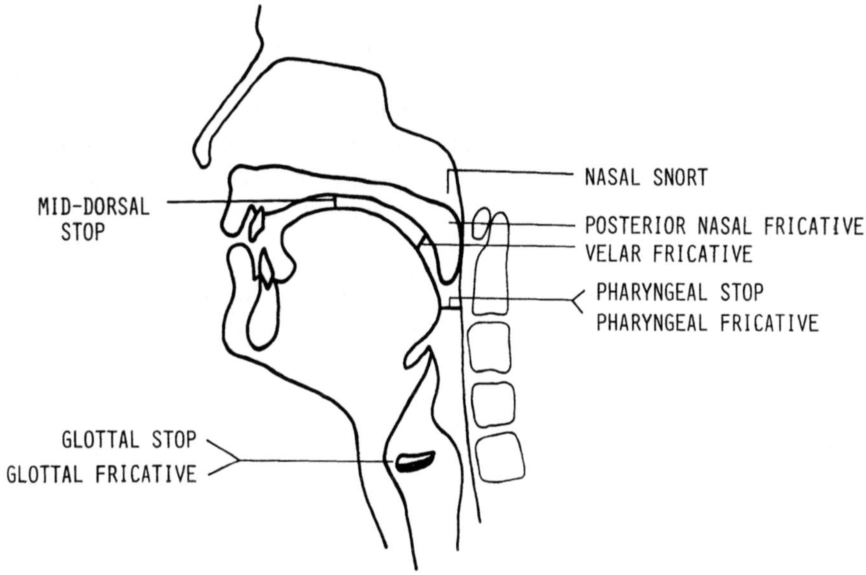

FIGURE 12.1. Sites of compensatory articulation within the vocal tract in an attempt to achieve oral-nasal valving.

neme-specific VPI and patients with dysarthric speech (10).

Middorsum Palatal Stop

Middorsum palatal stops are substitutions for the /t/, /d/, /k/, or /g/ sounds. Each substitution occurs in the place of production for the /j/ sound (as in "young"). With the middorsum palatal stop, voiced sound substitutions, like /d/ or /g/, are perceived equally, and unvoiced sound substitutions, like /t/ or /k/, sound similar (10). This is because the phoneme boundaries are lost between the /t/ and /k/, and /d/, and /g/.

Velar Fricative

The velar fricative involves the replacement of back velar stops /k/ and /g/ with a linguavelar articulation (10). This is seen often in dysarthric speakers who have neuromotor involvement of the posterior portion of the tongue. The sound produced has a fricative quality.

Pharyngeal Stop

The pharyngeal stop, described by Honjo and Isshiki (39) and by Trost (10) is a compensatory substitution for /k/ or /g/. This involves tongue positioning that is posterior and inferior to that required for /k/ or /g/ production. These stops can occur as high as the level of the greatest VP constriction or as low as the area slightly superior to the glottis.

Pharyngeal Fricative

The pharyngeal fricative described by Morris (38), Bzoch (37), and Morley (32) is an apparently compensatory frication between the dorsum of the tongue and pharyngeal wall.

Glottal Stop

The glottal stop, one of the earliest gross articulation errors to be described, is a plosive consonant produced by vocal-fold valving (10). Although still observed in cleft palate clinics, glottal stops are heard less frequently as care for cleft palate improves (40).

Glottal Fricative

Glottal fricative results from overly tense vocal cords, such that there is an increased frication between the two cords (1, 32).

Vocal Disorders

The diagnosis of voice disorders is complex. The speech-language pathologist must differentially assess several parameters of vocal function. These include vocal pitch, quality (e.g., presence of hoarseness), breathiness, harshness or loudness, and respiratory support for phonation. The speech-language pathologist must be knowledgeable about both organic and psychological contributions to a voice problem, and the interactions between the two.*

For the purposes of this text, we will limit our discussion to symptoms of vocal dysfunction that appear to be caused by VPI. These include hoarseness, soft voice, monotone, and strangled voice.

Chronic Hoarseness

McWilliams et al. (41, 42) first recognized that chronic hoarseness and the accompanying bilateral vocal cord nodules may be a result of velopharyngeal valving problems. In severe cases, patients may evidence periods of aphonia (loss of voice). These authors believe that the vocal hyperfunction represents the speaker's attempt to compensate for the VPI at the laryngeal level. In the diagnostic evaluation, hoarseness secondary to VPI must be differentiated from those vocal abuses such as screaming, yelling, and tense vocal production unrelated to velopharyngeal valving problems.

Soft Voice Syndrome

Bzoch (43) and McWilliams et al. (1) described another pattern of vocal behavior that may be symptomatic of velopharyngeal dysfunction. Patients may evidence a voice quality that has been variously described as soft, weak, or aspirate. This difficulty in achieving adequate loudness may be a result of nasal air loss through the inadequate VP valve.

Monotone

McWilliams et al. (1) note that patients with soft voice syndrome may also have excessively monotone speech and limited vocal-pitch range.

Strangled Voice

McWilliams et al. (1) report that another less frequently encountered voice problem is that of the "strangled voice," characterized by excessive abdominal and laryngeal tension such that the patient seems to be "trying to hold his or her breath and talk at the same time" (1). Their clinical experience has shown that therapy to increase relaxation in the laryngeal and abdominal structures in these patients often results in an increase in hypernasality.

*For further information in these areas, readers are referred to: A.E. Aronson, *Clinical Voice Disorders*, New York: BC Decker, 1980; D. Boone, *The Voice and Voice Therapy*, Englewood Cliffs, NJ: Prentice Hall, 1977, or latest edition. For reading about voice disorders in cleft-palate speakers, we recommend McWilliams BJ, Morris HL, Shelton RL, *Cleft Palate Speech*, Philadelphia: BC. Decker, 1984.

Velopharyngeal Function and Speech

In this section, we will outline and describe the speech characteristics of velopharyngeal competence and borderline, mild, moderate, and severe VPI. Although we would like to provide our readers with the exact size of the velopharyngeal gap that is characteristic of each of these speech groups, it would not be valid to do so because individuals vary in their adaptation to VPI. For example, we recall one speaker with a 1-2 mm^2 gap, who not only required a pharyngeal flap, but sounded worse than a second patient with a 3-4 mm^2 gap. A third patient with a 1-2 mm^2 gap had socially acceptable speech. The reasons for differences in adaptation have not been clearly documented, but are probably due to one or more of the following factors: coordination and timing; auditory, tactile and kinesthetic perception; intelligence; and motivation, to name but a few.

Thus, the following descriptions are only meant for use as a general guide.

Velopharyngeal Competence

An individual with velopharyngeal competence will generally not evidence speech stigmata of VPI. Exceptions to this include patients who adopt speech stigmata of VPI because of faulty learning, or patients who previously had VPI and, although capable of achieving velopharyngeal competency, retain one or more speech stigmata of VPI. However, the majority of patients with velopharyngeal competence will have normal resonance and voice quality, appropriate nasal emission, and an absence of facial grimacing. Articulation errors may be present due to dental problems or functional causes (e.g., faulty learning), but are not caused by VPI in this patient group (1).

Borderline VPI

Borderline VPI is a term generally used to describe very small degrees of VPI. In order to fully appreciate the speech symptoms of VPI, there must be a recognition of the variations of borderline VPI. The VPI may be present all, most, or some of the time. The consistency of the VPI often varies according to the speech task, the effort exerted by the speaker, and fatigue. Although we have heard clinicians use this term to designate velopharyngeal contact that does not appear to be tight, we prefer to reserve its use to indicate an opening in the portal.

Morris (44) differentiated between the "almost-but-not-quite" (ABNQ) and the "sometimes-but-not-always" (SBNA) groups.

The typical ABNQ patient has a consistent, small VP gap, very mild hypernasality, and no constriction of the nares. ABNQ patients show no evidence of timing difficulties. Instead, their hypernasality results from a small structural deficit in the velopharyngeal valve.

In contrast, the SBNA patient achieves VP competence on some tasks, usually on short utterances such as single words and consonant-vowel combinations. When task complexity increases to include long strings of words or words containing nasals, VP inadequacy and hypernasality surfaces. The SBNA patient appears to have difficulty in some aspect of velopharyngeal timing or coordination.

Borderline velopharyngeal closure is often difficult to recognize on the x-ray. Similarly, it is often difficult to perceive auditorially. We find ourselves "straining our ears" to hear excess nasality in the "borderline" speaker, who may be very mildly hypernasal only part of the time. Variation often appears to relate to the nature of the speech task, with speech sounds requiring high intraoral pressure such as /s/ most affected.

Observation of air escape from the nares will usually confirm that there is a faulty seal between the oral and nasal cavities. This is determined by holding a cold, small mirror under each nostril and noting whether condensation appears on the mirror while the person repeats words and sentences that do not contain nasal consonants. Sample sentences employed at the University of Pittsburgh Cleft Palate Center include: "Sissy sees the sky," "Patty paid Pat," and "Give Kate the chocolate cake." Nasal consonants may also be included to determine the effects of coarticulation, as in the word "technical," or "Put the baby in the buggy." (Discussed further in Chapter 14.)

Despite the presence of nasal air escape on the non-nasal consonants, if the speaker does not have evidence of audible symptoms of VPI, the speech end-product is considered normal. Inaudible nasal emission alone is not, in our opinion, sufficient cause to perform a videofluoroscopic x-ray, unless the x-ray is being performed to anticipate the effects of adenoidectomy or maxillary advancement.

Some patients with borderline VPI evidence mild hoarseness (42). This may occur either with or without hypernasality.

Mild VPI

Mild VPI is characterized by a small opening in the velopharyngeal valve. The patient may have mild hypernasality, and mild nasal escape will be seen on a mirror test of nasal emission. Articulation is generally not markedly influenced by mild degrees of VPI. Voice quality may be slightly hoarse. The type and severity of speech symptoms of mild VPI vary from patient to patient.

Moderate VPI

The moderately incompetent velopharyngeal valve will have a significant impact upon speech production. Resonance is moderately hypernasal and there is clearly nasal escape on a mirror test. The patient may or may not have nasal turbulence. Articulation distortions of sibilant or affricate sounds are apt to result from reduced intraoral pressure. Voice quality is often but not always hoarse.

Severe VPI

The speaker with a grossly incompetent velopharyngeal valve characteristically has severe hypernasality and nasal escape on the mirror test. Nasal turbulence may be present. Plosive sounds are often distorted as well as sibilant and affricate sounds. These articulation problems are a consequence of the reduced intraoral pressure subsequent to air escape through the valve. Patients may evidence compensatory articulation patterns (described in the previous section) that severely stigmatize speech.

Special Communication Considerations

Hearing

It is now well recognized that an individual born with a cleft palate is at an increased risk for developing otitis media and associated hearing loss. It has been hypothesized that the otitis media occurs because the poorly developed tensor veli palatini muscle fails to efficiently open the Eustachian tube, one function of which is to ventilate the middle ear. The lack of ventilation is thought to result in the accumulation of fluid and inflammation, allowing secondary bacterial infection to occur. The presence of the fluid in the middle ear, as well as a change in the compliance of the tympanic membrane, may impair the conductive function of the middle ear and result in a mild to moderate hearing loss. Otitis media decreases in frequency: (1) after palatal repair; (2) with increased age; and (3) with close monitoring and aggressive treatment.

Patients with velopharyngeal dysfunction in the absence of overt palatal clefts (i.e., submucous cleft palate) also experience a higher incidence of hearing loss and ear disease (1) although data are not as readily available on the exact incidence of these problems.

Radiologists should recognize that a great many of the patients they will examine with VPI may have chronic conductive hearing losses. The hearing impairment may be unilateral or bilateral and, if a result of otitis media, may fluctuate between 5 dB and 55 dB. This may result in symptoms as mild as subtle auditory dysfunction and inattention, or, if persistent and severe, speech and language retardation and handicapping hearing loss (1).

Patients who have very severe hearing problems may evidence hypernasality. Kuehn (28) hypothesized that this may be a result of difficulties in achieving normal timing of velopharyngeal movements. Research by Stevens et al. (45) suggests that the deaf may open the VP port when it should be closed and close the port when it should be open. Thus, they often interchange the nasal consonants /m/, /n/, /ng/ for /b/, /d/, and /g/, respectively.

Intelligence

The presence of a cleft *per se* does not signal the existence of intellectual impairment. Thus, it is important to evaluate the intellectual abilities of each child on an individual basis (47). However, the presence of congenital abnormalities *in addition* to a cleft seems to be associated with a higher probability of intellectual impairment. In general, patients with the isolated cleft palate type are more likely to have a congenital abnormality, and of these patients, problems in mental development occur more than 50% of the time (2). An excellent review of the literature that has investigated the intellectual status of patients with clefts is provided by McWilliams et al. (1).

As noted in Chapter 11, cognitive abilities are closely tied to speech and language competency and development. Typically, studies of mentally retarded children have shown them to have a high incidence of language delay, and language skills in these children are generally commensurate with their level of mental development (46).

Occasionally, the cleft palate team may be asked to evaluate the resonance and palatal function of an individual with Down's Syndrome. Daly (48), in fact, has reported an inexplicably high incidence of hypernasality in the trainable mentally retarded group.

Language

Children with clefts appear to have a slower onset of expressive language than noncleft speakers. In the preschool years, they have an increased risk of language impairment. These language delays appear to improve with age (1).

Neurologic Problems

Hypernasality has been diagnosed in many patients with dysarthria and/or apraxia of speech. Conditions with neurologic consequences in which hypernasality has been reported include amyotrophic lateral sclerosis, bulbar palsy (49), pseudobulbar palsy, chorea (50), multifocal eosinophilic granuloma (Hand Schuller-Christian disease) (51), myasthenia gravis (52, 53), and myotonic dystrophy (54).

Dentition

Dental abnormalities and malocclusion can have a profound impact upon articulation in speakers with or without clefting. However, the correspondence between structure and function is complex and imperfect. For example, the presence of a dental abnormality does not guarantee the co-occurrance of a speech problem. And, a speech problem in a patient with a dental abnormality may not be caused, totally or in part, by the structural difference. Consider, for instance, some of the possible etiologies of an /s/ distortion: (1) a functional, learned problem; (2) hearing loss; (3) developmental factors; (4) insufficient intraoral pressure secondary to VPI; (5) neurologic problems; (6) missing tooth; (7) anterior open bite; (8) crossbite; (9) ankyloglossia; (10) oronasal fistula; (11) recent acquisition of braces or retainer; and (12) palatal expander. The speech-language pathologist is trained to determine the cause or causes of such a speech error.*

The effect of dental malocclusion on speech is particularly pertinent to patients with cleft palate who have an increased incidence of maxillary collapse on the side of the cleft, midface deficiency, oronasal fistula, missing or supernumerary teeth, and/or crossbite. Sibilant articulation distortions are a common result of these problems and often relate to tongue protrusion into an open space (2).

It is important for the speech-language pathologist to differentiate a dentally related articulation problem from one caused by VPI alone. In our experience, the solution to dentally related articulation errors is improvement of the oral structure rather than speech therapy.

Speech Assessment Considerations

Recognizing "Normal Speech" Versus a "Speech Problem"

Why include a description of normal speech in a chapter proclaiming to do the opposite? Because the recognition of aberrant speech is very much based upon the ability to discern speech that is not normal from speech that is socially acceptable. What constitutes a speech problem?

Charles Van Riper (55), considered by many to be the "father" of speech-language pathology, defined three characteristics of a speech problem. Dr. Van Riper stated that speech is defective when it:

1. Deviates so far from the speech of other people that it calls attention to itself.
2. Interferes with communication.
3. Causes its possessor to be maladjusted.

Therefore, normal speech is that which is *understandable, socially acceptable,* and *does not cause the speaker concern.* Even though these concepts emerged in the 1950s, they are still widely accepted today.

In general, one does not require training to recognize normal speech. Sometimes, however, it helps simply to listen to the person without watching him speak. Facial scarring, dental or skeletal malocclusion, inappropriate tongue protrusion, or other atypical oral behaviors may influence the listener to more harshly judge speech (56). It is also important to recognize that as we have more exposure to a speaker, we may adapt to his speech pattern and consider it to be less deviant than it actually is.

When evaluating a patient's speech, it is helpful to consider whether his speech would "call attention to itself" outside of the clinic setting. In the case of a child, would he sound different from other children on the playground? In the case of an adult, would his speech sound "different" in the workplace or in the grocery store?

The significance of *context* cannot be overemphasized. The standards of speech acceptable for a television broadcaster are necessarily different from those for an individual who is not required to be a model communicator.

The issue of social, or cultural dialect, is also important to consider. We have known well-intentioned nonspeech professionals to mistakenly attribute speech that is characteristic of a foreign language (i.e., a Spanish accent) or cultural factors (i.e., black dialect) to the presence of a repaired cleft palate. These examples of different but not deviant speech represent rule-governed, nonpathologic systems of speech that bear no relationship to a cleft of the palate.

In the event that a patient or his family holds extremely perfectionistic or clinically unattainable standards for speech, it is far better to provide counseling to the patient or family than to conduct unnecessary x-ray evaluations or management procedures.

Occasionally the "trained-ear" of a speech-language pathologist may become "overtrained," and it is necessary to recognize that socially acceptable but not necessarily "perfect" speech is our treatment goal. It is important to remember that speech assessment is a personal decision that we make based upon our hearing acuity, perceptual abilities,

*For a comprehensive discussion of dentally related speech errors, the reader is referred to Bloomer (5).

and internal standards of "good speech." The latter may vary according to our training, the influence of colleagues, or even our emotional state.

One advantage of the "team approach" to treatment is that many different professionals evaluate each patient, an approach that helps to recognize and avoid perceptual errors. Team members can also periodically conduct reliability studies so that they systematically compare their perceptions to those of others.

The radiologist is a vital member of this "checks and balances" system. It is particularly important for him to question the need for an x-ray in the patient whom he believes has normal speech. In fact, the radiologist should expect to be informed as to the purpose of each videofluoroscopic examination he is asked to perform. The x-ray requisition form presented in Chapter 15 is one way to facilitate this communication. However, it is rare that the radiologist will be requested to conduct a videofluoroscopic x-ray on a normally speaking patient. Exceptions to this include the patient for whom a proposed surgical procedure may result in hypernasality, as in the case of maxillary advancement or adenoid removal.

Thus, the radiologist has an obligation to question and to refuse to perform an x-ray for an individual he feels has normal speech and for whom the x-ray will serve no useful purpose. Post-hoc sessions in which the obtained x-rays are related to speech descriptions are helpful in sensitizing team members to these issues.

Factors Affecting Speech Performance

The successful formulation of a representative speech assessment protocol depends upon an appreciation of the many factors which can affect speech performance. In the following section selected aspects of speech production are considered.

Length of Utterance

There is a difference of opinion in the literature as to whether hypernasal speakers will exhibit more (57) or less (58) hypernasality when uttering short phrases versus longer sentences. Therefore, the patient should utter a variety of utterance lengths during the videofluoroscopic study.

Phonetic Considerations

Fricative sounds, particularly sibilants, become defective as a result of mild degrees of VPI, whereas plosive sounds are produced correctly even with much greater degrees of VPI (2). This is because the demands for VP closure are greater for fricative sounds such as /s/ in which the airstream must be prolonged, versus the plosive /p/ in which air pressure is very briefly impounded, then released.

Voiceless sounds (e.g., /p/, /t/, /k/) are also associated with greater velopharyngeal defectiveness than voiced sounds (e.g., /b/, /d/, /g/). This is presumably because voiceless sounds are associated with higher intraoral pressure and, therefore, require greater degrees of velopharyngeal competence (2).

The phonetic composition of a speech task must also be considered. Evidence suggests that patients with VPI show increased hypernasality on nonnasal sounds when they are adjacent to nasal sounds (e.g., /m/, /n/, /ng/) (1).

Loudness

One characteristic often associated with VPI is the "soft voice" syndrome in which patients speak with reduced loudness. There are two possible explanations for this: one, that the patient is unable to generate sufficient loudness, and the other, that reduced loudness is used as a compensatory strategy (1, 2). In the latter instance, it is hypothesized that when speaking at low intensities, there is less respiratory effort, and less opportunity for air to be lost through the velopharyngeal valve (2). Thus, it is often helpful to ask a patient to increase his loudness in order to reveal greater degrees of hypernasality.

Rate

Patients with cleft palates may speak more slowly than those without clefts (59). Furthermore, in one study, when the rate of syllable production was increased, the /s/ phoneme was articulated less precisely (60). Thus, one way in which to unearth VPI in borderline patients is to increase the demands for rapid speech. This will disallow the opportunity that slow speech provides for compensatory accuracy.

Stimulus Conditions

The stimulus used to evoke a speech response is also important to consider. For instance, speech may be elicited by asking the patient to read, identify a picture, imitate an examiner's model, answer a question, complete a phrase or sentence, or by stimulating spontaneous conversation.

Although there is limited research in this area for cleft palate speakers, McDermott (60) reported that superior results for /s/ production occurred for the imitative rather than the spontaneous speech task.

These findings should be kept in mind when considering the imitative nature of the videofluoroscopic speech protocol. In order to unearth VPI in borderline speakers, it might be wise to include spontaneous conversation in the speech protocol.

References

1. McWilliams BJ, Morris HL, Shelton RL: Cleft Palate Speech. Philadelphia: BC Decker, 1984
2. McWilliams BJ, Philips BJ: Audio Seminars in Speech Pathology: Velopharyngeal Incompetence. Philadelphia: WB Saunders, 1979
3. Kelly AB: Congenital insufficiency of the palate. J Laryngol Rhin Otol 25:281, 1910
4. Stewart JM, Ott JE, Lagace R: Submucous cleft palate: prevalence in a school population. Cleft Palate J 9:246, 1972
5. Bloomer HH: Speech defects associated with dental malocclusions and related abnormalities. In Handbook of Speech Pathology and Audiology. Travis LE (ed). New York: Appleton-Century-Crofts, 1971
6. Morris H: Etiological bases for speech problems. In Cleft Palate and Communication. Spriestersbach DC, Sherman D (eds). New York: Academic Press, 1968
7. Van Demark DR: A factor analysis of the speech of children with cleft palate. Cleft Palate J 3:159, 1966
8. Fletcher SG: Diagnosing Speech Disorders from Cleft Palate. New York: Grune and Stratton, 1978
9. Shprintzen RJ, McCall GN, Skolnick ML: A new therapeutic technique for the treatment of velopharyngeal incompetence. J Speech Hear Disord 40:69, 1975
10. Trost JE: Articulatory additions to the classical descriptions of the speech of persons with cleft palate. Cleft Palate J 18:193, 1981
11. McDonald ET, Koepp-Baker H: Cleft palate speech: An integration of research and clinical observation. J Speech Hear Disord 16:9, 1951
12. Falk ML, Kopp GA: Tongue position and hypernasality in cleft palate speech. Cleft Palate J 5:228, 1968
13. McWilliams BJ: Cleft palate. In Human Communication Disorders. Shames G, Wiig E (eds). Columbus: CE Merrill, 1982
14. Peterson SJ: Nasal emission as a component of the misarticulation of sibilants and affricates. J Speech Hear Disord 40:106, 1975
15. Trost-Cardomone JE: Phone-specific nasal emission: A cleft palate soundalike. Paper presented at American Cleft Palate Association Meeting, Williamsburg, Virginia, April 26-29, 1988
16. Glaser(-Cohn) ER, Skolnick ML, McWilliams BJ, Shprintzen RJ: The dynamics of Passavant's ridge in subjects with and without velopharyngeal insufficiency—A multiview videofluoroscopic study. Cleft Palate J 16:24, 1979
17. McWilliams BJ: Speech and language problems in children with cleft palate. J Am Med Women's Assoc 21:1005, 1966
18. Dickson DR: A radiographic study of nasality. Cleft Palate J 6:60, 1969
19. Lawrence CW, Philips BJ: A telefluoroscopic study of lingual contacts made by persons with palatal defects. Cleft Palate J 12:85, 1975
20. McWilliams BJ: An experimental study of some of the components of intelligibility of speech of adult cleft palate patients. Doctoral dissertation, University of Pittsburgh, 1953
21. McWilliams BJ: Articulation problems of a group of cleft palate adults. J Speech Hear Res 1:68, 1958
22. Spriestersbach DC, Darley FL, Rouse V: Articulation of a group of children with cleft lips and palates. J Speech Hear Disord 21:436, 1956
23. Fletcher SG: Diagnosing Speech Disorders From Cleft Palate. New York: Grune and Stratton, 1978
24. Van Demark DR, Morris HL, VandeHaar C: Patterns of articulation abilities in speakers with cleft palate. Cleft Palate J 16:23, 1979
25. Counihan DT: A clinical study of speech efficiency and structural adequacy of operated adolescent and adult cleft palate persons. Doctoral thesis. Northwestern University, 1956
26. Moll KL: Speech characteristics of individuals with cleft palate. In Cleft Palate and Communication. Spriestersbach DC, Sherman D (eds). New York: Academic Press, 1968
27. Cullinan WL, Counihan DT: Ratings of vowel representatives. Percept Mot Skills 32:395, 1971
28. Kuehn DP: Assessment of resonance disorders in speech, language, and hearing. In Pathologies of Speech and Language. Lass NJ, McReynolds LV,

Northern JL, Yoder DE (eds). Philadelphia: WB Saunders, 1982
29. McWilliams BJ, Musgrave RH: Diagnosis of speech problems in patients with cleft palate. Br J Disord Commun 26: 31, 1971
30. Philips BJ, Harrison RJ: Articulation patterns of preschool cleft palate children. Cleft Palate J 6:245, 1969
31. Van Demark DR: The relationship between remedial speech instruction activities and articulation change. Cleft Palate J 6:57, 1969
32. Morley ME: Cleft Palate Speech. Baltimore: Williams and Wilkins, 1970
33. Spriestersbach DC, Moll KL, Morris HL: Subject classification and articulation of speakers with cleft palates. J Speech Hear Res 4:362, 1961
34. Powers GR: Cinefluorographic investigation of articulatory movements of selected individuals with cleft palate. J Speech Hear Res 5:59, 1962
35. Brooks AR, Shelton RL, Youngstrom KA: Compensatory tongue-palate-posterior pharyngeal wall relationships in cleft palate. J Speech Hear Disord 30:166-173, 1965
36. Brooks AR, Shelton RL, Youngstrom KA: Tongue palate contacts in persons with palatal defects. J Speech Hear Res 31:14-25, 1966
37. Bzoch KR: Etiologic factors related to cleft palate speech. In Communicative Disorders Related to Cleft Lip and Palate. Bzoch KR (ed). Boston: Little, Brown and Company, 1971
38. Morris HL: Abnormal articulation patterns. In Communicative Disorders Related to Cleft Lip and Palate. Bzoch KR (ed). Boston: Little, Brown and Company, 1971
39. Honjo I, Isshiki N: Pharyngeal stop in cleft palate speech. Folia Phoniatr 23:347, 1971
40. Cohn ER, McWilliams BJ: Letter to the Editor. J Plast Reconstr Surg 71:442, 1983
41. McWilliams BJ, Bluestone CD, Musgrave RH: Diagnostic implications of vocal cord nodules in children with cleft palate. Laryngoscope 79:2072, 1969
42. McWilliams BJ, Lavorato AS, Bluestone CD: Vocal cord abnormalities in children with velopharyngeal valving problems. Laryngoscope 83:1745, 1973
43. Bzoch KR: Etiological factors related to cleft palate speech. In Communicative Disorders Related to Cleft Lip and Palate. Bzoch KR (ed). Boston: Little, Brown and Company, 1979
44. Morris HL: Cleft palate. In Communicative Disorders. Weston AJ (ed). Springfield, IL: CC Thomas, 1972, p. 128
45. Stevens KN, Nickerson RS, Boothroyd A, Rollins A: Assessment of nasality in the speech of deaf children. Bolt, Beranek, and Newman, Report No. 2902, 1974
46. Edwards ML, Shriberg LD: Phonology: Applications in Communicative Disorders. San Diego: College Hill Press, 1983
47. Richman LC: The effects of facial disfigurement on teachers' perception of ability in cleft palate children. Cleft Palate J 15:155-160, 1978
48. Daly DA: Bioelectronic measurement of nasality in trainable mentally retarded children. J Speech Hear Disord 42: 436, 1977
49. Darley F, Aronson A, Brown J: Differential diagnostic patterns of dysarthria. J Speech Hear Res 12:246, 1969
50. Darley F, Aronson A, Brown J: Clusters of deviant speech dimensions in dysarthria. J Speech Hear Res 12:462, 1969
51. Cohn ER, Garver KL, Metz HC, McWilliams BJ, Skolnick ML, Garrett WS: Velopharyngeal incompetence in a patient with multifocal eosinophilic granuloma (Hand-Schuller-Christian Disease). J Speech Hear Disord 47:320, 1982
52. Wolski W: Hypernasality as the presenting symptom of myasthenia gravis. J Speech Hear Disord 32:36, 1967
53. Hagstrom W, Parsons R, Landa S, Robson M: Familial velo-pharyngeal incompetence caused by myasthenia gravis. Ann Plast Surg 3:155, 1979
54. Salomonson J, Kawamoto H, Wilson L: Velopharyngeal incompetence as the presenting symptom of myotonic dystrophy. Cleft Palate J 25:296, 1988
55. Van Riper C: Speech Correction: Principles and Methods. New York: Prentice Hall, 1954
56. Podol J, Salvia J: Effects of visibility of a prepalatal cleft on the evaluation of speech. Cleft Palate J 13: 361, 1976
57. Carney PJ, Sherman D: Severity of nasality in three selected speech tasks. J Speech Hear Res 14:396, 1971
58. Van Hattum RJ: Articulation and nasality in cleft palate speakers. J Speech Hear Res 1:383, 1958
59. Lass NL, Noll JD: A comparative study of rate characteristics in cleft palate and noncleft palate speakers. Cleft Palate J 7:275, 1970
60. McDermott RD: A study of /s/ production by individuals with cleft palates. Doctoral dissertation, University of Iowa, 1962

13
Preparing the Patient and Obtaining Cooperation

Honesty Is the Best Policy

It is important that patients and their families receive as complete an orientation as they can appreciate to the procedures and purposes of videofluoroscopy. Such education will help to achieve cooperation and lessen any anxiety surrounding the procedure. The extent of the orientation will, of course, depend upon the age, intellectual capabilities, and emotional status of the patient and accompanying adult.

A general "rule of thumb" is to be forthright and honest regarding any potentially unpleasant aspects of the examination. For example, a child 5 years old might be told: "Today we are going to take some pictures of your mouth. They will help us see how you talk. The pictures won't hurt. Before we take the pictures we need you to sniff this pink stuff in your nose, so we can take good pictures of your mouth." (Show barium solution.) "This tastes a little bit like a strawberry milkshake." (Note that we did not state that it tastes as good as a strawberry milkshake.) "We'll drip it in your nose with this squirter." (Show syringe or dropper.) "Here's where the drops come out. They may tickle a little bit.

"I'll have you rest your head on your neck like this so that the drops help us take a good picture. Then we'll ask you to say some words, like 'kitty cat' and 'popeye.' Let's practice some of the words. Say them real loud and clear, okay?

"After we take the pictures, would you like to see yourself on TV? Ok, let's go!"

Notice that in this passage, we presented information about a potentially unpleasant aspect of the procedure, the nosedrops. Honesty on the part of the examiners is important to establish trust and credibility. A failure by any one member of the cleft-palate team to maintain this trust may adversely affect the patient's faith in other team professionals.

There is also the factor of previous interactions with health professionals. If other professionals in the child's experience have not been forthright about the discomfort associated with medical procedures, the child may be rightfully suspicious and apprehensive. Furthermore, a child who has undergone surgical procedures may associate the hospital with unpleasantness, and cry in anticipation of even a relatively painless procedure.

While children primarily want to know "if it will hurt" and "if it is scary,",adults may have other concerns about the x-ray examination. These may include the risks of radiation exposure; the length of the procedure; the expected time in the waiting room; and the time needed to interpret the examination. Again, it is always best to provide realistic answers to these and other questions.

Age-Specific Considerations

One of the most important aspects of the successful videofluoroscopic examination is the recording of a representative speech sample. If a child is upset by the nosedrops, intimidated by the unfamiliar surroundings, or scared of the radiologist, the recording session may be doomed to failure by noncooperation and/or crying behavior. It is important to recognize that, unlike many other diagnostic medical procedures, patient cooperation is an essential component of the videofluoroscopic x-ray. This is because representative palatal function for speech cannot be obtained during a temper tantrum, whimpering, or crying. Even extremely reticent and uncharacteristically quiet speech may not demonstrate the patient's potential for velopharyngeal movement.

Occasionally a young child may be so uncooperative that it is impossible to perform the videofluoroscopic study. Such a child is most likely frightened and overwhelmed by the strangeness of the fluoroscopic room and its contents. In this situation we just show the child the equipment, briefly explain its use, and reschedule the patient for a second attempt at a study in 3 to 6 months, at which time we almost always succeed in performing the procedure.

The child's ability to adapt to this new situation will depend upon his general developmental status, intellectual capabilities, emotional health, and maturity. The following section presents special concerns related to specific-age groups.

Preschool-Aged Children

Preschool-aged children in the 3- to 5-year range are often literal and concrete in their interpretation of language and events according to Chambers and Abrams (1). For example, when asked if "the cat got your tongue," the preschool child may actually search for the "offending feline."

Chambers and Abrams also point out that the preschool population may invest inanimate objects with lifelike characteristics and feelings. A noisy piece of machinery may be thought of as being "angry." It is easy to imagine why the preschool child might consider the radiographic equipment to be alive and quite frightening.

The 3- to 5-year old's perceptions of the world may also be influenced by their own stature, which is small relative to that of the equipment and radiologist.

Young children are particularly concerned about their bodily sensations and are anxious to maintain control over their bodies. Thus, the instillation of the barium nosedrops can be a source of anxiety. Verbally explaining the procedure in understandable, neutral terms may help alleviate this concern. Focusing on "sniffing" the "strawberry liquid" so that we can take a "pretty picture" is preferable to discussing "nosedrops" and "the syringe."

Some young children are also more apt to respond with "no" than "yes" when they are asked whether they would like to do something. It is better to be quietly firm and optimistic, saying, "Now it is time for you to say some sentences," versus "How would you like to say some sentences for me now?"

As Chambers and Abrams so aptly note, "Preschool children are active, expressive and forgetful." Their fidgety behavior, inability to remember directions, and need for constant reminders may be age-appropriate behavior rather than a sign of disobedience. It may be unrealistic to expect to reason logically or negotiate with many preschoolers. They also may not remember or be able to follow a direction administered a short time ago. It is thus best to continually prompt the preschool-aged child.

The preschool child, and children in general, respond more favorably to positive reinforcement than to punishment or negative reinforcement. The positive reinforcement will be most powerful if it is delivered immediately following the commendable act, so that intervening stimuli are not confused with the desirable behavior.

The child should be told exactly what it is he is doing that you like (e.g., "I really like the way you are standing so very still while I take your picture," or "I can see that you're trying real hard to stand still—that's terrific." Less effective is praise of the child in a global fashion (e.g., "good boy") because not only does this not allude to the desirable behavior, but it also links performance of a task with the child's "goodness" or "badness."

Young School-Aged Children

Young school-aged children in the 5- to 7-year-old range are newly proud of their bodies, possessions, and abilities. However, despite their outward confidence, they are still vulnerable to and fearful of humiliation and loss of self-control (1).

Thus, it is important to avoid embarrassing, or diminishing the confidence of children this age. Their pride is easily wounded, and it is much more effective to appeal to their maturity and pride (e.g.,

"I'd like you to say these words using your big, grown-up voice," versus "I don't want you to cry like a little baby."). Patients in the 5- to 7-year range may feel "grown up" if they are able to briefly "assist" the radiologist or technician in some small way.

Chambers and Abrams advise that crying in this age group be handled with empathy and patience. For example: "I can see that all of the equipment in this room is scary to you, but I'll show you everything we are going to do. Here's where we take the picture of your mouth . . . "

It should be noted that young school-aged children are quite changeable. They may be calm and cooperative one moment, and agitated, withdrawn, and uncooperative the next. Thus, it is important to "strike while the iron is hot," by completing the examination during the maximal period of cooperation.

Preadolescents

Preadolescents in the 8-year to puberty range are usually more cooperative and pose fewer management problems to the radiologist. However, it has been noted that their comprehension is often overestimated by professionals. Although these children appear to understand all that is said to them, such may not be the case, as their cognitive space is inhabited by both real and imaginary concepts. Their concept of the word "x-ray" for example may be "blurred" and "imprecise." Their own conversation may be marked by rapid jumps in topic.

The professional should strive to give complete explanations to children in this age range. All questions should be treated seriously and addressed completely (1).

Adolescents

Adolescence is marked by the transition from childhood to adult behavior. At their best, teenagers will be cooperative with the radiologist. At their worst, they might be noncommunicative, or overly emotional. However, in general, the teenager will cooperate well during the videofluoroscopic procedure.

It is important to be frank and honest with the adolescent (i.e., "It is really important that you speak up so we can get a good recording of your voice"; or "I know you didn't like the nose drops, but we have to put them in the other side of your nose to get a good x-ray.").

Older Patients

It is rare that adult patients of normal intelligence will provide management difficulties for the radiologist. And it is highly unusual for the radiologist to perform a videofluoroscopic examination of the velopharyngeal mechanism on geriatric patients. However, it should be noted that some of the latter group may have diminished sight, memory, or hearing.

Special Populations

It is beyond the scope of this book to provide management techniques for the wide variety of special populations. These may include the mentally retarded, hard of hearing, emotionally disturbed, or hyperactive, to name a few. However, patients who fit into these or other "special" groups may pose added difficulties for the radiologist.

It is recommended that the radiologist, the speech-language pathologist, and the parent(s) work together to accomplish the best x-ray examination. Advance preparation and rehearsal of an individualized speech sample may be required.

In the case of a hard-of-hearing individual, a list of speech stimuli to read may be provided. This would be more effective than asking the hearing-impaired patient to repeat after the radiologist.

The speech-language pathologist will, by virtue of his training, be able to adapt the standard speech stimuli and behavioral management techniques to special populations.

The Family

It is very important for the parents of the children and the teenagers to receive adequate information and preparation. If family members are anxious about the videofluoroscopic x-ray, it is likely that the children will be similarly concerned.

Parents will usually want to know the length of the x-ray, the benefits and risks, and treatment options should VPI be absent or present.

After the x-ray is completed, parents who are anxiously sitting in the waiting room will appreciate a few words from the radiologist. The radiologist might relate the child's response to the x-ray (e.g., "Johnny didn't like the nose drops at all, but he really cooperated well with the examination. We got a good, clear study."). The radiologist should also tell the family when and from whom they should expect to receive a report of the examination. Usually the report comes from a member of the cleft-palate team as part of its total evaluation of the patient.

Finally a note about timeliness. One of the surest ways to antagonize patients and their families is to keep them waiting for long periods of time. Generally, waits in excess of 20 to 30 minutes create resentment and annoyance. Similarly, the radiologist should make every effort to provide a timely interpretation of the x-ray study, as the family may be anxiously awaiting the results.

Reference

1. Chambers DW, Abrams RG. Dental Communication. Norwalk, CT: Appleton-Century-Crofts, 1986

14
The Speech Protocol

Much of this text has described technical aspects of the videofluoroscopic examination, velopharyngeal anatomy, and the characteristics of normal and disordered speech. However, this chapter, "The Speech Protocol," is perhaps the most crucial to the conduct of the velopharyngeal examination.

We will provide examples of speech tasks and refer to the underlying reasons for their inclusion in the x-ray examination.

However, it is important to note that the speech-language pathologist, by virtue of his training, is the professional best qualified to formulate an individualized speech protocol for every patient. It is also helpful if the speech-language pathologist can be present during the x-ray examination so that the optimal speech protocol is devised, and if necessary, revised for each patient.

A speech-protocol "menu" is included, which the speech-language pathologist can use to select and include in each patient's x-ray protocol.

The "Ideal" Speech Protocol

Formulating the "ideal" speech protocol is difficult because of the varying needs and characteristics of individual patients.

The ideal speech protocol:

1. Incorporates speech tasks that demonstrate a patient's best VP performance, worst VP performance, and habitual VP performance.
2. Incorporates a representative sample of different phonemes.
3. Obtains the needed information efficiently in as short a time as possible.
4. Takes into account a speaker's age, interests, intellectual and emotional status.
5. Enables the clinician to differentiate between palatal dysfunction that is phoneme specific and that which is secondary to articulation problems, versus that which is a result of VPI.
6. Includes speech tasks of varying lengths (i.e., isolated vowel; consonant-vowel syllable; word; phrase; sentence; connected speech).
7. Demonstrates palatal timing and coordination, and the ability to accomplish transitions between nasal to non-nasal consonants (e.g., <u>m</u>o<u>p</u>) and non-nasal to nasal consonants (e.g., <u>p</u>e<u>nn</u>y).
8. Includes some standard speech tasks that enable the clinician to compare VP performance among patients.
9. Incorporates special tasks as needed: tasks that fatigue or stress the VP mechanism; barium swallow; alternative head positions; and playing of musical instrument.

Speech Tasks

A variety of speech task types are presented below.

Isolated Vowels

Many vowels produced in isolation do not require VP closure. Thus, the absence of VP closure for many vowels produced in isolation is not diagnostic of VP dysfunction.

However, we do find this isolated vowel task helpful in demonstrating that the patient is properly positioned, especially for the base view. Patients can be asked to produce isolated vowels in succession [e.g., i (eat), a (bottle), u (boot)].

Consonant-Vowel Combinations

Consonant-vowel pairs, such as "ma," "pa," "ka," and consonant-vowel strings, such as "ma-ma-ma," "pa-pa-pa," "ka-ka-ka," and "sa-sa-sa," are useful in demonstrating palatal movement in uncomplicated phonetic contexts. Although dramatic palatal movement is not expected with nasal sounds (e.g., ma), they are included to insure abnormal movement is not present. Nasal-non-nasal consonant-vowel combinations (e.g., ma-pa; pa-ma) demonstrate the patient's ability to accomplish these transitions.

Words and Phrases

Words and short phrases are particularly useful for young children or older patients who have difficulty repeating longer sentences.

Selection of these words and sentences should reflect patients' interests and be consistent with their vocabulary.

In addition, phoneme selections should be carefully planned to uncover abnormal palatal function during the production of various phoneme classes or combinations. Examples are as follows:

nasals	lemon jam
	ninety-nine
plosives	pretty puppy
	baby bottle
velars	cookie
	kitty cat
fricatives	silly Sue
	ship shape
affricates	chocolate chips
	George
nasal–non-nasal consonant transitions	technical
	bread and butter
	Donald Duck
	Mickey Mouse
	Kermit the Frog

In the rare event that patients are unwilling to repeat a specific word, a forced-choice task is useful, providing the individual is able to made a rapid decision. The patient is asked, for example:

Which do you like better:

nasals	milk or meat?
plosives	apple pie or pepper?
velars	coca-cola or chocolate candy?
fricative	fish or fries?
affricates	chili or chicken?
nasal–non-nasal consonant transitions	Michael Jackson or Janet Jackson?

Sentences

Sentences, "loaded" with specific phonemes or nasal–non-nasal consonant combinations, are perhaps the most efficient means to demonstrate differential aspects of velopharyngeal function.

We have found the following sentences to be extremely useful [starred sentences were reported by McWilliams and Philips (1)]:

nasals	Mama made lemon jam*
glides	We love yellow light
plosives	Pull the baby buggy
plosive-nasal	Put the baby in the buggy*
velar	Give Kate the cake
velar-nasal	Kindly give Kate the cake*
	Go get the wagon for the girl*
fricative	Silly Sissy sees the sky*
fricative-nasal	Sissy sees the sun in the sky*
	The ship goes in the shallow water*
	Please tie the stamps with string*
affricates	Charge the check
affricate-nasal	Jim and Charlie chew gum*

Conversation

It is unusual for us to employ conversation in the videofluoroscopic x-ray because if the patient's response time is slow, it could prolong the examination time and increase exposure to the x-ray. If the patient's speech is difficult to understand, a further disadvantage is that the clinicians may not be able to determine what the patient is saying. Conversation also typically includes many phoneme types and does not allow the examiners to view classes of phonemes in isolation. However, because much of speech occurs in conversation, and conversation is the most demanding of speech tasks, this is often useful in assessing habitual performance.

We find it helpful to: (1) direct the conversation toward a topic within the patient's realm of experience and; (2) avoid yes-no questions. Sample questions might include:

1. Tell us about your family.
2. What do you do in your spare time (or after school)?
3. Tell me about school.
4. Tell me about your job.

Because velopharyngeal function may be influenced by rate and volume, the patient might be asked to vary these parameters.

Memorized Speech

Memorized or learned speech tasks such as counting and nursery rhymes, provide easily available speech stimuli. However, it must be remembered that they generally contain nasal as well as non-nasal consonants. Examples are as follows:

Prayer
Alphabet
Nursery rhyme
Pledge of Allegiance
Counting:
  90–91 (nasally loaded)
  60–69 (sibilant loaded, with nasals on 61, 67, and 69)

Special Tasks

In the following section we will describe various types of tasks performed for special purposes.

Quiet Oral Breathing

Although not a speech task per se, it is important to view the velopharyngeal portal during quiet oral breathing to determine whether the airway space is sufficient. This demonstrates the distance between a pharyngeal flap and the lateral and posterior walls during breathing (see Chapter 10).

Barium Swallow

This maneuver is useful in demonstrating nasopharyngeal reflux, borderline VP closure, and the presence of an oronasal fistula (2). The barium swallow is described in detail in Chapter 4.

Fatigue Testing

In order to assess changes in a patient's speech as a result of fatigue, the patient is first x-rayed during routine tasks. The x-ray is then turned off, and the patient asked to engage in prolonged and strenuous speech, as in rapid counting (1 to 400+). The x-ray is turned on at variable points to demonstrate the absence or emergence of fatigue.

Stimulability Testing/Trial Therapy

When examining a patient with inconsistent or defective articulation, it is often helpful to stimulate correct production of the phoneme. This can be done by simply asking the patient to repeat what you say. Another strategy, best performed by the speech-language pathologist, is to provide the patient with verbal instructions regarding tongue placement, etc., while the x-ray is off, and quickly turning on the x-ray to record the patient's performance.

Length of Protocol

We recommend that unless the radiologist and family are committed to a research protocol, an effort should be made to obtain the needed information about VP function in the shortest possible time.

If, for example, the early part of the x-ray consistently and unequivocably reveals gross velopharyngeal insufficiency, or completely competent VP function, then we recommend that fewer speech stimuli be incorporated into the remainder of the examination.

Furthermore, it is also possible to eliminate the use of the hyperextended lateral head position, if it is clear that gross velopharyngeal insufficiency exists on the nonhyperextended view.

Appendix 1. Menu of Suggested Speech Tasks*

Isolated Vowels:

i (beat)
e (bait)
æ (bat)
ʊ (but)
u (boot)
o (boat)
a (bottle)

Consonant-Vowel Combinations:

ma
pa
ka
sa
ma-ma-ma
pa-pa-pa
ka-ka-ka
sa-sa-sa

Words and Phrases:

lemon jam
ninety-nine
pretty puppy
baby buggy
cookie
kitty cat
silly Sue
ship shape
chocolate chips
George

Forced Choice:

"Do you like":
  milk or meat
  apple pie or pepper
  coca-cola or chocolate candy
  fish or fries
  chili or chicken
  Michael Jackson or Janet Jackson

Sentences:

Many of the following sentences were derived from sentences utilized at the University of Pittsburgh (1).

Mama made lemon jam
We love yellow light
Pull the baby buggy
Give Kate the cake
Go get the wagon for the girl
Silly sissy sees the sky
The ship goes in the shallow water
Please tie the stamps with the string
Charge the check
Jim and Charlie chew gum

Conversation:

Specify topics

Memorized Speech Tasks:

Alphabet
Pledge of Allegiance
Counting (specify)
Other

Special Tasks:

Quiet oral breathing
Barium swallow
Fatigue testing (specify)
Stimulability testing/trial therapy (specify)

Appendix 2. Sample Speech Protocols

I. Three-Year-Old
  (normal intelligence; reticent)
  ma-ma-ma
  pa-pa-pa
  ka-ka-ka
  sa-sa-sa
  mommy (nasal)
  daddy (non-nasal)
  puppy dog (non-nasal)
  chocolate chip cookie (non-nasal)
  gummy bears (nasal–non-nasal transition)
  I want candy (nasal–non-nasal transition)
  Alphabet

II. Four-Year-Old
  (normal intelligence; cooperates easily)
  Quiet oral breathing
  i (beat) (high vowel)
  a (bottle) (low vowel)
  o (boat) (mid vowel)

*All tasks are not needed for each patient.

ma-ma-ma (nasal)
pa-pa-pa (plosive)
ka-ka-ka (velar)
sa-sa-sa (fricative)
count 1-10 (nasal–non-nasal transitions)
Pull the baby buggy (plosive)
Give Kate cake (velar)
Sissy sees the sky (fricative)
Donald Duck is nice (nasal–non-nasal transitions)

III. Twelve-Year-Old
(normal intelligence)
quiet oral breathing
i (beat) (high vowel)
a (bottle) (low vowel)
o (boat) (mid vowel)
ma-ma-ma (nasal)
pa-pa-pa (plosive)
ka-ka-ka (velar)
sa-sa-sa (fricative)
Mama made lemon jam (nasal)
Pull the baby buggy (plosive)
Give Kate the cake (velar)
Silly sissy sees the sky (high pressure–fricative)
Jim and Charlie chew gum (nasals in affricate environment)
counting 1–200 (film 1–20, 80–90, and 180–190 to unearth fatigue)

References

1. McWilliams BJ, Philips BJ: Audio seminars in speech pathology: velopharyngeal incompetence. Philadelphia: WB Saunders, 1979
2. Skolnick, ML, Glaser(-Cohn) ER, McWilliams BJ: The use and limitations of the barium pharyngogram in the detection of velopharyngeal insufficiency. Radiology 135(2):301, 1980

15
Ratings and Report Writing

Purpose

The purpose of this chapter is to provide the reader with alternative systems of developing reports of the videofluoroscopic studies ranging from a highly structured rating system for clinicians who are just beginning to perform videofluoroscopic studies or who wish to gather data for research purposes, to a brief report for the experienced interpreter that summarizes how various aspects of the velopharyngeal valving mechanism functionally relate to one another.

We wish to point out that despite our inclusion of detailed rating scales, the clinical interpretation of the videofluoroscopic study is truly a subjective process. "The art" of interpreting the x-rays is honed over time, and is best accomplished when the radiologist, speech-language pathologist, and plastic surgeon read the x-rays together, and relate the radiologic observations to clinical findings. Toward this end, we have included an x-ray requisition form and described clinical review procedures that have been helpful in stimulating these types of interactions at the University of Pittsburgh Cleft Palate Center.

Clinical Reporting

Clinicians responsible for interpreting the videofluoroscopic examination have a variety of methods by which to record their observations and impressions. Each method or combination of methods has strengths and weaknesses. We will, therefore, provide the reader with several alternative strategies. Specifically, we will discuss what we term the "bottom-line" report, the "narrative report," and the "rating sheet." These may be used singly or in combination.

The "Bottom-Line" Report

The "bottom-line" report consists of a brief, one- or two-sentence summarization of the videofluoroscopic x-ray study and its application to clinical treatment. Typical "bottom-line" reports are as follows:

Example 1:
The patient has mild velopharyngeal insufficiency, consistently seen on lateral and base views. Approximately 70% movement of the lateral pharyngeal walls toward the midline is present on frontal projection.
Example 2:
There is gross velopharyngeal insufficiency affecting velar, lateral, and posterior pharyngeal wall movement. There appears to be a neurologic component to the VPI.
Example 3:
There is evidence of a submucous cleft palate but the patient consistently achieves velopharyngeal closure.
Example 4:
The pharyngeal flap is placed too low in the pharynx to achieve maximum benefit from the active lateral pharyngeal wall movement. Barium blowing through the velopharyngeal portal suggests mild to moderate VPI.

Example 5:
The velopharyngeal mechanism appears competent, but an oronasal fistula appears to be the cause of the hypernasal speech.

The advantages of the bottom-line report are that it is concise and states the *clinically significant* material. Less important observations that might not be clinically symptomatic (i.e., the palate is short and thin) are not given undue emphasis. Thus, the unsophisticated reader of the radiologist's report is less likely to misinterpret the nonclinically significant observations.

The disadvantages of this approach lie in the loss of detail that might be useful in comparing the report to that of a future x-ray examination.

The bottom-line report is best used by the very experienced interpreter who is able to quickly perceive the "gestalt."

It is also useful as a summarization of a narrative report or the rating sheet.

The Narrative Report

The narrative report is a concise way to describe velopharyngeal structure and function. We recommend organizing the report as follows:

1. Brief statement of clinical problem
2. Type of videofluoroscopic views performed
3. Brief description of the findings for each of the views performed
4. Conclusions and possible recommendations

Examples of Narrative Reports

1. Clinical Information: 7-year-old patient with repaired posterior cleft palate. Speech evidence of mild velopharyngeal insufficiency, nasal turbulance.

Lateral, frontal, and base views of barium-coated nasopharynx were recorded with speech on videotape.

Lateral: Adenoid mass, consistent with patient's age, is noted. The palate appears of approximately normal length. During phonation no distinct eminence is seen. Approximately a 2-mm midsagittal gap between the vertical portion of the velum and surface of the adenoid is noted during speech. During hyperextension, a small Passavant's ridge appears below adenoid level but no widening of the velopharyngeal gap occurs.

Frontal: Approximately 90% medial movement of the pharyngeal walls. Maximum medial movement is at the nasopharyngeal level. Movement is bilaterally symmetric.

Base: Approximately a 2-mm gap within a circular closure pattern. Barium blows through the gap.

Conclusion: Passavant's ridge with minimal VPI. Circular closure pattern.

2. Clinical Information: 8-year-old patient with hyponasality was referred to assess potential effects of planned adenoidectomy on speech.

Lateral, frontal, oblique, and base views of barium-coated nasopharynx were recorded with speech on videotape.

Lateral: Massive adenoid and tonsillar masses are present. During phonation, the velum closes against the inferior surface of the adenoid mass. If, however, the adenoid mass was removed, the velar length would probably not be adequate to reach the posterior pharyngeal wall. I am hedging on this observation because with "s" in isolation the velum does exhibit some stretching. How much further stretching it could have if the adenoids were removed would be difficult to say. There is very little movement of the velum because the space between it and the adenoid mass during quiet breathing is only several millimeters.

Frontal: Approximately 20% to 30% medial movement of the lateral pharyngeal walls occurs. Wall movement is limited because of the large adenoid mass. In the oral pharyngeal region, the lateral pharyngeal walls are laterally displaced during phonation by large tonsils.

Oblique: Views confirm frontal and lateral views.

Base: View is technically inadequate.

Conclusion: Adequate velopharygneal closure between palate that appears slightly short and huge adenoids. Removal of adenoids may result in velopharyngeal insufficiency.

3. Clinical Information: 15-year-old patient with cerebral palsy who had pharyngeal flap surgery 5 years ago. Moderate-to-severe hypernasality, mild hoarseness, nasal turbulance and plosive, sibilant, and affricate distortions characterize speech.

Lateral, frontal, and base views of barium-coated nasopharynx were recorded with speech on videotape.

Lateral: A horizontally positioned flap between the vertical portion of the velum and posterior pharyngeal wall is present. During phonation, the palate shows almost no motion except with "s" and "k," where slight flickering is noted. The tongue exhibits loss of its normal fine and rapid movements. Palatal length is approximately normal.

Frontal: Approximately only 10% medial movement of the lateral pharyngeal walls occurs, and only rarely. "K" and "s" are the sounds mainly producing this motion. The movement, when it occurs, is equal and symmetric. Each lateral portal is approximately 1 cm wide. During swallowing, normal pharyngeal wall movement is seen.

Base: Confirms frontal and lateral views. Bilaterally wide and incompetent lateral portals.

Conclusion: Status postpharyngeal flap. Almost no palatal, pharyngeal, or rapid tongue movement. Bilateral portal incompetence.

4. Clinical Information: 8-year-old patient with repaired unilateral cleft of the upper anterior palate. Pharyngeal flap surgery 5 years ago. Mild-to-moderate hypernasality, mild hoarseness, and sibilant distortions.

Lateral, frontal, and base views of barium-coated nasopharynx recorded with speech on videotape.

Lateral: A pharyngeal flap at the oral pharyngeal level is present. No midsagittal closure occurs above the flap. No barium blowing is seen.

Frontal: There is approximately a 1-2 mm gap between the medial surface of each lateral pharyngeal wall at its maximum medial excursion and the lateral edge of each side of the flap. The pharyngeal walls show greater medial movement about 1 cm cephalad to the flap than they do at the flap level. No barium blowing through the lateral portals is appreciated.

Base: The pharyngeal flap occupies approximately one third of the width of the portal. Each lateral portal appears to close completely during phonation. No barium blowing through the portal is noted. The discrepancy between the apparent closure of the portals in the base view, and the small gaps between the pharyngeal walls and edges of the flap on the frontal view is explained by the observation that the pharyngeal walls show slightly greater medial movement above the level of the flap than they do at the flap level. Images projected on the base view are a summation of structures moving at various levels, and therefore, the narrower area of the portal is that projected on the base view.

Conclusion: Status postpharyngeal flap with minimal velopharyngeal gap involving both lateral portals during phonation.

Rating Forms

Rating forms (Table 15.1), organized by the videofluoroscopic projection (e.g., frontal, lateral, or base) and subdivided by structure (e.g., lateral pharyngeal walls; Passavant's ridge; velum) are yet another way to record interpretations of the videofluoroscopic study.* This information management approach is especially useful for the inexperienced reader. The form provides a structured approach to the review of the videofluoroscopic study so that one does not omit important information. Clinical alternatives are provided in the answer portion of each question. These stimulate the interpreter to consider many possible scenarios. The rating form is also advantageous because the interpreter judges the same parameters from patient to patient.

Rating systems also provide a disciplined and complete approach to interpretation that is useful for research. Although the experienced viewer may find the rating process lengthy and redundant, even the most experienced clinician may have difficulty perceiving all aspects of VP function during one review of the tape and may need to view a tape repeatedly. We recommend that novices limit their attention to a few parameters during a single viewing of the tape.

Finally, the rating system is helpful in that it underscores the unique contributions that each videofluoroscopic view makes to the study of the velopharyngeal mechanism. For example, while reviewing the rating scales it becomes obvious that the lateral view does not provide information about the presence or configuration of lateral pharygneal wall motion. However, the lateral view provides a unique opportunity to observe the height of velar move-

*To our knowledge, the first published rating scale for the evaluation of videofluoroscopic data was developed by McWilliams and Bradley (1) for the lateral projection. Subsequent additions to the rating scales were made by B.J. McWilliams, M.L. Skolnick, E.R. Cohn, Q.C. Beery, A.S. Lavorato, S.R. Rood, and other members of the University of Pittsburgh Cleft Palate Center. The contributions of these individuals to the rating scales published in this manual cannot be overemphasized.

TABLE 15.1. Rating system

Identification Data

Patient:
Date of Birth:
Date of Study:
Cassette No.:

Technical Aspects of Study

Technical quality (diagnostic/nondiagnostic)
- Lateral
  - With barium _____
    - Neutral head position _____
    - Hyperextended head position _____
  - Without barium _____
- Oblique (with barium) _____
- Frontal (with barium) _____
- Base (with barium) _____

Barium swallow—lateral _____

Lateral View: Erect, Neutral Head Position
(Primarily for postbarium study but can be used where applicable on nonbarium lateral view.)

Velum—Structure/Position

1. Velar length relative to pharyngeal dimensions (evaluate during quiet nasal breathing)
   a. normal
   b. short
   c. very short
2. Velar thickness (evaluate during quiet nasal breathing)
   a. very thick (swelling/postoperative change)
   b. normal
   c. moderately thin
   d. very thin
3. Submucous cleft palate suggested by
   a. bifid uvula
   b. double line of barium on palate
4. Distance between velum (during quiet nasal breathing) at rest and posterior pharyngeal wall (to estimate depth of nasopharynx)
   a. short
   b. normal
   c. long

Velar Function:

1. Superiormost point of velar elevation relative to palatal plane
   a. above plane
   b. at level of plane
   c. below plane
   d. well below plane
2. Prominence of velar eminence
   a. prominent
   b. observable
   c. flicker only
   d. not present
3. Position of velar eminence
   a. more anterior than usual
   b. appropriate position
   c. more posterior than usual
4. Evidence of palatal asymmetry or hemiparesis (see double barium coating during phonation indicating that one side of palate elevates more than other side)
   a. no asymmetry
   b. possible asymmetry
   c. definite asymmetry resulting in unilaterally reduced movement
5. Velar stretch
   a. absent
   b. minimal
   c. marked
6. Speed of movement
   a. movement is barely observable or absent
   b. sluggish
   c. movement is within normal limits
   d. movement is more active than usual
7. Coordination (timing) of velar movement with speech tasks
   a. severe discoordination
   b. mild discoordination
   c. adequate coordination
8. Presence of midsagittal velopharyngeal closure or velopharyngeal gap with non-nasal consonants should be evaluated at both the level of maximum effort and the most consistent effort using the following ratings:
   a. competent—no gap
   b. minimally incompetent, 0–5 mm
   c. moderately incompetent, 6–10 mm
   d. grossly incompetent, > 10 mm
9. Estimated vertical length of velar-PPW contact*
   a. 0 mm
   b. 1–2 mm
   c. 3–4 mm
   d. 5–6 mm
10. Consistency of degree of velar contact during speech
    a. consistent
    b. inconsistent
11. Ability of patient to increase closure or narrowing with stimulation (being prompted by speech-language pathologist to repeat specific speech tasks as videofluoroscopic examination is in progress) (specify speech tasks).
    a. not applicable
    b. good—substantial improvement
    c. fair
    d. poor—little or no improvement
12. Size of adenoidal tissue
    a. not present
    b. small for age
    c. age appropriate
    d. large for age
13. Role of adenoidal tissue in VP constriction/closure
    a. above site
    b. at site
14. Anterior motion of PPW during speech
    a. generalized motion
    b. localized motion—Passavant's ridge
    c. no motion
15. Passavant's ridge
    a. Pattern of closure/constriction involving Passavant's ridge
       1. ridge is primary structure for VP narrowing or closure
       2. ridge is a noncontinuous, second point of narrowing/closure

TABLE 15.1 continued

3. ridge is a contiguous, second point of narrowing/ constriction that increases the extent of the posterior pharyngeal wall, which approximates/ contacts the velum
4. ridge is below the uvula

b. Palatal region closest to Passavant's ridge
1. velar eminence
2. vertical length of the velum
3. the uvula
4. region below the uvula

c. Orientation of Passavant's ridge to velum
1. upward toward the velum
2. perpendicular to the velum
3. downward relative to the velum

d. Ridge size
1. small
2. moderate
3. large

e. Consistency of ridge movement across speech tasks
1. consistent
2. inconsistent—specify:

16. Barium blowing into nasal cavity during speech
a. Site—if present
1. anterior hard palate
2. posterior hard palate
3. velopharyngeal portal (no flap)
4. velopharyngeal portal (flap present)
b. Sounds that produce barium blowing—define:

17. Pharyngeal flap
a. vertical level of attachment on posterior pharyngeal wall at:
1. nasopharynx
2. oropharynx
b. velopharyngeal closure above flap level—yes/no
c. flap effect on velar elevation:
1. limits it
2. does not limit it

18. Abnormal vessels in nasopharynx (identified by vessel pulsations)—yes/no

19. Visibility of tonsillar tissue on x-ray
a. not present
b. small
c. moderate
d. large

20. Prosthetic appliance—type:

21. Prosthetic appliance relationship to palate or pharyngeal flap—if present (state relationship to which)
a. assists in closure
b. only partially assists in closure
c. does not assist in velopharyngeal closure
d. obstructs palatal movement

22. Pharyngeal wall implants (e.g. teflon)
a. Position/size on pharyngeal wall. Diagram:

b. Position relative to site of maximum velar movement
1. above site of constriction
2. below site of constriction
3. at site of constriction

c. Most consistent state of velopharyngeal contact on non-nasal consonants throughout exam:
1. competent—no gap
2. minimally incompetent, 0–5 mm gap
3. moderately incompetent, 6–10 mm gap
4. grossly incompetent, > 10 mm gap

d. Estimated vertical length of velar-PPW contact*
1. 0 mm
2. 1–2 mm
3. 3–4 mm
4. 5–6 mm

Lateral View: Head Hyperextended (with nasal barium)

Change of velopharyngeal relationships versus head in neutral position
1. No change
2. Increase in VP gap (use same criteria for rating gap as in lateral view in neutral position)
3. Appearance of or increase in size of Passavant's ridge (use same criteria for rating Passavant's ridge as for lateral view in neutral position)
4. Increase in velar length (use same criteria for rating Passavant's ridge as for lateral view in neutral position)
5. Increase in velar elevation (use same criteria for rating Passavant's ridge as for lateral view in neutral position)

Frontal View

1. Timing of LPW movement relative to speech task

	Right	Left
Properly timed, brisk	a	a
Adequate timing, but not brisk	b	b
Poor timing	c	c

2. Percentage of LPW movement toward midline during maximum displacement versus LPWs at rest
Right______% Left______%

3. Level of LPWs in relationship to velum at maximum medial position
a. above palate during maximum palatal elevation
b. at level of palate during maximum palatal elevation
c. below palate during maximum palatal elevation

4. Barium blowing during speech (specify speech tasks)
a. through portal (no flap)
b. through flap
c. through portal(s) on either side of flap (R–L)

5. Level of LPW motion relative to pharyngeal flap

	Right	Left
a. above flap level	a	a
b. at level of flap	b	b
c. below level of flap	c	c

6. Width of velopharyngeal portals with pharyngeal flap (during quiet nasal breathing)

	Right	Left
a. no opening, 0 mm	a	a
b. small, 1–5 mm	b	b
c. moderate, 6–10 mm	c	c
d. large, > 10 mm	d	d

7. Width of pharyngeal flap during quiet nasal breathing
a. narrow, < 1/3 width of total portal
b. average, 1/3–1/2 width of total portal

c. wide, > 1/2 portal width but not completely occluding portal
d. totally occludes portal
8. Position of flap during quiet nasal breathing
a. central
b. asymmetric to right side
c. asymmetric to left side
9. Width of lateral portals during phonation (specify speech tasks)
a. no opening, 0 mm
b. small, 1–5 mm
c. moderate, 6–10 mm
d. large, > 10 mm

Base View

1. Barium blowing (specify speech tasks)
a. through portal (no flap)
b. through flap
c. through portal(s) on either side of flap (R–L)
2. Orifice shape at maximum effort (specify speech tasks)
a. coronal
b. circular
c. sagittal
d. oval
3. Passavant's ridge visible
a. yes
b. no
4. Size of VP orifice during non-nasal consonant production
a. no visible opening
b. tiny opening, 0–10 mm^2
c. small, 10–20 mm^2
d. moderate, 20–40 mm^2
e. gross opening, 40+ mm^2
5. Width of velopharyngeal portals with pharyngeal flap (during quiet nasal breathing)
a. no opening, 0 mm
b. small, 1–5 mm
c. moderate, 6–10 mm
d. large, > 10 mm
6. Width of pharyngeal flap during quiet nasal breathing
a. narrow, < 1/3 width of total portal
b. average, 1/3–1/2 width of total portal
c. wide, > 1/2 portal width but not completely occluding portal
d. totally occluding portal
7. Position of flap during quiet nasal breathing
a. central
b. asymmetric to right side
c. asymmetric to left side
8. Width of lateral portals during phonation (specify speech tasks)
a. no opening, 0 mm
b. small, 1–5 mm
c. moderate, 6–10 mm
d. large, > 10 mm

Overall Rating

a. competent
b. borderline
c. inadequate
d. grossly inadequate

*Refers to demagnified dimensions (see Chapter 6 for details of measurement).

TABLE 15.2. Comparison of information obtained from lateral, frontal, and base views

	Lateral	Frontal	Base
Velar structure			
Length	x		
Thickness	x		
Submucous cleft palate	x		x
Fistula	x	Possible	Possible
Tissue sufficiency	x		Possible
Velar function			
Elevation	x		
Velar eminence			
prominence	x		
position	x	Possible	
Asymmetric elevation or unilaterla paresis	x	Rarely	Possible
Velar stretch	x		
Speed of movement	x	Possible	x
Coordination of movement	x	Possible	x
Consistency of movement	x	Possible	x
Elevation relative to palatal plane	x	Possible	
Elevation relative to adenoid tissue	x		
Elevation relative to site of lateral pharyngeal wall movement	Rarely	x	
Adenoid tissue			
Presence/absence	x		
Size	x		
Role in velopharyngeal closure	x		

TABLE 15.2 continued

	Lateral	Frontal	Base
Generalized posterior pharyngeal wall activity	x		x
Maxillary oro-nasal fistula			
Position	x		
Size	x	Possible	
Vascular Tissue			
Aberrant carotid position (carotid pulsations)		x	
Tonsillar tissue			
Presence/absence	x	x	Possible
Size	x	x	
Role in VP closure	x	x	
Effectiveness of palatal lift/or obturator	x	x	x
Passavant's Ridge			
Presence/absence in speech	x	Possible	x
Level relative to velum	x	Possible	
Size	x		Possible
Orientation	x		
Consistency	x		x
Role in VP constriction/closure coordination	x		x
Velopharyngeal closure			
Presence or size of gap	x	Possible–flap	x
Consistency	x	Possible–flap	x
Vertical length of contact—velum to pharyngeal wall	x	x	
Pharyngeal flap			
Presence/absence	x	x	x
Thickness	x		
Insertion points	x		
Level	x	x	
Palatal-flap movement complex	x		
Interaction with lateral pharyngeal walls			
Symmetry		x	x
Incomplete portal closure		x	x
Level		x	
Presence of fistula	x	Possible	Possible
Barium reflux on swallow	x	Possible	Possible
Barium-coated mucus in portal	x	Possible	x
Lateral pharyngeal walls			
Length of constriction		x	
Timing relative to speech		x	x
% traveled toward midline		x	x
Symmetry of movement		x	x
Major site of movement relative to level of pharyngeal flap		x	
Major site of movement relative to obturator		x	
Major site of movement relative to palatal eminence		Possible	
Lingual movement			
Appropriateness of site of articulation	x		
Lingual-palatal contact	x		
Coordination			
Tip	x		
Blade	x		
Back	x		
Size	x		
Pattern of constriction			
Circular			x
Sagittal			x
Coronal			x
Oval			x
Consistency			x

TABLE 15.3. Videofluoroscopy requisition form

Patient: ______________________________

Clinical History:

Date of exam:

Date of birth:

Cleft type: Unilateral _____

Bilateral _____

Hard palate _____

Soft palate _____

Submucous cleft palate _____

_____ bifid uvula

_____ blue raphe

_____ palatal notching

No cleft _____

Pharyngeal flap: ______________________________
date: ______________________________

Prosthetic device: ______________________________

T % A: date: ______________________________

Teflon implant: date: ______________________________

Oronasal fistula:
anterior ______________________________
posterior ______________________________

Symptomatic:
speech ______________________________
food ______________________________

Swallow (milk: other fluids, solids)

A. Nasal reflux
1. No reflux
2. Little reflux
3. Substantial, noticeable reflux

B. Ease in swallowing
1. Normal
2. Labored

Neuromotor problems ______________________________

Mental retardation ______________________________

Speech Evaluation:

Nasal emission:

	Non-nasal Sounds		Nasal Sounds	
	Left	Right	Left	Right
None	_____	_____	_____	_____
Mild	_____	_____	_____	_____
Moderate	_____	_____	_____	_____
Severe	_____	_____	_____	_____
Consistent	_____	_____	_____	_____
Inconsistent	_____	_____	_____	_____

Facial grimace:	None	Mild	Mod.	Severe
	1	2-3	4-5	6-7
Hypernasality:	None	Mild	Mod.	Severe
	1	2-3	4-5	6-7
Hyponasality:	1	2-3	4-5	6-7
Hoarseness:	1	2-3	4-5	6-7
Nasal turbulence:	1	2-3	4-5	6-7

Articulation errors due to:
(specify)

VPI ______________________________
Dentition ______________________________
Functional ______________________________
Neuromotor ______________________________
ONF ______________________________
Developmental Factors ______________________________
Other ______________________________

Characterized by:

Vowel distortions ______________________________
Sibilant distortions ______________________________
Plosive distortions ______________________________
Affricative distortions ______________________________
Pharyngeal fricatives ______________________________
Other ______________________________

Intelligibility:

Single words:	100%		50%			0%	
	1	2	3	4	5	6	7
Connected Speech:							
	1	2	3	4	5	6	7

Specific questions to be resolved from videofluoroscopic study:

Special requests for videofluoroscopic evaluation in addition to standard speech sample

Please circle: conversation; examination of fistula; speech with fatigue; evaluation with prosthesis, evaluation without prosthesis; specific tasks (specify)

Key: (Plosives: p, b, t, d, k, g; Fricatives: f, v, voiced th, unvoiced th, s, z, sh, zh, h; Affricates: tch, dzh; Nasals: m, n, ng; Glides: hw, w, r, j; Lateral: l)

Remarks:

ment. See Table 15.2 for a summary and comparison of the specific contributions of the lateral, frontal, and base projections.

The Videofluoroscopy Requisition Form

At the University of Pittsburgh Cleft Palate Center, the authors, in conjunction with Betty Jane McWilliams, Ph.D., developed a "videofluoroscopy requisition form."* This was done to provide the radiologist with background information, the reason the videofluoroscopic study was ordered, and to request any special procedures (Table 15.3).

Reliability and Validity

Intrarater reliability can be established by rating a study on two or more occasions. The clinician can determine what percentage of responses are judged the same way on repeated trials. Interrater reliability is established when two or more raters independently rate a study and compute their percentage of agreement. A minimum of 90% agreement is desirable. Reliability should be considered on a question-by-question basis and not simply on the result of an entire rating instrument.

Validity refers to the veracity or accuracy of the data. Are videofluoroscopic ratings of velopharyngeal function consistent with the presence or absence and degree of velopharyngeal dysfunction seen on nasopharyngoscopy? If a very large velopharyngeal gap is seen on videofluoroscopy, is speech production also highly reflective of gross velopharyngeal iadequacy? One must not only be consistent but be accurate as well.

Reference

1. McWilliams-Neely BJ, Bradley DP: A rating scale for evaluation of video tape recorded x-ray studies. Cleft Palate J 1: 88–94, 1964

*We are also indebted to other staff members of the University of Pittsburgh (Alice Brown, Bernice Dressler, Michelle Ferketic, Linda Vallino) for suggested amendments to this form.

16
The Cleft Palate Team

The purpose of this chapter is twofold. First, it is to describe the team approach to the radiological study of the velopharyngeal portal. This portion of the chapter is included for the benefit of the radiologist who has had little or no exposure to the cleft palate team approach. For as noted by Skolnick in 1977 (1):

> To most radiologists, the roentgen examination of the velopharyngeal portal is an entirely new procedure. This is not a technique learned during one's residency, and it is not a technique discussed at radiological meetings or in the radiological journals. There are probably too few needs for such studies to teach the procedure as a routine one to the radiological community. Furthermore, the radiologist is not accustomed to working with other than roentgenological data. The recording of speech and the understanding of the types of speech tasks that will produce specific movements in the structures of the velopharyngeal portal complicate the examination and put additional burdens on the radiologist. Also, for this examination to be successful, it has to be an interactive procedure in which the speech tasks may be modified as the study is being performed because of particular events that are observed. Thus, the radiologist is now being asked to perform as a speech pathologist as well. In addition, except for specialized pediatric radiologists, most radiologists are not accustomed to working with small children, especially in situations where the full cooperation of an unsedated patient is required in order to perform the study.

A second purpose of this chapter is to describe the composition and functions of the cleft palate team. In doing so, we will focus on the cooperative nature of the videofluoroscopic examination.

Team Approach to Management

Team management of cleft lip and palate was originated by H.K. Cooper, founder of the Lancaster Cleft Palate Clinic in the early 1930s (2). Cooper recognized that a cooperative approach to cleft-palate management yielded better patient care than could be provided by professionals working in isolation.

The number and kinds of specialties associated with a cleft palate team may vary, depending upon both clinical needs and local talents. The American Cleft Palate Association specified that a cleft palate team be minimally staffed by a plastic surgeon, speech-language pathologist, and dentist. However, it is not uncommon for cleft palate teams to include many more than only these core specialists. The expanded cleft palate team may contain an orthodontist, otolaryngologist, prosthodontist, pediatrician, psychologist, radiologist, geneticist, social worker, and pediatric nurse practitioner, to name but a few. When craniofacial surgery is performed, a neurologist, neurosurgeon, and ophthalmologist may also be among those serving on the team.

The advantages of a cooperative approach to cleft-palate habilitation are widely recognized, primarily because optimal care results when specialists develop a coordinated approach to treatment. In contrast, cleft palate habilitation performed in isolation is often characterized by incomplete information and discoordination such that aspects of treatment may be ill-timed and counterproductive.

In addition to improved clinical care, the cleft palate team approach offers conveniences for patients and their families. As McWilliams et al. (3) note:

The major advantages of a team, in addition to coordinated care, are that the family can often see a number of specialists at one appointment; follow-up can be managed in a controlled system; the family associates all aspects of care with one program, and only one contact has to be made if emergencies arise.

Team Interaction Models

Cleft palate teams may employ one or more of a variety of interaction models. The choice of the model may be determined by such factors as the history of team interaction, the size and composition of the team, the physical characteristics of the clinic, and thc complexity of the patient's problems.

One of the most common approaches to team decision-making occurs when the cleft palate team meets as a group in one setting at a prescribed time. The team members examine the patient and then together as a group formulate a treatment strategy.

Another type of interaction model incorporates private meetings between the patient and each specialist throughout the course of a scheduled clinic session. During this "clinic," team members work separately and simultaneously in a common building or suite of offices. When all of the scheduled patients have been examined, the team members convene to summarize findings and coordinate treatment.

A variation on this theme occurs when team members read each others' notes on the patient's chart in lieu of a concluding team meeting. This interaction pattern works well when team members have worked with each other in the past, share a common philosophy of treatment, and do not need to discuss "routine" clinical scenarios. More complex management questions are resolved either in brief conversations held between two or more specialists, or in more formal patient staffings if the need arises.

Team Recommendation for Videofluoroscopy

We suggest that two or more team members agree that a videofluoroscopic x-ray is indeed necessary for a particular patient's diagnosis and management. The authorizing team members should minimally include a speech-language pathologist and the plastic surgeon. Representatives of these specialities, and the team leader, comprise what we consider to be the minimal "videofluoroscopic subteam." A cleft-palate team that utilizes prosthetic management of VPI might include a maxillofacial prosthodontist on this subteam.

It is important to include the speech-language pathologist in this decision-making process. Otherwise, patients may be x-rayed for whom there is no speech evidence of velopharyngeal insufficiency. We recall one patient from a distant clinic who was referred for an x-ray because of "speech indicative of VPI." However, this patient had severe hyponasality with no speech evidence of VPI, and did not require x-ray examination. Because of the potential differences in examiners' perception of speech, we recommend that when a patient with the diagnosis of velopharyngeal insufficiency is referred to the cleft-palate team with a request for a videofluoroscopic study, the diagnosis should first be confirmed by the team speech-language pathologist before ordering the examination. It is also important that the plastic surgeon sanction the x-ray requisition because this will guard against x-raying patients who are not candidates for surgical or other management.

The radiologist generally serves as a final source of approval or disapproval of the requisitioned x-ray study. As described in Chapter 17, it is important that the radioloigst assess the clinical and ethical implications of the x-ray for each patient.

X-Ray Interpretation

We feel it is important that the videofluoroscopic evaluation be viewed and interpreted by a radiologist experienced with the technique. One reason is the possibility that a pathological condition unrelated to velopharyngeal insufficiency may be present on the study. We recall one patient, for example, who was found on videofluoroscopy to have an aberrant carotid artery lying close to the pharyngeal wall that could have been injured if a pharyngeal flap were to be performed. This is obviously not an isolated case, as the presence of abnormal carotid arteries in velocardiofacial syndrome has been documented elsewhere (4). While nonphysician team members may be well experienced with the structure and function of the valving mechanism or dental structures, they may not be as familiar with other regions of the head and neck.

Among the cleft palate team members, the radiologist will generally have the most extensive training in radiologic interpretation. Furthermore, he or she is the professional most likely to carry the appropriate liability insurance. (It is interesting to note that a malpractice insurance policy recently purchased by the speech-language pathologist coauthor of this text specifically excluded coverage for procedures traditionally performed by physicians.)

Although we believe that the videofluoroscopic report should always be interpreted by a radiologist knowledgeable about the VP valve, we recommend that studies also be viewed by a speech-language pathologist and plastic surgeon. This combined viewing will produce a broader understanding of the patient's problems as a result of a pooling of the three specialists' knowledge.

The speech-language pathologist is uniquely qualified to assess the functional and acoustic aspects of speech production. Factors to be considered include: the position and movement of the oral articulators, the adequacy of the speech-sample, and phoneme-specific velopharyngeal dysfunction. As Skolnick (1) pointed out, it is helpful for the speech-language pathologist to be present during the performance of the radiological examination:

> He conducts the speech portion of the examination as the radiologist performs or supervises the radiological portion. The speech pathologist should also be monitoring the radiological images, and if necessary, modify the speech tasks to emphasize abnormalities suggested by roentgen images. For example, if the patient can close with certain isolated sounds but not with the sounds of connected speech, the patient should be briefly stimulated using words with these sounds (with the x-ray off) and then momentarily re-examined to see if closure can now occur.

The plastic surgeon will be particularly interested in the structural characteristics of the velopharyngeal valve, the site of velopharyngeal movement, and the consistency and degree of movement. Such information will determine the need for surgical management of velopharyngeal insufficiency. Results of the videofluoroscopy may also suggest the optimal placement and construction of a proposed pharyngeal flap. In patients with an ineffective pharyngeal flap, the videofluoroscopy will assist the surgeon in visualizing the problem and planning for treatment.

The videofluoroscopy study will also provide information to the dentist interested in determining a patient's need for prosthetic management of velopharyngeal insufficiency.

In order to improve the value of the videofluoroscopic procedure, we suggest that the cleft palate team periodically review patient charts to see how the information provided by these studies affected patient management. In this manner, the team would obtain a better understanding of the benefits and limitations of the videofluoroscopic examination and learn how to improve their utilization of the procedure.

Team Communication

The traditional style of medical communication, wherein the physician unilaterally makes diagnoses and directs others to follow a prescribed treatment plan, is not the optimal model for communication pertaining to the evaluation of cleft palate patients because of the need for multispecialty care of the patient.

Knowledge of the velopharyngeal valve is generally acquired outside of formal existing medical training. Because members of different disciplines may offer unique perspectives on the same x-ray study, members of the cleft palate team should have the opportunity to present and support their opinions. Furthermore, it is important that a person's expertise in this procedure, rather than power derived from longevity on the team or professional title, determine his or her decision-making function on the team. In our experience, patient care improves when open and creative communication prevails over traditional styles of medical communication.

Communication with the Radiologist

Communication within a team is generally facilitated and enhanced by the face-to-face contact that occurs when a "small group" of people convene for a common purpose. Indeed, to fully appreciate the communicative dynamics within a cleft palate team [insightfully documented by Koepp-Baker (5)], the team's unique mission and composition must be viewed within the larger context of "small-group communication" principles.

In our experience, a key member of this team, the radiologist, can be inadvertently deprived of the full enrichment afforded by this "small-group" experience, because of his geographic separation from the team. Since the radiologist's office is often physically removed from that of other team members, special efforts must be made to avoid relating to this team member as a technician who simply performs the x-ray studies that others have ordered.

In order to increase the radiologist's commitment and contribution to the team, he should be encouraged to attend team clinical, administrative, and educational meetings. Like other clinicians the radiologist should also have the opportunity to examine patients. At clinical meetings he should present the radiologic findings to the other members. These activities promote both team functioning and foster collegiality. They also provide team members with opportunities to educate the radiologist about the functions and structures of the velopharyngeal portal under conditions of both normal and pathological speech. And, face-to-face interaction enables the radiologist to impart knowledge about the radiologic technique, the views used and what each of these show. As a result of this mutual interaction, the radiologist gains a better understanding of velopharyngeal behavior while others on the team acquire an increased appreciation of the limitations and advantages of videofluoroscopy.

Patient Counseling

While the diverse flow of ideas and opinions regarding treatment may be stimulating to team members, patients may perceive these differences as evidence of uncertainty and confusion. Therefore, whenever possible, it is important for patients and their families to receive consistent and noncontradictory information about diagnosis and treatment. Individual members of the team, especially when the patient has a complex problem, should refrain from providing "off the cuff" evaluations to the patient and family until the entire cleft palate team has reviewed the case and arrived at its diagnosis and treatment plan.

References

1. Skolnick ML: A plea of an interdisciplinary approach to the radiological study of the velopharyngeal portal. Cleft Palate J 1977;14:4
2. Krogman W: The cleft palate team in action. In: Cooper HK, Harding RL, Krogman WM, Mazaheri M, Millard R, eds. Cleft lip and palate: A team approach to clinical management and rehabilitation of the patient. Philadelphia: WB Saunders, 1979
3. McWilliams BJ, Morris HL, Shelton RL. Cleft palate speech. Philadelphia: BC Decker, 1984
4. Stepner-MacKenzie K, Witzel MA, Stringer DA, Lindsay WK, Munro IR, Hughes H: Abnormal carotid arteries in the velocardiofacial syndrome: A report of three cases. Paper presented at the Annual Meeting of the American Cleft Palate Association, New York, 1986
5. Koepp-Baker H. The cleft palate team. In: Bzock KR, ed. Communicative disorders related to cleft lip and palate. Boston: Little, Brown and Company, 1972

17
Ethical Concerns

Shortly after roentgen rays were discovered, x-ray technology was recognized as a means to study the structure and function of the speech mechanism (1). Since then, x-ray studies, some of which were static (e.g., cephalometry), and others dynamic (e.g., cine- and videofluoroscopy), have greatly contributed to our understanding of normal and pathologic palatopharyngeal function.

Over the years, we have come to appreciate the dangers of radiographic imaging technologies, and the need to exercise restraint in their use.

As a result of concerns about the long-term hazards of radiation exposure, increasing numbers of researchers and institutional review boards in the United Stated are applying more stringent standards for the use of x-rays. Clinicians are now less likely to employ radiographic imaging to study patients with speech disorders for whom the obtained information would not be clinically useful. Indeed, the use of x-rays is best limited to those patients with suspected pathologic palatopharyngeal function for whom there is a need to document valving abnormalities and for whom the obtained information would significantly contribute to their treatment. A case in point would be the patient who, despite having a repaired cleft palate, has socially acceptable speech production. Barring specific circumstances, such as an impending adenoidectomy or maxillary advancement, the aforementioned patient would not be a candidate for a videofluoroscopic study.

The purposes of this chapter are thus to address ethical considerations surrounding the use of the videofluoroscopic studies; consider the applicability of alternative techniques for studying the velopharyngeal valve; and discuss ways to encourage members of the cleft palate team to judiciously use videofluoroscopy.

Ethical Issues

The casual reader of this book may wonder why the issue of ethics is being addressed in a "how-to" manual, and may be even more puzzled as to how topics like "length of the studies" and "repeated studies" relate to ethical concerns. We have included these and other topics because the decision to refer a patient for videofluoroscopy or alternative methodologies, and the ways in which the physicians regard patients and their families are very much affected by the clinician's values and goals.

Is the clinician basically patient oriented or research oriented? Does the radiologist feel it is worth his effort to provide a period of orientation for the patient? Is the radiologist committed to conducting the shortest evaluation possible, and would he even refuse to perform a seemingly unnecessary study? Can the clinician-researcher place the needs of the patient before his needs to obtain subjects and data? The ethical orientation and values of the radiologist and other members of the cleft palate team will influence the answers to these questions.

Treatment versus Research

With increasing awareness of the long-range effects of radiation exposure, limitations have been placed upon the use of x-rays for research purposes. Many institutional research review boards in the United

Stated now restrict radiographic research studies to adults past childbearing age, or to patients for whom there is a clinical need for the x-ray study.

In addressing the issue of clinical need, there is often a fine distinction between clinical studies that will potentially alter the course of treatment, and those that will simply provide interesting, but not clinically useful information. We thus recommend that the following questions be addressed prior to ordering a "clinically necessary" videofluoroscopic study:

The *Clinical Need* Question: "Is it strongly anticipated that the data supplied by the videofluoroscopic examination will potentially alter the course of clinical treatment?

The *Alternative Methods* Question: "Are there equally effective means by which to obtain the needed information which involve less or no radiation?"

The *"So What"* Question: "What will be done if velopharyngeal insufficiency is detected?"

An example of a patient for whom there would be a clear need for a videofluoroscopic study would be a hypernasal candidate being considered for a pharyngeal flap. In addition to confirming the presence of VPI, the videofluoroscopic exam would provide useful information about the valving mechanism for the purposes of surgical planning (e.g., flap width and positioning) and prognosis formation.

On the other hand it would be inappropriate to perform a videofluoroscopic study in a patient with bilateral inaudible nasal emission, normal speech, and no other VPI-related complaints. Although it would be interesting for the cleft palate team to know the status of velopharyngeal valving, such information might not influence clinical treatment. If the same patient were to be a candidate for an adenoidectomy, a stronger case could be made for the usefulness of videofluoroscopy. In such an instance, it would be important to know if adenoidal tissue is used to achieve velopharyngeal closure because removal of the adenoids could result in velopharyngeal insufficiency and hypernasal speech.

Repeated Studies

Under typical circumstances, it is not often necessary or desirable to perform repeated videofluoroscopic studies. Exceptions to this would include the following:

VPI is unresolved following pharyngeal flap surgery, and there is the possibility of surgical revision of the flap.

The pharyngeal flap appears to be creating sleep apnea, and assessment is required to determine management.

Speech symptoms of VPI worsen with age and require intervention.

Speech has improved consequent to speech therapy, or maturation but still demonstrates evidence of VPI. It is now felt that the repeat videofluoroscopic x-ray will yield a more valid and representative picture of velopharyngeal valve capability than was the case during the initial study.

The patient was initially too young to cooperate with examiners during the initial videofluoroscopic study and poor results were obtained. Now the patient is older and it is likely that cooperation will be obtained.

The same criteria that govern the administration of the first videofluoroscopic study apply to subsequent studies. In fact, it might be argued that repeat studies should be judged with even more stringent standards because of the added radiation exposure. Clearly, the videofluoroscopic study should not be repeated to routinely document growth, nor should it serve to assess VP function following surgery or other treatment unless a clinical problem exists, or is strongly anticipated.

Length of the Studies

A videofluoroscopic study should be performed in the minimum time necessary to obtain the needed diagnostic information. Experienced clinicians can often shorten the length of the evaluation by eliminating aspects of the exam that would not be likely to yield new information. For example:

If a patient demonstrates gross VPI, it is not necessary to assess competence in the hyperextended head position. Hyperextension is primarily useful in unearthing borderline VPI, and gross VPI is readily recognizable with the head in an upright position.

When testing for fatigue of the speech mechanism, fluoroscopy can be briefly turned on at the beginning and end of a long and continuous task such as counting from 1 to 400, to see if there has been any change in the closure mechanism during the exercise trial.

If the role of the adenoids is the issue, it is not necessary to perform a prolonged examination of the lateral pharyngeal walls.

Give instructions to a patient when the fluoroscopy is off, rather than on.

Do not persist with the examination if patient positioning (e.g., hyperextension for base view) is inadequate. React quickly to improve positioning. The same applies relative to the quality of the barium coating. Repeat the barium coating if the pharyngeal walls are poorly seen.

Explaining the Procedure

It is the responsibility of the cleft palate team members to fully inform patients and their families why the videofluoroscopic study is being recommended, the benefits and risks of the study, and how the results might affect treatment.

Clear, nontechnical explanations should be offered. It should be recognized that patients and their families may differ both in their desire for detail and their ability to comprehend it. Thus, professionals must strive to convey both the appropriate level and amount of information. Questions and concerns should be addressed in a thorough, but not intimidating manner. Further suggestions about patient preparation are included in Chapter 13.

If the videofluoroscopic study is being performed for research purposes with no expected clinical benefits, the consenting adult should be so informed. The researcher should stress that a decision not to participate in the study will not result in the denial of future treatment. It should also be clear that neither the patient nor their family will be expected to pay for a study conducted for research purposes only.

Patient Protection

It is ultimately the radiologist's responsibility to exercise the precautions necessary for patient protection. These may include the placing of a lead apron, thyroid and ocular shields on the patient if the collimator of the x-ray tube doesn't sufficiently limit the beam to the region of interest. The radiologist should always be aware of the possibility of pregnancy in females of child-bearing age. When parents or other professionals are enlisted to help obtain the cooperation of a young child, the adults should also receive protective garb.

Alternative Methodologies

It is the ethical responsibility of cleft palate team members to consider the use of alternative methods of studying the velopharyngeal valve. These are discussed in Chapter 2. If two techniques can provide comparable information, the technique that has the least risk, is least invasive, least unpleasant, and least uncomfortable should be employed.

The applicability of alternative methodologies will vary according to the specific diagnostic questions to be answered, the patient's age and his ability to cooperate. For example, when a pharyngeal flap is suspected of obstructing the airway in an adult patient, nasopharyngoscopy may be a useful technique. It has the added advantage of being radiation free. However, videofluoroscopy might be preferable for a similar problem in a 4-year-old patient unwilling to cooperate during the nasopharyngoscopic procedure.

Examiner Experience

It is well recognized that the utility of a specific methodology will greatly depend upon the clinician's experience with the technique. It is important that a physician not perform a technique unless he has adequate training and experience. If such is not the case, consultation with a more experienced individual, or referral of the patient to another team is ethically desirable.

Equipment

If it is known that the same results can be achieved in another setting with equipment that emits less radiation, is it ethical to perform videofluoroscopy using higher radiation exposure equipment? On the other hand, what if you suspect that despite superior equipment, the staff in the other facility does not perform good videofluoroscopic studies, and makes errors in interpretation? What if you are totally unacquainted with the other staff's competence?

These ethical problems are at best, difficult to answer, but deserve our concern and attention. We must consider our patients' interests, safety and comfort in the same way we would those of our own children or adult family members.

Reference

1. Scheier M: Über die Verwertung der Röntgenstrahlen in der Rhino- und Laryngologie. Arch Laryngol 6:57, 1897

Appendix Story Orientation to Videofluoroscopy for Young Children

The following is intended to help orient the young patient to videofluoroscopy. In our experience, children who are familiar with the purpose and procedures of the x-ray are generally less anxious and better able to cooperate than children with no prior orientation.

Reproduction of this portion of this book is permitted, in limited quantities, for clinical purposes, so that it can be distributed to and illustrated by young patients.

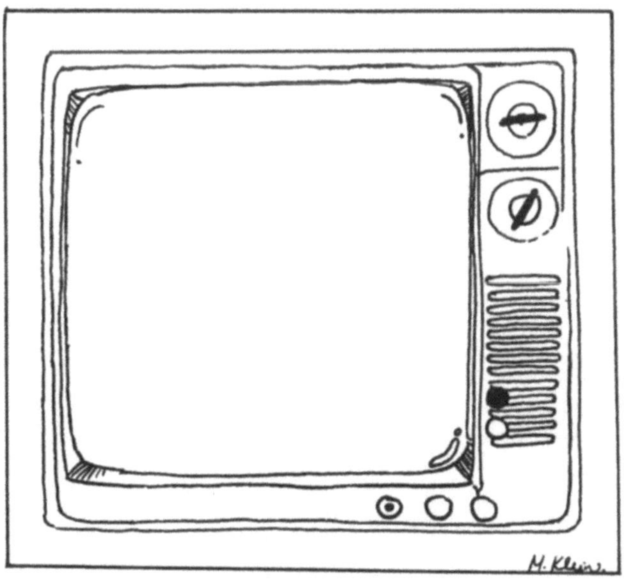

Do you like to watch TV? What is your favorite television show? You can draw a picture of your television show right here.

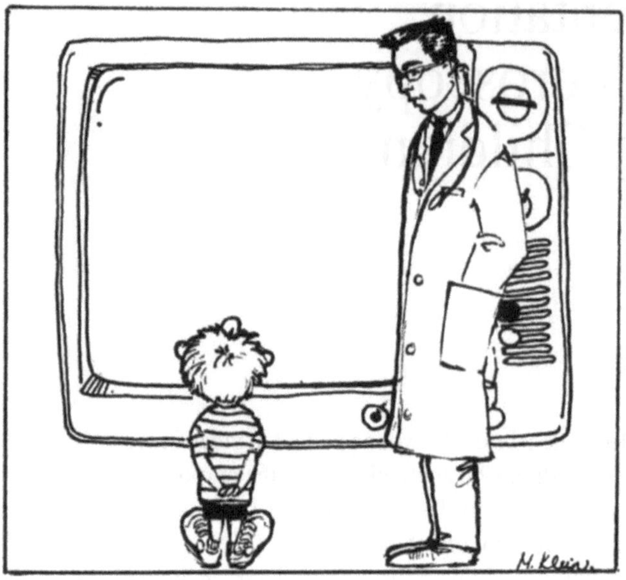

This is a story to tell you how a special doctor, called a radiologist, is going to make a special television picture. Dr. ______________ is going to take an x-ray so we can see how your mouth moves when you talk. So YOU are going to be on TV! But only the people who work on the cleft-palate team will see you on TV. After we finish the x-ray, if you like, you can watch how your mouth moves when you talk.

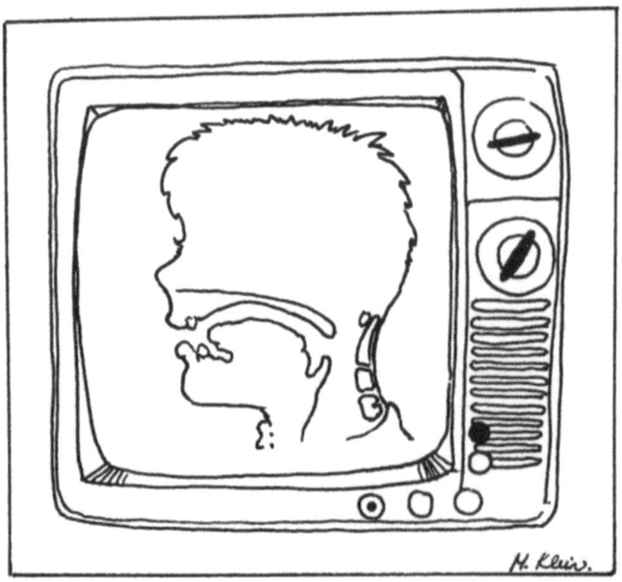

We will look at how your palate moves. Here is a picture of the palate and the tongue. They move when you talk. Can you point to your palate? How about your tongue?

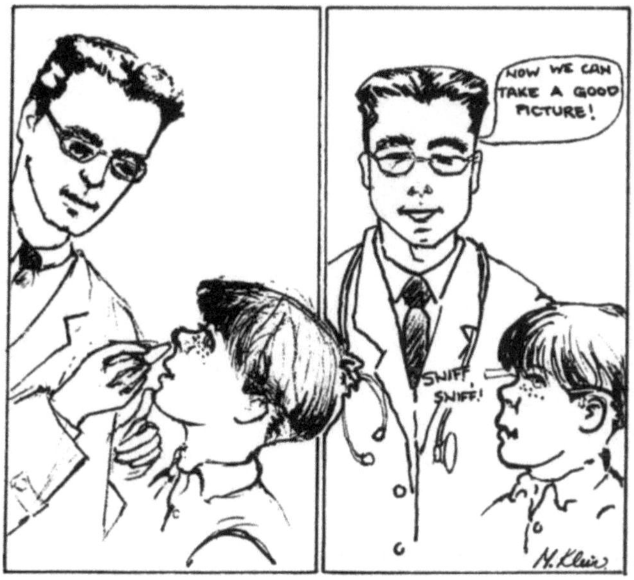

When the doctor takes an x-ray, it's sort of like taking a picture. It doesn't hurt! The doctor will have you sniff some special nosedrops, so he can take a real good picture. The drops taste a little bit like a strawberry milkshake.

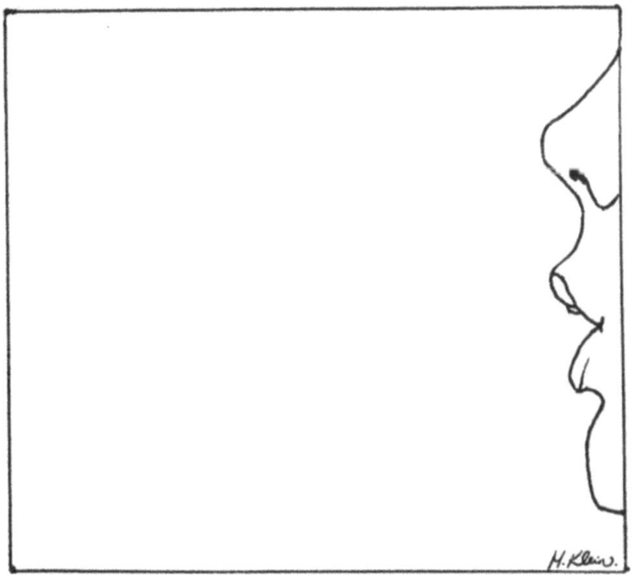

The doctor and his helper are going to take a "talking picture," just like on TV. You will wear a microphone, and say some words, nice and loud. Let's practice. . . .

(Space can be filled with a printed list of the speech protocol for readers, or pictures or stickers for nonreaders.)

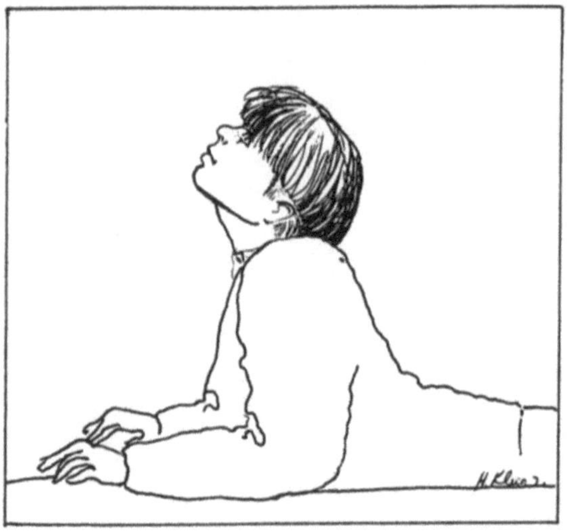

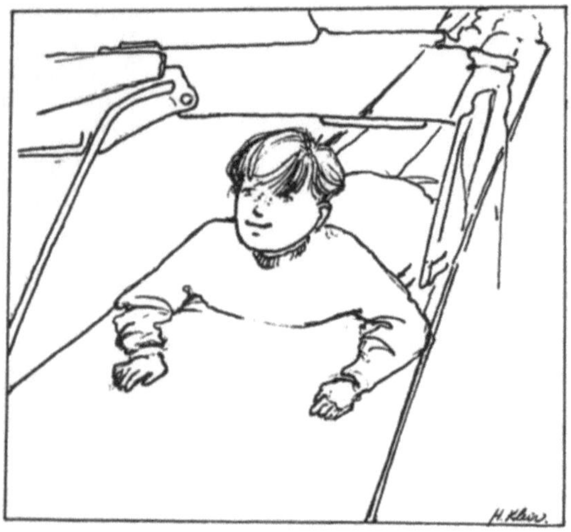

While you are saying the words, the doctor may ask you to hold very still. He may even ask you to tilt your head back, or lie on your stomach. That way he can take a really good movie of how you talk.

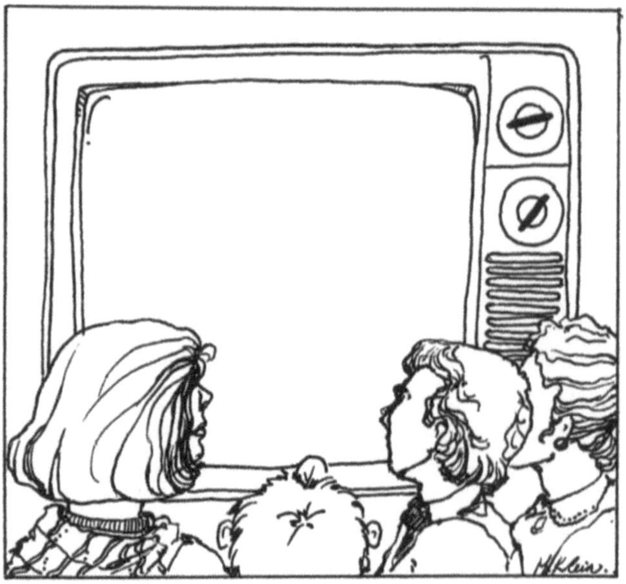

When you're all done, if you'd like, we can all watch you on TV.

Index

MIX
Papier aus verantwortungsvollen Quellen
Paper from responsible sources
FSC® C105338

If you have any concerns about our products,
you can contact us on
ProductSafety@springernature.com

In case Publisher is established outside the EU,
the EU authorized representative is:
Springer Nature Customer Service Center GmbH
Europaplatz 3, 69115 Heidelberg, Germany

Printed by Libri Plureos GmbH
in Hamburg, Germany